Nonlinear Assignment Problems

T0205632

COMBINATORIAL OPTIMIZATION

VOLUME 7

Through monographs and contributed works the objective of the series is to publish state of the art expository research covering all topics in the field of combinatorial optimization. In addition, the series will include books which are suitable for graduate level courses in computer science, engineering, business, applied mathematics, and operations research.

Combinatorial (or discrete) optimization problems arise in various applications, including communications network design, VLSI design, machine vision, airline crew scheduling, corporate planning, computer-aided design and manufacturing, database query design, cellular telephone frequency assignment, constraint directed reasoning, and computational biology. The topics of the books will cover complexity analysis and algorithm design (parallel and serial), computational experiments and applications in science and engineering.

The titles published in this series are listed at the end of this volume.

Nonlinear Assignment Problems

Algorithms and Applications

edited by

Panos M. Pardalos
University of Florida

and

Leonidas S. Pitsoulis
Princeton University

KLUWER ACADEMIC PUBLISHERS
DORDRECHT / BOSTON / LONDON

A C.I.P. Catalogue record for this book is available from the Library of Congress.

ISBN 978-1-4419-4841-0

Published by Kluwer Academic Publishers,
P.O. Box 17, 3300 AA Dordrecht, The Netherlands.

Sold and distributed in North, Central and South America
by Kluwer Academic Publishers,
101 Philip Drive, Norwell, MA 02061, U.S.A.

In all other countries, sold and distributed
by Kluwer Academic Publishers,
P.O. Box 322, 3300 AH Dordrecht, The Netherlands.

Printed on acid-free paper

To Our Parents

Contents

List of Figures

List of Tables

Preface

Nonlinear Assignment Problems (NAPs) are natural extensions of the classic Linear Assignment Problem, and despite the effort of many researchers over the past three decades, they still remain some of the hardest combinatorial optimization problems to solve exactly. The purpose of this book is to provide in a single volume, major algorithmic aspects and applications of NAPs as contributed by leading international experts.

The chapters included in this book are concerned with major applications and the latest algorithmic solution approaches for NAPs. Approximation algorithms, polyhedral methods, semidefinite programming approaches and heuristic procedures for NAPs are included, while applications of this problem class in the areas of multiple-target tracking in the contexts of military surveillance systems and experimental high energy physics, and parallel processing are presented.

We would like to thank all the authors of this book for cooperating with us and providing excellent chapters, and also John Martindale and the staff members at Kluwer Academic Publishers for their help with the manuscript.

The intended audience includes researchers and graduate students in the areas of combinatorial optimization, mathematical programming, operations research, physics, and computer science.

PANOS PARDALOS AND LEONIDAS PITSOULIS

Contributing Authors

Ioannis T. Christou
Delta Technology Inc.
Dept. of Operations Research
1001 International Blvd.
Atlanta, GA 30354, USA
ioannis.christou@delta-air.com

Michel Cosnard
LORIA INRIA Lorraine
615, rue du Jardin Botanique
54602 Villers les Nancy, France
Michel.Cosnard@loria.fr

Tao Yang
CS Dept. UCSB
Engr Building I
Santa Barbara, CA 93106, USA
tyang@cs.ucsb.edu

Emmanuel Jeannot
LaBRI, University of Bordeaux I
351, cours de la Liberation
33405 Talence Cedex, France
ejeannot@labri.u-bordeaux.fr

Volker Kaibel
Fachbereich Mathematik, Sekr. 7–1
Technische Universität Berlin
Straße des 17. Juni 136
10623 Berlin, Germany
kaibel@math.TU-Berlin.DE

Robert R. Meyer
University of Wisconsin - Madison
Computer Sciences Dept.
1201 W. Dayton Str.
Madison, WI 53706, USA
rrm@cs.wisc.edu

Rob A. Murphey
Air Force Research Laboratory, Munitions Directorate,
101 W. Eglin Blvd., Ste 330
Eglin AFB, FL 32542-6810, USA
Murphey@eglin.af.mil

Aubrey B. Poore

Department of Mathematics
Colorado State University
Fort Collins, Colorado 80523, USA
poore@math.colostate.edu

Jean-Francois Pusztaszeri

Logistics Division,
Sabre, Inc.
pusztaszeri@sabre.com

Liqun Qi

School Of Mathematics
The University of New South Wales
Sydney 2052, Australia
L.Qi@unsw.edu.au

Frits C.R. Spieksma

Department of Mathematics
Maastricht University
P.O. Box 616
NL-6200 MD Maastricht, The Netherlands
spieksma@math.unimaas.nl

Defeng Sun

School Of Mathematics
The University of New South Wales
Sydney 2052, Australia
D.Sun@unsw.edu.au

Stefan Voss

Technische Universität Braunschweig
Allgemeine Betriebswirtschaftslehre
Wirtschaftsinformatik und Informationsman-
agement
Abt-Jerusalem-Str. 7, D-38106 Braunschweig,
Germany
voss@tu-bs.de

Henry Wolkowicz

Department of Combinatorics and Optimiza-
tion
Waterloo, Ontario N2L 3G1, Canada
hwolkowi@orion.math.uwaterloo.ca

Introduction

Panos Pardalos and Leonidas Pitsoulis

Consider that we are given n facilities and n locations, and a matrix $\mathcal{Z}^{n \times n} \ni C := (c_{ij})$ where c_{ij} represents the cost of placing facility i to location j. Our objective is to find an optimal assignment of facilities to locations such that the total placement cost is minimized. Using the decision variables $x_{ij} \in \{0, 1\}$ as

$$x_{ij} = \begin{cases} 1 & \text{if facility } i \text{ is assigned to location } j, \\ 0 & \text{otherwise,} \end{cases}$$

we can formulate the Linear Assignment Problem (LAP) as an integer $0 - 1$ program

$$\min \quad \sum_{i,j=1}^{n} c_{ij} x_{ij} \tag{I.1}$$

$$\text{s.t.} \quad \sum_{i=1}^{n} x_{ij} = 1, \quad j = 1, 2, \dots, n, \tag{I.2}$$

$$\sum_{j=1}^{n} x_{ij} = 1, \quad i = 1, 2, \dots, n, \tag{I.3}$$

$$x_{ij} \in \{0, 1\}, \quad i, j = 1, 2, \dots, n. \tag{I.4}$$

The constraints (I.2), (I.3) and (I.4) are the so called *assignment constraints* which define the vertices of the well known assignment polytope (other names include: the Birkhoff polytope, the perfect matching polytope of $K_{n,n}$, and the transportation polytope). For each solution matrix $X = (x_{ij}) \in \{0, 1\}^{n \times n}$ to the LAP defined in (I.1)–(I.4), we can associate a permutation $p : N \to N$, where $N = \{1, 2, \dots, n\}$, by assigning

$$p(i) = j \quad \text{iff} \quad x_{ij} = 1, \quad i, j = 1, 2, \dots, n.$$

Therefore, we can equivalently formulate the LAP using permutations as follows

$$\min_{p \in \Pi_N} \sum_{i=1}^{n} c_{ip(i)}, \tag{I.5}$$

where Π_N is the set of all permutations p. The LAP is a classical optimization problem and can be solved very efficiently. Nonlinear Assignment Problems (NAPs) are extensions of the LAP, and they can be divided into two major classes, M-adic assignment problems, and Multidimensional Assignment Problems (MAPs).

In the M-adic assignment problem class, the dimension of the cost array in (I.5) becomes $2M$, $M \geq 2$, and we permute over M indices of C in the objective function. For example, the 3-adic assignment problem can be formulated as

$$\min_{p \in \Pi_N} \sum_{i=1}^{n} \sum_{j=1}^{n} \sum_{k=1}^{n} c_{ijkp(i)p(j)p(k)}, \tag{I.6}$$

where $C := (c_{ijklms}) \in \mathcal{Z}^{n^{2M}}$. In a similar way we can define the Quadratic ($M = 2$), and the BiQuadratic ($M = 4$) assignment problems. In the MAP class, the dimension of the cost array in (I.5) becomes M, and we sum over n values of C. However, C is permuted over $M - 1$ indices by different permutations. For example, the three dimensional assignment problem can be formulated as

$$\min_{p,q \in \Pi_N} \sum_{i=1}^{n} c_{ip(i)q(i)}, \tag{I.7}$$

where $C := (c_{ijk}) \in \mathcal{Z}^{n^M}$. All of the problems presented in the chapters that this volume contains can be classified as either M-adic or multidimensional assignment problems.

The first four chapters of this book deal with multidimensional assignment problems. Chapter 1 presents an overview of complexity and approximability issues regarding MAPs, as well as various applications. Chapters 2 and 3, deal exclusively with an important military application of MAPs, and specifically that of multiple-target tracking. Chapter 4 presents an application of MAPs arising in experimental high energy physics, which also involves solving the data association phase of a multiple-target tracking problem where the targets are elementary particles. Chapters 5 through 8 present algorithmic approaches for solving NAPs. Chapters 5 and 6 present polyhedral methods for solving the 3-index assignment problem, and the QAP respectively. Chapter 7 presents a Semidefinite Programming approach for solving the QAP. Here it has to be noted that although the QAP is probably the most celebrated of the NAPs in terms of applications and amount of research devoted to it, we have decided not to include a separate chapter for it, but rather include two chapters with the most recent solution approaches for solving the QAP. The reason for not including a separate chapter is twofold. First, the amount of literature for the QAP is so vast that a survey chapter for the subject would just be too large to be included in this book. Secondly, there are two recent publications, specifically the survey chapter by Burkard, Çela, Pardalos, and Pitsoulis [Burkard et al., 1998] and the book by Çela [Çela, 1998], which cover every aspect of the QAP, and an additional chapter would be redundant. Chapter 8 is a survey of heuristic approaches for solving NAPs such as Local Search, GRASP, Simulated Annealing, Tabu Search, Genetic Algorithms and Ant Systems. Finally the last two chapters present applications of NAPs in the area of parallel processing.

References

[Burkard et al., 1998] Burkard, R., Çela, E., Pardalos, P., and Pitsoulis, L. (1998). The quadratic assignment problem. In Pardalos, P. and Du, D.-Z., editors, *Handbook of Combinatorial Optimization*, pages 241–338, Boston. Kluwer.

[Çela, 1998] Çela, E. (1998). *The Quadratic Assignment Problem: Theory and Algorithms*. Kluwer, Dordrecht.

Chapter 1

MULTI INDEX ASSIGNMENT PROBLEMS: COMPLEXITY, APPROXIMATION, APPLICATIONS

Frits C.R. Spieksma
Department of Mathematics
Maastricht University
P.O. Box 616
NL-6200 MD Maastricht, The Netherlands
spieksma@math.unimaas.nl

1. INTRODUCTION

This chapter deals with approximation algorithms for and applications of multi index assignment problems (MIAPs). MIAPs and relatives of it have a relatively long history both in applications as well as in theoretical results, starting at least in the fifties (see e.g. [Motzkin, 1952], [Schell, 1955] and [Koopmans and Beckmann, 1957]). Here we intend to give the reader i) an idea of the range and diversity of practical problems that have been formulated as an MIAP, and ii) an overview on what is known on theoretical aspects of solving instances of MIAPs. In particular, we will discuss complexity and approximability issues for *special cases* of MIAPs. We feel that investigating special cases of MIAPs is an important topic since real-world instances almost always posses a certain structure that can be exploited when it comes to solving them.

Before doing so, let us first describe a somewhat frivolous, quite unrealistic situation that gives rise to an instance of an MIAP. Suppose that you have become mayor of a set of villages. Everything is running smoothly, except in one village where chaos seems to rule. You, as a responsible mayor should, start looking into this and after careful gathering of information you find that there 30 cats, 30 houses, 30 men and 30 women present in the village. Now, in order to establish some peace you impose that 30 "units" will have to be formed, each consisting of a cat, a house, a man, and a woman. It turns out that for each possible 4-tuple (and there are $30^4 = 810000$ of them), a number is

1

known that reflects the happiness of this particular unit. Since you as a mayor want to maximize total happiness, the problem becomes to find those 30 units that achieve total happiness.

This is an example of a so-called axial 4-index assignment problem. Different versions of the MIAP may arise by varying the number of indices, or by asking not for 30 4-tuples but instead for 30^2 4-tuples such that each pair consisting of a cat and a house is represented exactly once in a 4-tuple, or by considering other objective functions (e.g. bottleneck).

This chapter is organized as follows. In Section 1.1 we give some technical preliminaries. Section 2. deals with the two versions of the 3-index assignment problem; Section 3. treats the general case. Finally, in Section 4. we discuss extensions of the MIAP.

1.1   TECHNICAL PRELIMINARIES

An introduction to the issue of approximation and complexity can be found in [Papadimitriou, 1994] and [Crescenzi and Kann, 1999]. Here, we shortly describe two concepts we need.

- A *ρ-approximation algorithm* for a maximization (minimization) problem P is a polynomial time algorithm that, for all instances, outputs a solution with a value that is at least (at most) equal to ρ times the value of an optimal solution of P. Observe that for maximization problems $0 \leq \rho \leq 1$ and for minimization problems $\rho \geq 1$.

- A *polynomial time approximation scheme* (PTAS) for a maximization (minimization) problem is a family of polynomial time $(1 - \epsilon)$-approximation algorithms ($(1 + \epsilon)$-approximation algorithms) for all $\epsilon > 0$.

2.   THE 3-INDEX ASSIGNMENT PROBLEM

In this section we introduce two types of 3-index assignment problems, the so-called axial 3-index assignment problem and the so-called planar 3-index assignment problem (these names were first used by [Schell, 1955]). For each of these types we give a formulation and discuss complexity issues. We mention heuristic and exact approaches that have appeared in literature and sketch applications.

2.1   THE AXIAL 3IAP

Given are 3 n-sets A_1, A_2 and A_3. For each triple in $A_1 \times A_2 \times A_3$ a number is known (either a profit w_{ijk} or a cost c_{ijk}). The problem is now to find n triples such that each element in $A_1 \cup A_2 \cup A_3$ is in exactly one triple. So in the axial 3IAP one is asked to output n triples that either maximize total profit or minimize total cost.

2.1.1   Maximization.

$$(A3IAPMAX) \quad \text{maximize} \quad \sum_{i=1}^{n}\sum_{j=1}^{n}\sum_{k=1}^{n} w_{ijk}x_{ijk}$$

$$\text{subject to} \quad \sum_{i=1}^{n}\sum_{j=1}^{n} x_{ijk} \leq 1 \text{ for } k = 1, \ldots, n,$$

$$\sum_{i=1}^{n}\sum_{k=1}^{n} x_{ijk} \leq 1 \text{ for } j = 1, \ldots, n,$$

$$\sum_{j=1}^{n}\sum_{k=1}^{n} x_{ijk} \leq 1 \text{ for } i = 1, \ldots, n,$$

$$x_{ijk} \in \{0,1\} \text{ for } i, j, k = 1, \ldots, n.$$

Notice that this formulation generalizes the well-known (2-index) assignment problem; moreover in this generalization we lose the integrality property of the assignment problem. In other words, when relaxing the integrality constraints and solving the resulting Linear Programming formulation, fractional values for the variables may arise.

The axial 3IAP is a generalization of the 3-dimensional matching problem (3DM). Indeed: for $w_{ijk} \in \{0,1\}$ 3DM is *equivalent* to 3IAP. Thus, concerning complexity and approximation for the 3IAP with coefficients in $\{0,1\}$, we can state the following (see [Crescenzi and Kann, 1999]):

- It appears in the seminal work of [Karp, 1972], and thus has the honor of being one of the seven first problems to be proved to be NP-complete.

- Subsequently, [Kann, 1991] proves that even for a special case called *bounded* 3DM there is no PTAS (unless P=NP). Bounded means that for each element in $A_1 \cup A_2 \cup A_3$, the number of triples with weight 1 in which an element of $A_1 \cup A_2 \cup A_3$ occurs is bounded by a prespecified constant $B \geq 3$.

- When the coefficients are *planar* (admittedly, this may be confusing terminology) the problem remains NP-hard ([Dyer and Frieze, 1983]), but a PTAS exists ([Nishizeki and Chiba, 1988]). Planar means that the graph constructed as follows: have a node for each element of $A_1 \cup A_2 \cup A_3$ and for each triple with weight 1, and connect two nodes iff the corresponding triple contains the corresponding element, is planar.

- [Hurkens and Schrijver, 1989] give a $\frac{2}{3} - \epsilon$ approximation algorithm (see Section 3.).

When allowing arbitrary values for the coefficients the following is known:

- [Hausmann et al., 1980] show in a more general context that a simple greedy algorithm achieves a performance guarantee of $\frac{1}{3}$ (see Section 3.).

- [Arkin and Hassin, 1998] give a $\frac{1}{2} - \epsilon$ approximation algorithm.

- if there exist numbers a_i, b_j and c_k, $i, j, k = 1, \ldots, n$ such that $w_{ijk} = a_i \cdot b_j \cdot c_k$, for all i, j, k, the problem is easy ([Gilbert and Hofstra, 1988], [Burkard et al., 1996]).

- A recent result in [Barvinok et al., 1998] is as follows. Let the $3n$ points be points in R^l for some dimension l, and let distances d be computed according to some geometric norm. For the case of polyhedral norms (which includes the rectilinear norm and the sup norm) and with $w_{ijk} = d_{ij} + d_{ik} + d_{jk}$ for all i, j, k (see Section 3. for a general framework encompassing this cost-structure) it is claimed that the problem is polynomially solvable.

2.1.2   Minimization.

$(A3IAPMIN)$ minimize $\displaystyle\sum_{i=1}^{n}\sum_{j=1}^{n}\sum_{k=1}^{n} c_{ijk} x_{ijk}$

subject to $\displaystyle\sum_{i=1}^{n}\sum_{j=1}^{n} x_{ijk} = 1$ for $k = 1, \ldots, n,$

$\displaystyle\sum_{i=1}^{n}\sum_{k=1}^{n} x_{ijk} = 1$ for $j = 1, \ldots, n,$

$\displaystyle\sum_{j=1}^{n}\sum_{k=1}^{n} x_{ijk} = 1$ for $i = 1, \ldots, n,$

$x_{ijk} \in \{0, 1\}$ for $i, j, k = 1, \ldots, n.$

Concerning complexity and approximation we have the following:

- Again, the NP-hardness of this problem follows from [Karp, 1972].

- Even stronger, it is proven in [Crama and Spieksma, 1992] that no polynomial time algorithm can even achieve a constant performance ratio unless P=NP. In fact, this result already holds in each of the following two cases. Let d_{ij}, d_{ik} and d_{jk} be nonnegative numbers specified for each pair of elements in different sets. One can interpret these numbers as distances (not necessarily in a metric space). The result mentioned above holds when

  – $c_{ijk} = d_{ij} + d_{ik} + d_{jk}$ for all i, j, k, and when

– $c_{ijk} = \min\{d_{ij} + d_{ik}, d_{ij} + d_{jk}, d_{ik} + d_{jk}\}$ for all i, j, k.

However, when in addition to one of these cost-structures the d's satisfy the triangle inequality a $\frac{4}{3}$ approximation algorithm exists.

- If the $3n$ points are located in the plane and distances are computed according to the Euclidean norm, [Spieksma and Woeginger, 1996] prove that minimizing the total circumference ($c_{ijk} = d_{ij} + d_{ik} + d_{jk}$ for all i, j, k) as well as minimizing the total area is NP-hard.

- Also, if $c_{ijk} = a_i \cdot b_j \cdot c_k$, for all i, j, k the problem remains NP-hard and even hard to approximate ([Burkard et al., 1996]), which contrasts with the maximization case.

2.1.3   Applications and Solution Methods.   [Pierskalla, 1967, Pierskalla, 1968] mention a number of settings in which axial 3IAPs arise: capital investment, dynamic facility location, satellite launching. Further, in [Crama et al., 1996] a situation in the assembly of printed circuit boards is described that is modeled using formulation (A3IAPMIN). [Arbib et al., 1999] describe a problem in perishable production planning formulated as an axial 3IAP with $\{0, 1\}$ coefficients.

Exact approaches for axial 3IAPs are presented in [Leue, 1972], [Hansen and Kaufman, 1973], [Burkard and Fröhlich, 1980] and in [Balas and Saltzman, 1991]. [Burkard and Rudolf, 1993] study the effect of different branching rules in a branch and bound framework. A GRASP for the axial 3IAP is presented in [Lidstrom et al., 1999].

The polytope induced by the convex hull of the feasible solutions to (A3IAPMIN) is studied in [Balas and Saltzman, 1989], [Balas and Qi, 1993], [Qi et al., 1994] and [Gwan and Qi, 1992].

2.2   THE PLANAR 3IAP

Given are 3 n-sets A_1, A_2 and A_3. For each triple in $A_1 \times A_2 \times A_3$ a number p_{ijk} is known. The problem is now to find n^2 triples such that each pair of elements from $(A_1 \times A_2) \cup (A_1 \times A_3) \cup (A_2 \times A_3)$ is in exactly one triple. So in the planar 3IAP one is asked to output n^2 triples containing each pair of indices exactly once that either maximize or minimize the sum of the p_{ijk}'s corresponding to selected triples.

$$(P3IAP) \quad \text{maximize/minimize} \quad \sum_{i=1}^{n}\sum_{j=1}^{n}\sum_{k=1}^{n} p_{ijk}x_{ijk}$$

$$\text{subject to} \quad \sum_{i=1}^{n} x_{ijk} = 1 \text{ for } j, k = 1, \ldots, n,$$

$$\sum_{j=1}^{n} x_{ijk} = 1 \text{ for } i, k = 1, \ldots, n,$$

$$\sum_{k=1}^{n} x_{ijk} = 1 \text{ for } i, j = 1, \ldots, n,$$

$$x_{ijk} \in \{0, 1\} \text{ for } i, j, k = 1, \ldots, n.$$

The planar 3IAP is proved to be NP-complete in [Frieze, 1983] even when the $p_{ijk} \in \{0, 1\}$. When the coefficients are invariant with respect to one index, that is they satisfy $p_{ijk} = p_{ijl}$ for all $k \neq l$, [Gilbert and Hofstra, 1987] show that the problem becomes polynomially solvable.

2.2.1   Applications and Solution Methods.
Planar 3IAPs find applications in timetabling (see e.g. [Hilton, 1980]). [Gilbert and Hofstra, 1987] formulate a practical rostering problem as a planar 3IAP. In [Balas and Landweer, 1983] a problem in launching satellites is modeled using a planar 3IAP with a special objective function.

Branch and bound methods for planar 3IAP are proposed by [Vlach, 1967], [Burkard and Fröhlich, 1980] and by [Magos and Miliotis, 1994]. A heuristic procedure is proposed in [Evans, 1981]. A tabu search approach is presented in [Magos, 1996].

The polytope corresponding to the convex hull of feasible solutions of (P3IAP) has been studied in [Euler et al., 1986].

3.   MIAPS

Of course, one can straightforwardly generalize the formulation (A3IAP) and (P3IAP) to higher dimensions using multiple summations. Here, however we give a more compact formulation (see also [Queyranne and Spieksma, 1999]). Observe that in case of a k index assignment problem there are $k - 1$ variants possible (cf. the 3 index case). Thus, in order to specify a k index assignment problem another parameter, say q, needs to be given. We refer to the resulting problem as a q-fold kIAP.

Given are k n-sets $A_1, A_2, \ldots, A_k$, and let $A = \bigotimes_{i=1}^{k} A_i = A_1 \times A_2 \times \ldots \times A_k$. Thus, A is the set of all k-tuples $a = (a(1), a(2), \ldots, a(k)) \in A$.

There is a variable x_a for each $a \in A$. Given costs c_a for all $a \in A$ we can formulate the objective function as minimize $\sum_{a \in A} c_a x_a$.

Given an integer q, $1 \leq q \leq k - 1$, the constraints of a q-fold k index assignment problem are defined as follows. Let Q be the set of all subsets of $\{1, 2, \ldots, k\}$ consisting of $k - q$ elements. An element F of this set Q corresponds to a set of "fixed" indices. Given such an F, let $A_F = \otimes_{f \in F} A_f$. Further, given some $g \in A_F$, let $A(F, g) = \{a \in A \mid a(f) = g(f) \text{ for all } f \in F\}$ be the set of k-tuples that coincide with g on the fixed indices.

$$(q\text{-fold } kIAP) \quad \text{max/min} \quad \sum_{a \in A} c_a x_a$$

$$\text{subject to} \quad \sum_{a \in A(F,g)} x_a = 1 \text{ for all } g \in A_F, F \in Q,$$

$$x_a \in \{0, 1\} \text{ for all } a \in A.$$

Observe that the case $q = 1$ corresponds to a planar k index assignment problem, whereas the case $q = k - 1$ corresponds to an axial problem. Some of the approaches mentioned for the 3 index case can be generalized to the k index case:

- The results in [Hausmann et al., 1980] imply a $\frac{1}{k}$ approximation algorithm for the axial k index assignment problem (when maximizing).

- The approach in [Hurkens and Schrijver, 1989] implies a $\frac{2}{k} - \epsilon$ approximation algorithm for the axial k index assignment problem with $\{0, 1\}$ cost coefficients (when maximizing).

Further, [Kuipers, 1990] proves for the axial kIAP with $\{0, 1\}$ coefficients (when maximizing) that the ratio between the value of the LP-relaxation and an optimal solution is bounded by $k - 1$, $k \geq 2$.

An important special case of the (min) axial multi index assignment problem is the case where the cost coefficients are so-called *decomposable* (see [Bandelt et al., 1994]). Decomposable costs arise when a number d_{st} is specified for each pair of elements $(s, t) \in A_i \times A_j$, $(i, j = 1, \ldots, k)$ and a function h exists such that the costs of a k-tuple depend on the d's involved in the k-tuple. More formally, let $c_a = h(d_{a(1),a(2)}, d_{a(1),a(3)}, \ldots, d_{a(k-1),a(k)})$ for all $a \in A$. Different possibilities exist for specifying h; popular examples are sum costs, tour costs, diameter costs and others (see [Bandelt et al., 1994]). Depending upon the specific form of the function h there are results in approximation: for instance there is a $2 - \frac{2}{k}$ approximation algorithm for sum costs (see [Bandelt et al., 1994]).

3.0.2   Applications and Solution Methods. An important application that is modeled using an axial multi index assignment problem occurs in the field of target tracking and plays a central role in operating surveillance systems. A description is as follows. Consider a radar system used to monitor some area. At each of discrete time-units $t_1, t_2, \ldots$ a set of *measurements* (called a *scan*) induced by targets becomes available. For simplicity it is usually assumed that each target moves in a straight line in the plane with a constant velocity (but other movement patterns are conceivable). The problem is now to associate the measurements with the targets, or in other words to identify *tracks*, each consisting of a series of measurements (at most 1 from each scan) caused by the same target. Such a track reflects the trajectory of the target. This so-called *data-association* problem can be naturally formulated as an axial multi index assignment problem by defining the sets A_i as the set of measurements in scan i. The cost-coefficients then express how likely it is that a particular set of measurement (one from each scan) are caused by the same target.

Solution methods for this application based on Lagrangian relaxation are proposed in [Pattipatti et al., 1992], [Poore, 1994], [Poore and Rijavec, 1993]. Other approaches are described in [Murphey et al., 1998].

An application concerning routing in meshes that can be formulated as an axial MIAP is described in [Fortin and Tusera, 1994].

4.   EXTENSIONS

As in the 2-dimensional case, an important extension of multi index assignment problems is to multi index *transportation* problems. Observe that in the formulations considered so far all right-hand sides are 1. By relaxing this property (and by allowing sets A_i of possibly unequal size) a more general problem arises which we call multi index (integer) transportation problems (depending upon whether integrality requirements are present). See [Queyranne and Spieksma, 1999] for an overview of these problems.

MIAPs as discussed in this chapter can be extended in various other ways:

- The bottleneck case. Axial 3IAPs with a bottleneck objective are considered in [Malhotra et al., 1985] and [Geetha and Vartak, 1994]. Special cases of axial 3IAPs with a bottleneck objective function are considered in [Klinz and Woeginger, 1996].

- Multi criteria. An solution method for an axial 3IAP with two criteria can be found in [Geetha and Vartak, 1989].

References

[Arbib et al., 1999] Arbib, C., Pacciarelli, D., and Smriglio, S. (1999). A three-dimensional matching model for perishable production scheduling. *Discrete*

Applied Mathematics, 92:1–15.

[Arkin and Hassin, 1998] Arkin, E. and Hassin, R. (1998). On local search for weighted packing problems. *Mathematics of Operations Research*, 23:640–648.

[Balas and Landweer, 1983] Balas, E. and Landweer, P. (1983). Traffic assignment in communication satellites. *Operations Research Letters*, 2:141–147.

[Balas and Qi, 1993] Balas, E. and Qi, L. (1993). Linear time separation algorithms for the three-index assignment polytope. *Discrete Applied Mathematics*, 43:1–12.

[Balas and Saltzman, 1989] Balas, E. and Saltzman, M. (1989). Facets of the three-index assignment polytope. *Discrete Applied Mathematics*, 23:201–229.

[Balas and Saltzman, 1991] Balas, E. and Saltzman, M. (1991). An algorithm for the three-index assignment problem. *Operations Research*, 39:150–161.

[Bandelt et al., 1994] Bandelt, H.J., Crama, Y., and Spieksma, F. (1994). Approximation algorithms for multidimensional assignment problems with decomposable costs. *Discrete Applied Mathematics*, 49:25–50.

[Barvinok et al., 1998] Barvinok, A., Johnson, D., Woeginger, G., and Woodroofe, R. (1998). Finding maximum length tours under polyhedral norms. In *Report Woe-20*. TU Graz.

[Burkard and Fröhlich, 1980] Burkard, R. and Fröhlich, K. (1980). Some remarks on 3-dimensional assignment problems. *Methods of Operations Research*, 36:31–36.

[Burkard and Rudolf, 1993] Burkard, R. and Rudolf, R. (1993). Computational investigations on 3-dimensional axial assignment problems. *Belgian Journal of Operations Research, Statistics and Computer Science*, 32:85–98.

[Burkard et al., 1996] Burkard, R., Rudolf, R., and Woeginger, G. (1996). Three-dimensional axial assignment problems with decomposable cost-coefficients. *Discrete Applied Mathematics*, 65:123–140.

[Crama et al., 1996] Crama, Y., Oerlemans, A., and Spieksma, F. (1996). *Production planning in automated manufacturing*. Springer Verlag.

[Crama and Spieksma, 1992] Crama, Y. and Spieksma, F. (1992). Approximation algorithms for three-dimensional assignment problems with triangle inequalities. *European Journal of Operational Research*, 60:273–279.

[Crescenzi and Kann, 1999] Crescenzi, P. and Kann, V. (1999). A compendium of $\mathcal{NP}$ optimization problems. http://www.nada.kth.se/~viggo/wwwcompendium/.

[Dyer and Frieze, 1983] Dyer, M. and Frieze, A. (1983). Planar 3dm is np-complete. *Journal of Algorithms*, 7:174–184.

[Euler et al., 1986] Euler, R., Burkard, R., and Grommes, R. (1986). On latin squares and the facial structure of related polytopes. *Discrete Mathematics*, 62:155–181.

[Evans, 1981] Evans, J. (1981). The multicommodity assignment problem: a network aggregation heuristic. *Computers and Mathematics with Applications*, 7:187–194.

[Fortin and Tusera, 1994] Fortin, D. and Tusera, A. (1994). Routing in meshes using linear assignment. In A. Bachem, U. Derigs, M. J. and Schrader, R., editors, *Operations Research '93*, pages 169–171.

[Frieze, 1983] Frieze, A. (1983). Complexity of a 3-dimensional assignment problem. *European Journal of Operational Research*, 13:161–164.

[Geetha and Vartak, 1989] Geetha, S. and Vartak, M. (1989). Time-cost trade-off in a three dimensional assignment problem. *European Journal of Operational Research*, 38:255–258.

[Geetha and Vartak, 1994] Geetha, S. and Vartak, M. (1994). The three-dimensional bottleneck assignment problem with capacity constraints. *European Journal of Operational Research*, 73:562–568.

[Gilbert and Hofstra, 1987] Gilbert, K. and Hofstra, R. (1987). An algorithm for a class of three-dimensional assignment problems arising in scheduling applications. *IIE Transactions*, 8:29–33.

[Gilbert and Hofstra, 1988] Gilbert, K. and Hofstra, R. (1988). Multidimensional assignment problems. *Decision Sciences*, 19:306–321.

[Gwan and Qi, 1992] Gwan, G. and Qi, L. (1992). On facet of the three index assignment polytope. *Australasian Journal of Combinatorics*, 6:67–87.

[Hansen and Kaufman, 1973] Hansen, P. and Kaufman, L. (1973). A primal-dual algorithm for the three-dimensional assignment problem. *Cahiers du Centre d'Etudes de Recherche Opérationnell*, 15:327–336.

[Hausmann et al., 1980] Hausmann, D., Korte, B., and Jenkyns, T. (1980). Worst case analysis of greedy type algorithms for independence systems. *Mathematical Programming*, 12:120–131.

[Hilton, 1980] Hilton, A. (1980). The reconstruction of latin squares with applications to school timetabling and to experimental design. *Mathematical Programming Study*, 13:68–77.

[Hurkens and Schrijver, 1989] Hurkens, C. and Schrijver, A. (1989). On the size of systems of sets every t of which have an sdr, with an application to the worst-case ratio of heuristics for packing problems. *SIAM Journal on Discrete Mathematics*, 2:68–72.

[Kann, 1991] Kann, V. (1991). Maximum bounded 3-dimensional matching is max snp-complete. *Information Processing Letters*, 37:27–35.

[Karp, 1972] Karp, R. (1972). Reducibility among combinatorial problems. In Miller, R. and Thatcher, J., editors, *Complexity of Computer Computations*, pages 85–103. Plenum Press.

[Klinz and Woeginger, 1996] Klinz, B. and Woeginger, G. (1996). A new efficiently solvable case of the three-dimensional axial bottleneck assignment problem. *Lecture Notes in Computer Science*, 1120:150–162.

[Koopmans and Beckmann, 1957] Koopmans, T. and Beckmann, M. (1957). Assignment problems and the location of economic activities. *Econometrica*, 25:53–76.

[Kuipers, 1990] Kuipers, J. (1990). On the LP-relaxation of multi-dimensional assignment problems with applications to assignment games. In *Report M 90-10*. Maastricht University.

[Leue, 1972] Leue, O. (1972). Methoden zur lösung dreidimensionaler zuordnungsprobleme. *Angewandte Informatik*, 14:154–162.

[Lidstrom et al., 1999] Lidstrom, N., Pardalos, P., Pitsoulis, L., and Toraldo, G. (1999). An approximation algorithm for the three-index assignment problem. In Floudas, C. and Pardalos, P., editors, *Encyclopedia of Optimization (to appear)*.

[Magos, 1996] Magos, D. (1996). Tabu search for the planar three-index assignment problem. *Journal of Global Optimization*, 8:35–48.

[Magos and Miliotis, 1994] Magos, D. and Miliotis, P. (1994). An algorithm for the planar three-index assignment problem. *European Journal of Operational Research*, 77:141–153.

[Malhotra et al., 1985] Malhotra, R., Bhatia, H., and Puri, M. (1985). The three dimensional bottleneck assignment problem and its variants. *Optimization*, 16:245–256.

[Motzkin, 1952] Motzkin, T. (1952). The multi-index transportation problem. *Bulletin of the American Mathematical Society*, 58:494.

[Murphey et al., 1998] Murphey, R., Pardalos, P., and Pitsoulis, L. (1998). A parallel GRASP for the data association multidimensional assignment problem. In *IMA Volumes in Mathematics and its Applications, Vol. 106*, pages 159–180. Springer-Verlag.

[Nishizeki and Chiba, 1988] Nishizeki, T. and Chiba, N. (1988). Planar graphs: Theory and algorithms. In *Volume 32 of Annals of Discrete Mathematics*. Elsevier, Amsterdam.

[Papadimitriou, 1994] Papadimitriou, C. (1994). *Computational Complexity*. Addison-Wesley.

[Pattipatti et al., 1992] Pattipatti, K., Deb, S., Bar-Shalom, Y., and R.B. Washburn, J. (1992). A new relaxation algorithm passive sensor data association. *IEEE Transactions on Automatic Control*, 37:198–213.

[Pierskalla, 1967] Pierskalla, W. (1967). The tri-substitution method for the three-dimensional assignment problem. *Journal of the Canadian Operational Research Society*, 5:71–81.

[Pierskalla, 1968] Pierskalla, W. (1968). The multidimensional assignment problem. *Operations Research*, 16:422–431.

[Poore, 1994] Poore, A. (1994). Multidimensional assignment formulation of data-association problems arising from multitarget and multisensor tracking. *Computational Optimization and Applications*, 3:27–57.

[Poore and Rijavec, 1993] Poore, A. and Rijavec, N. (1993). A lagrangian relaxation algorithm for multidimensional assignment problems arising from multitarget tracking. *SIAM Journal on Optimization*, 3:544–563.

[Qi et al., 1994] Qi, L., Balas, E., and Gwan, G. (1994). A new facet class and a polyhedral method for the three-index assignment problem. In Du, D., editor, *Advances in Optimization*, pages 256–274.

[Queyranne and Spieksma, 1999] Queyranne, M. and Spieksma, F. (1999). Multi-index transportation problems. In Floudas, C. and Pardalos, P., editors, *Encyclopedia of Optimization (to appear)*.

[Schell, 1955] Schell, E. (1955). Distribution of a product by several properties. In of Management Analysis, D., editor, *Second Symposium in Linear Programming 2*, pages 615–642. DCS/Comptroller HQ, US Air Force, Washington DC.

[Spieksma and Woeginger, 1996] Spieksma, F. and Woeginger, G. (1996). Geometric three-dimensional assignment problems. *European Journal of Operational Research*, 91:611–618.

[Vlach, 1967] Vlach, M. (1967). Branch and bound method for the three-index assignment problem. *Ekonomicko-Matematicky Obzor*, 3:181–191.

Chapter 2

MULTIDIMENSIONAL ASSIGNMENT PROBLEMS ARISING IN MULTITARGET AND MULTISENSOR TRACKING

Aubrey B. Poore *

Department of Mathematics
Colorado State University
Fort Collins, Colorado 80523
poore@math.colostate.edu

Abstract   The ever-increasing demand in surveillance is to produce highly accurate target and track identification and estimation in real-time, even for dense target scenarios and in regions of high track contention. The use of multiple sensors, through more varied information, has the potential to greatly enhance target identification and state estimation. For multitarget tracking, the processing of multiple scans all at once yields high track identification. However, to achieve this accurate state estimation and track identification, one must solve an NP-hard data association problem of partitioning observations into tracks and false alarms in real-time. Over the last ten years a new approach to the problem formulation based on multidimensional assignment problems and near optimal solution in real-time by Lagrangian relaxation has evolved and is proving to be superior to all other approaches. This work reviews the problem formulation and algorithms with some suggested future directions.

1.   INTRODUCTION

The ever-increasing demand in surveillance is to produce highly accurate target identification and estimation in real-time, even for dense target scenarios and in regions of high track contention. Past surveillance sensor systems have relied on individual sensors to solve this problem; however, current and

---

*This work was partially supported by the Air Force Office of Scientific Research through AFOSR Grant Number F49620-97-1-0273 and by the Office of Naval Research through ONR Grant Number N00014-99-1-0118.

P.M. Pardalos and L.S. Pitsoulis (eds.), Nonlinear Assignment Problems, 13–38.

future needs far exceed single sensor capabilities. The use of multiple sensors, through more varied information, has the *potential* to greatly improve state estimation and track identification. Fusion of information from multiple sensors is part of a much broader subject called data or information fusion, which for surveillance applications is defined as "a multilevel, multifaceted process dealing with the detection, association, correlation, estimation and combination of data and information from multiple sources to achieve refined state and identity estimation, and complete and timely assessments of situation and threat" [Waltz and Llinas, 1990]. (A comprehensive discussion can be found in the book of Waltz and Llinas [Waltz and Llinas, 1990].) Level 1 deals with single and multi-source information involving tracking, correlation, alignment, and association by sampling the external environment with multiple sensors and exploiting other available sources. Numerical processes thus dominate Level 1; symbolic reasoning involving various techniques from artificial intelligence permeate Levels 2 and 3.

Given this level of fusion, the particular fusion architecture of interest in this work is called centralized in that all reports are sent to a central processing unit. For this architecture, current data association methods for multitarget tracking generally fall into two categories: sequential and deferred logic. Methods for the former include nearest neighbor, one-to-one or few-to-one assignments, and all-to-one assignments as in the joint probabilistic data association (JPDA) [Bar-Shalom and Li, 1995]. For track maintenance, the nearest neighbor method is valid in the absence of clutter when there is little or no track contention, i.e., when there is little chance of misassociation. Problems involving one-to-one or few-to-one assignments are generally formulated as (two dimensional) assignment or multi-assignment problems for which there are some excellent optimal algorithms [Bertsekas, 1998, Jonker and Volgenant, 1987]. This methodology is real-time but can result in a large number of partial and incorrect assignments in dense or high contention scenarios, and thus incorrect track identification. The difficulty is that decisions, once made, are irrevocable, so that there is no mechanism to correct misassociations. The use of all observations in a scan or within a neighborhood of a predicted track position (e.g., JPDA) to update a track moderates the misassociation problem and has been highly successful for tracking a few targets in dense clutter [Bar-Shalom and Li, 1995].

Deferred logic techniques consider several data sets or frames of data all at once in making data association decisions. At one extreme is batch processing in which all observations (from all time) are processed together, but this is computationally too intensive for real-time applications. The other extreme is sequential processing. Deferred logic methods between these two extremes are of primary interest in this work. The most popular deferred logic method used to track large numbers of targets in low to moderate clutter is called multiple hypothesis tracking (MHT) in which one builds a tree of possibilities, assigns a likelihood score to each track, develops an intricate pruning logic, and then solves the data association problem by an explicit enumeration scheme. The use of these enumeration schemes to solve this NP-hard combinatorial optimization problem in real-time is inevitably faulty in dense scenarios since the time required to solve the problem optimally can grow exponentially with the size of the problem.

Over the last ten years a new formulation and class of algorithms for data association have shown itself to be superior to all other deferred logic methods [Poore and Rijavec, 1991b, Poore and Rijavec, 1993, Poore and A.J. Robertson, 1997, Shea and Poore, 1998, Poore and Yan, 1999, Barker et al., 1995, Poore, 1996, Poore, 1999, Blackman and Popoli, 1999]. This formulation is based on multidimensional assignment problems and the algorithms, on Lagrangian relaxation. We stress however that the use of combinatorial optimization in multitarget tracking is not new and dates back to the pioneering work of Morefield [Morefield, 1977] who used integer programming to solve set packing and covering problems arising from a data association problem. Multiple hypothesis tracking has been popularized by the fundamental work of Reid [Reid, 1996]. These works are further discussed in the books of Blackman and Popoli [Blackman and Popoli, 1999], Bar-Shalom and Li [Bar-Shalom and Li, 1995], and Waltz and Llinas [Waltz and Llinas, 1990], which also serve as excellent introductions to the field of multitarget tracking and multisensor data fusion. Bar-Shalom, Deb, Kirubarajan, and Pattipati [Deb et al., 1993, Deb et al., 1997, Kirubarajan et al., 1997, Kirubarajan et al., 1998, Popp et al., 1998, Chummun et al., 1999] have also formulated sensor fusion problems in terms of these multidimensional assignment problems and have developed algorithms as discussed in Section 4.

The performance of any tracking system is dependent on a large number of components and having one component superior to all others does not guarantee a superior tracking system. To address some of these other issues, we give a brief overview of the many issues in the design of a tracking system and a placement of the problem within the context of the more general surveillance and fusion problem in Section 2. The formulation of the problem is presented in Section 3 and an overview of the Lagrangian relaxation based methods in Section 4. The concluding Section 5 contains a summary of some of the problems that should be investigated in the future.

2. PROBLEM BACKGROUND

A general surveillance problem involves the use of multiple platforms such as ships, planes, or stationary ground based radar systems, on which there are located one or more sensors to track multiple objects. The term target is used instead of the term object since the objects being tracked generally do not identify themselves. As one might imagine, optimization problems permeate this field of surveillance, particularly in the collection and fusion of information. First, there are the problems of routing and scheduling surveillance platforms and then dynamically retasking these platforms as more information becomes available. For each platform, one must allocate and manage the scarce sensor resources in order to maximize the information returned to each platform. The second area of information fusion is the subject of this work.

There are many issues in the design of a fusion system for multiple surveillance platforms such as fusion architectures, communication links between sensor platforms, misalignment problems, tracking coordinate systems, motion models, likelihood ratios, filtering and estimation, and the data association problem of partitioning reports into tracks and false alarms. The recent book by Blackman and Popoli [Blackman and Popoli, 1999] presents an excellent overview of these topics along with an extensive list of references.

The targets under consideration are classified as point or small targets and measurements or observations of these targets are in the form of kinematic information such as range, azimuth, elevation, and range rate. Future sensor systems will provide additional feature or attribute information [Blackman and Popoli, 1999].

A question that generally arises is that of the difference between the air traffic control problem and the surveillance problem. In the former, planes, through their beacon codes, generally identify themselves so that observations are associated with the correct plane. For the surveillance problem, the objects being tracked do not identify themselves, so that one must derive a figure of merit for the association of a sequence of observations to a particular target. In this work, the aforementioned likelihood ratios are used for this purpose.

One aspect of this problem that is seldom discussed in the literature is the collection of data structures that put all of this information together efficiently. In reality, we generally consider a tracking system to be composed of a dynamic search tree which organizes this information and recycles memory for real-time processing. However, *the central problem is the data association problem.*

To place the current data association problem within the context of the different architectures of multiplatform tracking, a brief review of the architectures is helpful. The first architecture is *centralized fusion* in which raw observations are sent from the multiple platforms to a central processing unit where they can be combined to give superior state estimation [Blackman and Popoli, 1999] (compared to sensor level fusion). At the other extreme is *track fusion*, wherein each sensor forms tracks along with the corresponding statistical information from its own reports and then sends this preprocessed information to a processing unit where one correlates the tracks. Once the correlation is complete, one then combines the tracks with appropriate modification in the statistics. In reality, many sensor systems are *hybrids* of these two architectures in which one sends some preprocessed data and some raw data with possibly a switch between these two. A discussion of the advantages and disadvantages of these architectures is presented in the book of Blackman and Popoli [Blackman and Popoli, 1999]. The current data association problem is most applicable to the centralized or hybrid architecture.

3. ASSIGNMENT FORMULATION OF SOME GENERAL DATA ASSOCIATION PROBLEMS

The goal of this section is to formulate the data association problem for a large class of multiple target tracking and sensor fusion problems as a multidimensional assignment problem. This development extracts the salient features and assumptions that occur in a large class of these problems and is a brief update to earlier work [Poore, 1994]. A general class of data association problems was posed as set packing problems by Morefield [Morefield, 1977] in 1977. We begin with an abstracted view of Morefield's work to include set coverings, packings, and partitionings and then proceed to a formulation of the assignment problem.

In tracking, a common surveillance problem is to estimate the past, current, or future state of a collection of targets (e.g., airplanes in the air, ships on the sea, or cars and trucks on the ground) from a sequence of observations made of the surveillance region by one or more sensors. This work addresses specifically what are called "small" targets [Drummond, 1999] for which the sensors generally supply kinematic information such as range, azimuth, elevation, range rate and some limited attribute or feature information.

Suppose now that one or more sensors, either co-located or distributed, survey the surveillance region and produce a stream of observations (or measurements), each with a distinct time tag. These observations are then arranged into sets of observations called frames of data. Mathematically, let the $Z(k) = \{z_{i_k}^k\}_{i_k=1}^{M_k}$ denote the kth frame of data where each $z_{i_k}^k$ is a vector of noise contaminated observations with an associated time tag $t_{i_k}^k$. The index k represents the frame number and i_k, the i_k^{th} observation in frame k. An observation in the frame of data $Z(k)$ may be a false report or may emanate from a true target.

In this work, we ultimately assume that each frame of data is a "proper frame" in which each target is seen no more than once. For a rotating radar, one sweep or scan of the field of view generally constitutes a proper frame. For sensors such as electronically scanning phased array radar wherein the sensor switches from surveillance to tracking mode, the partitioning of the data into proper frames of data is more interesting with several choices. More efficient methods will be addressed in forthcoming work.

Given this data, the data association problem to be solved is to correctly partition the data into observations emanating from targets and false alarms.

The combinatorial optimization problem that governs a large number of data association problems in multitarget tracking and multisensor data fusion [Poore and Rijavec, 1991b, Bar-Shalom and Li, 1995, Deb et al., 1993, Blackman and Popoli, 1999, Morefield, 1977, Barker et al., 1995, Poore, 1996, Poore, 1999, Kirubarajan et al., 1997, Kirubarajan et al., 1998, Popp et al., 1998, Chummun et al., 1999, Sittler, 1964, Waltz and Llinas, 1990] is generally posed as

$$\text{Maximize} \left\{ P \left(\Gamma = \gamma \mid Z^N \right) \mid \gamma \in \Gamma^* \right\}, \tag{2.1}$$

where Z^N represents N data sets (2.2), γ is a partition of indices of the data ((2.3) and (2.4a)-(2.4d)), Γ^* is the finite collection of all such partitions (2.4a)-(2.4d), Γ is a discrete random element defined on Γ^*, γ^0 is a reference partition, and $P(\Gamma = \gamma \mid Z^N)$ is the posterior probability of a partition γ being true given the data Z^N. Each of these terms must be defined. Having done so, the objective will be to formulate a reasonably general class of these data association problems (2.1) as multidimensional assignment problems (2.15).

In the above surveillance example, the data sets were observations of the objects in the surveillance region, including false reports. To allow for more general types of data such as tracks and track-observation combinations as well as observations, we follow [Reid, 1996], who used the term *reports* to include each of these. Thus, let $Z(k)$ denote a data set of M_k reports $\{z_{i_k}^k\}_{i_k=1}^{M_k}$ and let Z^N denote the cumulative data set of N such sets defined by

$$Z(k) = \left\{ z_{i_k}^k \right\}_{i_k=1}^{M_k} \quad \text{and} \quad Z^N = \{Z(1), \ldots, Z(N)\}, \tag{2.2}$$

respectively. In multisensor data fusion and multitarget tracking the data sets $Z(k)$ may represent different classes of objects. For track initiation in multitarget tracking the objects are observations that must be partitioned into tracks and false alarms. In our formulation of track maintenance (Section 3.3) one data set will be tracks and remaining data sets will be observations which are assigned to existing tracks, as false observations, or to initiating tracks. In sensor level tracking, the objects to be fused are tracks [Waltz and Llinas, 1990]. In centralized fusion [Blackman and Popoli, 1999, Waltz and Llinas, 1990], the objects may all be observations that represent targets or false reports, and the problem is to determine which observations emanate from a common source.

The next task is to define what is meant by a partition of the cumulative data set Z^N in (2.2). Since this definition is to be independent of the actual data in the cumulative data set Z^N, we first define a partition of the indices in Z^N. Let

$$I^N = \{I(1), I(2), \ldots, I(N)\}, \quad \text{where } I(k) = \{i_k\}_{i_k=1}^{M_k} \tag{2.3}$$

denotes the indices in the data sets (2.2). A partition γ of I^N and the collection of all such partitions Γ^* is defined by

$$\gamma = \{\gamma_1, \ldots, \gamma_{n(\gamma)} \mid \gamma_i \neq \emptyset \text{ for each } i\}, \tag{2.4a}$$
$$\gamma_i \cap \gamma_j = \emptyset \text{ for } i \neq j, \tag{2.4b}$$
$$I^N = \cup_{j=1}^{n(\gamma)} \gamma_j, \tag{2.4c}$$
$$\Gamma^* = \{\gamma \mid \gamma \text{ satisfies } (2.4a) - (2.4c)\}. \tag{2.4d}$$

Here, $\gamma_i \subset I^N$ in (2.4a) will be called a track, so that $n(\gamma)$ denotes the number of tracks (or elements) in the partition γ. A $\gamma \in \Gamma^*$ is called a *set partitioning* of the indices I^N if properties (2.4a) - (2.4c) are valid, a *set covering* of I^N if property (2.4b) is omitted but the other two properties (2.4a) and (2.4c) are retained, and a *set packing* if (2.4c) is omitted but (2.4a) and (2.4b) are retained [Nemhauser and Wolsey, 1988]. A partition $\gamma \in \Gamma^*$ of the index set I^N induces a partition of the data Z^N via

$$Z_\gamma = \{Z_{\gamma_1}, \ldots, Z_{\gamma_{n(\gamma)}}\} \text{ where } Z_{\gamma_i} \subset Z^N. \tag{2.5}$$

Clearly, $Z_{\gamma_i} \cap Z_{\gamma_j} = \emptyset$ for $i \neq j$ and $Z^N = \cup_{j=1}^{n(\gamma)} Z_{\gamma_j}$. Each Z_{γ_i} will be called a track of data. Note that a Z_{γ_i} need not have observations from each frame of data $Z(k)$, but it must by definition have at least one observation.

Under several independence assumptions between tracks [Poore, 1994], one can establish a probabilistic framework in which

$$P(\Gamma = \gamma \mid Z^N) = \frac{C}{p(Z^N)} \prod_{\gamma_i \in \gamma} p(Z_{\gamma_i}) G(\gamma_i), \tag{2.6}$$

where C is a constant. This completes the formulation of the general data association problem as presented in the work of Poore [Poore, 1994] and Morefield [Morefield, 1977].

The next objective is to refine this formulation in a way that is amenable to the assignment problem. For notational convenience in representing tracks, we add a *zero index* to each of the index sets $I(k)$ ($k = 1, \ldots, N$) in (2.3), a *dummy report* z_0^k to each of the data sets $Z(k)$ in (2.2), and require that each

$$\gamma_i = (i_1, \ldots, i_N), \tag{2.7}$$
$$Z_{\gamma_i} = Z_{i_1 \cdots i_n} \equiv (z_{i_1}^1, \ldots, z_{i_N}^N),$$

where i_k and $z_{i_k}^k$ can now assume the values of 0 and z_0^k, respectively. The dummy report z_0^k serves several purposes in the representation of missing data, false reports, initiating tracks, and terminating tracks. If Z_{γ_i} is missing an actual report from the data set $Z(k)$, then $\gamma_i = (i_1, \ldots, i_{k-1}, 0, i_{k+1}, \ldots, i_N)$

and $Z_{\gamma_i} = \{z_{i_1}^1, \ldots, z_{i_{k-1}}^{k-1}, z_0^k, z_{i_{k+1}}^{k+1}, \ldots, z_{i_N}^N\}$. A false report $z_{i_k}^k$ ($i_k > 0$) is represented by $\gamma_i = (0, \ldots, 0, i_k, 0, \ldots, 0)$ and $Z_{\gamma_i} = \{z_0^1, \ldots, z_0^{k-1}, z_{i_k}^k, z_0^{k+1}, \ldots, z_0^N\}$ in which there is but one actual report. The partition γ^0 of the data in which all reports are declared to be false reports is defined by

$$Z_{\gamma^0} = \left\{ Z_{0 \cdots 0 i_k 0 \cdots 0} \equiv \left(z_0^1, \ldots, z_0^{k-1}, z_{i_k}^k, z_0^{k+1}, \ldots, z_0^N\right) \right. \tag{2.8}$$
$$\left. \mid i_k = 1, \ldots, M_k; k = 1, \ldots, N \right\}.$$

If each data set $Z(k)$ represents a "proper frame" of observations, a track that initiates on frame $m > 1$ will contain only the dummy report z_0^k from each of the data sets $Z(k)$ for each $k = 1, \ldots, m-1$. Likewise, a track that terminates on frame m would have only the dummy report from each of the data sets for $k > m$. These representations are discussed further in Section 3 for both track initiation and track maintenance.

The use of the 0-1 variable

$$z_{i_1 \cdots i_N} = \begin{cases} 1 & \text{if } (i_1, \ldots, i_N) \in \gamma, \\ 0 & \text{otherwise,} \end{cases} \tag{2.9}$$

yields an equivalent characterization of a partition (2.4a)-(2.4d) and (2.7) as a solution of the equations

$$\sum_{i_1=0}^{M_1} \cdots \sum_{i_{k-1}=0}^{M_{k-1}} \sum_{i_{k+1}=0}^{M_{k+1}} \cdots \sum_{i_N=0}^{M_N} z_{i_1 \cdots i_N} = 1 \tag{2.10}$$
$$\text{for } i_k = 1, \ldots, M_k \text{ and } k = 1, \ldots, N.$$

With this characterization of a partition of the cumulative data set Z^N as a set of equality constraints (2.10), we now proceed to the formulation of the multidimensional assignment problem.

Observe that for $\gamma_i = (i_1 \cdots i_n)$ as in (2.7) and the reference partition (2.8),

$$\frac{P(\Gamma = \gamma \mid Z^N)}{P(\Gamma = \gamma^0 \mid Z^N)} \equiv L_\gamma \equiv \prod_{(i_1 \cdots i_N) \in \gamma} L_{i_1 \cdots i_N} \tag{2.11}$$

where

$$L_{i_1 \cdots i_N} = \frac{p(Z_{i_1 \cdots i_N}) G(Z_{i_1 \cdots i_N})}{\prod_{k=1, i_k \neq 0}^N p(Z_{0 \cdots 0 i_k 0 \cdots 0}) G(Z_{0 \cdots 0 i_k 0 \cdots 0})}. \tag{2.12}$$

Here the index i_k in the denominator corresponds to the kth index of $Z_{i_1\cdots i_N}$ in the numerator. Next define

$$c_{i_1\cdots i_N} = -\ln L_{i_1\cdots i_N},\qquad(2.13)$$

so that

$$-\ln\left[\frac{P(\gamma|Z^N)}{P(\gamma^0|Z^N)}\right] = \sum_{(i_1,\ldots,i_N)\in\gamma} c_{i_1\cdots i_N}.\qquad(2.14)$$

Thus in view of the characterization of a partition (2.4a)-(2.4d) and (2.5) specialized by (2.7) as a solution of the equation (2.10), the independence assumptions [Poore, 1994] and the expansion (2.6), problem (2.1) is equivalently characterized as the following N-dimensional assignment problem:

$$\text{Minimize}\quad \sum_{i_1=0}^{M_1}\cdots\sum_{i_N=0}^{M_N} c_{i_1\cdots i_N} z_{i_1\cdots i_N}\qquad(2.15)$$

$$\text{Subject To:}\quad \sum_{i_2=0}^{M_2}\cdots\sum_{i_N=0}^{M_N} z_{i_1\cdots i_N} = 1,\ \ i_1 = 1,\ldots,M_1,\qquad(2.16)$$

$$\sum_{i_1=0}^{M_1}\cdots\sum_{i_{k-1}=0}^{M_{k-1}}\sum_{i_{k+1}=0}^{M_{k+1}}\cdots\sum_{i_N=0}^{M_N} z_{i_1\cdots i_N} = 1,\qquad(2.17)$$

$$\text{for}\ \ i_k = 1,\ldots,M_k\ \text{and}\ k = 2,\ldots,N-1,$$

$$\sum_{i_1=0}^{M_1}\cdots\sum_{i_{N-1}=0}^{M_{N-1}} z_{i_1\cdots i_N} = 1,\ \ i_N = 1,\ldots,M_N,\qquad(2.18)$$

$$z_{i_1\cdots i_N} \in \{0,1\}\ \text{for all}\ i_1,\ldots,i_N,\qquad(2.19)$$

where $c_{0\cdots 0}$ is arbitrarily defined to be zero. Note that the definition of a partition (2.4a)-(2.4d) and (2.7) and the 0-1 variable $z_{i_1\cdots i_N}$ in (2.9) imply $z_{0\cdots 0} = 0$. (If $z_{0\cdots 0}$ is not preassigned to zero and $c_{0\cdots 0}$ is defined arbitrarily, then $z_{0\cdots 0}$ is determined directly from the value of $c_{0\cdots 0}$, since it does not enter the constraints other than being a zero-one variable.) Also, each cost coefficient with exactly one nonzero index is zero (i.e., $c_{0\cdots 0 i_k 0\cdots 0} = 0$ for all $i_k = 1,\ldots,M_k$ and $k = 1,\ldots,N$) due to use of the normalizing partition γ^0 in the likelihood ratio in (2.1) and (2.12). It is entirely possible to derive the same problem formulation not assuming that the cost coefficients with exactly one nonzero index are zero [Poore, 1994]; however, these other formulations can be reduced to the one above using the following invariance theorem.

Having completed the derivation of the assignment problem (2.15), several remarks are in order. The definition of a partition in (2.4a)-(2.4d) and (2.5) implies that each actual report belongs to at most one track of reports Z_{γ_i} in a partition Z_γ of the cumulative data set. One can modify this to allow multi-assignments of one, some, or all the actual reports. The assignment problem (2.15) is changed accordingly. For example, if $z_{i_k}^k$ is to be assigned no more than, exactly, or no less than $n_{i_k}^k$ times, then the " $= 1$" in the constraint (2.15) is changed to " $\leq, =, \geq n_{i_k}^k$," respectively. (This allows both set coverings and packings in the formulation.) In making these changes, one must pay careful attention to the independence assumptions [Poore, 1994]. Such problems with inequality constraints arise in multiresolution problems in sensor fusion as well as for the problem of unresolved closely spaced objects.

The likelihood ratio $L_{i_1\cdots i_N}$ is a complicated expression containing probabilities for detection, termination, model maneuvers as well as the density functions for and expected number of false alarms and initiating targets. Also included is the likelihood that an observation arises from a particular target, which requires that one estimates the target dynamics through the corresponding sequence of observations $\{z_{i_1}^1, \ldots, z_{i_k}^k, \ldots, z_{i_N}^N\}$. Filtering such sequences is the most time consuming part of the problem formulation and far exceeds the time to solve the data association problem.

4.   MULTIPLE FRAME TRACK INITIATION AND TRACK MAINTENANCE

We have developed a general expression for the data association problem arising from tracking. The underlying tracking problem is a dynamic one in that information from one or more sensors continually arrives at the processing unit where one partitions the data into frames of data for the assignment problem. Thus the dimension of the assignment problem grows with the number of frames of data N. Processing all of the data at once is called batch processing, which eventually becomes computationally unacceptable as the dimension N increases. To circumvent this problem, we have introduced a sliding window in which data association decisions are hard prior to the window and soft within the window. The first sliding window (single pane) formulation was presented in 1992 [Poore et al., 1992] and refined in [Poore and Drummond, 1997] to include a dual pane window. We briefly discuss the single pane sliding window here and then refinements in the last subsection of this section.

The moving window and resulting "search tree" are truly the heart of a tracking system. One cannot overemphasize the importance of the underlying data structures to the efficiency of a tracking system; however, these data structures are not discussed here.

4.1   TRACK INITIATION

The pure track initiation problem is to formulate and solve the assignment problem described in the previous section with an appropriate number of frames of data N. The choice of the number of frames N is itself a nontrivial problem. Choices of $N = 4, 5, 6$ have worked well in many problems; however, a good research problem is to develop a method that adaptively chooses the number of frames based on the problem complexity.

4.2   TRACK MAINTENANCE USING A SLIDING WINDOW

The term "track maintenance" as used in this section includes three functions: extending and terminating existing tracks and initiating new ones. Suppose now that the observations on P frames of observations have been partitioned into tracks and false alarms and that K *new* frames of observations are to be added. One approach to solving the resulting data association problem is to formulate the problem as a track initiation problem with $P + K$ frames. This is the previously mentioned *batch* approach.

The *deferred logic approach* adopted here is to treat the track extension problem within the framework of a window sliding over the frames of observations. The P frames are partitioned into two components: the first H frames in which *hard* data association decisions are made and the next S frames in which *soft* decisions are made. Then, one adds the K *new* frames of observations. The number of frames in the sliding window is $N = S + K$ while the number of frames in which data association decisions are hard is $H = P - S$. There are various sliding windows that one can develop, including "single pane," "double pane," and "multiple pane" windows. The intent of each of these is efficiency in solving the underlying tracking problem. It is perhaps easier to describe the simplest single pane window with $K = 1$.

4.2.1   A Single Pane Sliding Window.   Let M_0 denote the number of confirmed tracks (i.e., that arise from the solution of the data association problem) on frame k constructed from a solution of the data association problem utilizing frames up to $k + N - 1$. Data association decisions are fixed on frames up to k. Now a new frame of data is added. Thus, frame k denotes a list of tracks and frames $k + 1$ to $k + N$ denote observations. For $i_0 = 1, \ldots, M_0$ the i_0^{th} such track is denoted by T_{i_0} and the $(N + 1)$-tuple $\{T_{i_0}, z_{i_1}^1, \ldots, z_{i_N}^N\}$ will denote a track T_{i_0} plus a set of observations $\{z_{i_1}^1, \ldots, z_{i_N}^N\}$, actual or dummy, that are feasible with the track T_{i_0}. The $(N + 1)$-tuple $\{T_0, z_{i_1}^1, \ldots, z_{i_N}^N\}$ will denote a track that initiates in the sliding window. A false report in the sliding window is one with all but one non-zero index in the $(N + 1)$-tuple $\{T_0, z_{i_1}^1, \ldots, z_{i_N}^N\}$.

The hypothesis about a partition $\gamma \in \Gamma^*$ being true is now conditioned on the truth of the M_0 tracks entering the N-scan window. (Thus the assignments prior to this sliding window are fixed.) The likelihood function is given by $L_\gamma = \prod_{\{T_{i_0}, z_{i_1}, \ldots, z_{i_N}\} \in \gamma} L_{i_0 i_1 \cdots i_N}$, where $L_{i_0 i_1 \cdots i_N} = L_{T_{i_0}} L_{i_1 \cdots i_N}$, $L_{T_{i_0}}$ is the composite likelihood from the discarded frames just prior to the first scan in the window for $i_0 > 0$, $L_{T_0} = 1$, and $L_{i_1 \cdots i_N}$ is defined as in (3.8) for the N-scan window. ($L_{T_0} = 1$ is used for any tracks that initiate in the sliding window.) Thus the track extension problem can be formulated as Maximize $\{L_\gamma \mid \gamma \in \Gamma^*\}$. With the same convention as in Section 2, a feasible partition is one which is defined by the properties in (2.4a)-(2.4d) and (2.7). Analogously, the definition of the zero-one variable

$$z_{i_0 i_1 \cdots i_N} = \begin{cases} 1 & \text{if } \{T_{i_0}, z_{i_1}^1, \ldots, z_{i_N}^N\} \text{ is assigned as a unit,} \\ 0 & \text{otherwise,} \end{cases}$$

and the corresponding cost for the assignment of the sequence $\{T_{i_0}, z_{i_1}, \ldots, z_{i_N}\}$ to a track by $c_{i_0 i_1 \cdots i_N} = -\ln L_{i_0 i_1 \cdots i_N}$ yield the following multidimensional assignment formulation of the data association problem for track maintenance:

$$\text{Minimize} \quad \sum_{i_0=0}^{M_0} \cdots \sum_{i_N=0}^{M_N} c_{i_0 \cdots i_N} z_{i_0 \cdots i_N} \tag{2.20}$$

$$\text{Subj. To} \quad \sum_{i_1=0}^{M_1} \cdots \sum_{i_M=0}^{M_N} z_{i_0 \cdots i_N} = 1, \quad i_0 = 1, \ldots, M_0, \tag{2.21}$$

$$\sum_{i_0=0}^{M_0} \sum_{i_2=0}^{M_2} \cdots \sum_{i_M=0}^{M_N} z_{i_0 i_1 \cdots i_N} = 1, \quad i_1 = 1, \ldots, M_1, \tag{2.22}$$

$$\sum_{i_0=0}^{M_0} \cdots \sum_{i_{k-1}=0}^{M_{k-1}} \sum_{i_{k+1}=0}^{M_{k+1}} \cdots \sum_{i_M=0}^{M_N} z_{i_0 \cdots i_N} = 1, \tag{2.23}$$

$$\text{for} \quad i_k = 1, \ldots, M_N \text{ and } k = 2, \ldots, N-1,$$

$$\sum_{i_0=0}^{M_0} \cdots \sum_{i_{N-1}=0}^{M_{N-1}} z_{i_0 \cdots i_N} = 1, \quad i_N = 1, \ldots, M_N, \tag{2.24}$$

$$z_{i_0 \cdots i_N} \in \{0, 1\} \text{ for all } i_0, \ldots, i_N. \tag{2.25}$$

Note that the association problem involving N frames of observations is an N-dimensional assignment problem for track initiation and an $(N+1)$-dimensional one for track maintenance.

4.2.2   Double and Multiple Pane Window.   In the single pane window, the first frame contains a list of tracks and the remaining N frames, observations. The assignment problem is of dimension $N+1$ where N is the number of frames of observations in front of the existing tracks. The same window is being used to initiate new tracks and continue existing ones. A newer approach is based on the belief that once a track is well established, only a few frames are needed to benefit from the soft decisions on track continuation while a longer window is needed for track initiation. Thus, if the current frame is numbered k with frames $k+1,\ldots,k+N$ being for track continuation, one can go back to frames $k - M,\ldots,k$ and allow observations not attached to tracks that exist through frame k to be used to initiate new tracks. Indeed, this concept works extremely well in practice and was initially posed in the work of Poore and Drummond [Poore and Drummond, 1997]. The next approach to evolve is that of a multiple pane window in which the position of hard data association decisions of observations to tracks can vary within the frames $k - M,\ldots,k,\ldots,k+N$ depending on the difficulty of the problem as measured by track contention. The efficiency of this approach has yet to be determined.

5.   ALGORITHMS

The multidimensional assignment problem for data association is one of combinatorial optimization and NP-hard [Garvey and Johnson, 1979], even for the case $N = 3$. The only known method for solving this problem optimally is branch and bound. The problems arising in tracking are generally sparse, large scale, and noisy with real-time needs. Due to the noise in the problem, the objective is generally to solve the problem to within the noise level. This section presents some of the Lagrangian relaxation algorithms that have been particularly successful in solving complex tracking problems. An advantage is that they generally provide both a lower and upper bound on the optimal solution and thus some measure of closeness to optimality. (This is obviously limited by the duality gap.) There are however many potential algorithms that could be used. For example, GRASP [Murphey et al., 1998a, Murphey et al., 1998b, Robertson, 2000] has also been used successfully.

5.1   PREPROCESSING

Two often used pre–processing techniques, namely *fine gating* and *problem decomposition*, are presented in this section. These two methods can substantially reduce the complexity of the multidimensional assignment problem.

5.1.1   Fine Gating.   The term *fine gating* is used since it is the last in a sequence of techniques used to reduce the unlikely paths in the layered graph or pairings of combinations of reports. This method is based on the following theorem [Poore, 1994].

Theorem 1 (Invariance Property) *Let $N > 1$ and $M_k > 0$ for $k = 1, \ldots, N$, and assume $\hat{c}_{0\cdots0} = 0$ and $u_0^k = 0$ for $k = 1, \ldots, N$. Then the minimizing solution and objective function value of the following multidimensional assignment problem are independent of any choice of $u_{i_k}^k$ for $i_k = 1, \ldots, M_k$ and $k = 1, \ldots, N$.*

$$\textit{Minimize} \quad \sum_{i_1=0}^{M_1} \cdots \sum_{i_N=0}^{M_N} \left(\hat{c}_{i_1 \cdots i_N} - \sum_{k=1}^{N} u_{i_k}^k \right) z_{i_1 \cdots i_N}$$
$$+ \sum_{k=1}^{N} \sum_{i_k=0}^{M_k} u_{i_k}^k \tag{2.26}$$

$$\textit{Subject To:} \quad \sum_{i_2=0}^{M_2} \cdots \sum_{i_N=0}^{M_N} z_{i_1 \cdots i_N} = 1, \quad i_1 = 1, \ldots, M_1, \tag{2.27}$$

$$\sum_{i_1=0}^{M_1} \cdots \sum_{i_{k-1}=0}^{M_{k-1}} \sum_{i_{k+1}=0}^{M_{k+1}} \cdots \sum_{i_N=0}^{M_N} z_{i_1 \cdots i_N} = 1, \tag{2.28}$$
$$\text{for } i_k = 1, \ldots, M_k \text{ and } k = 2, \ldots, N - 1,$$

$$\sum_{i_1=0}^{M_1} \cdots \sum_{i_{N-1}=0}^{M_{N-1}} z_{i_1 \cdots i_N} = 1, \; i_N = 1, \ldots, M_N, \tag{2.29}$$

$$z_{i_1 \cdots i_N} \in \{0, 1\} \textit{ for all } i_1, \ldots, i_N. \tag{2.30}$$

If one identifies $\hat{c}_{i_1 \cdots i_N} = c_{i_1 \cdots i_N}$ and $u_{i_k}^k = c_{0 \cdots 0 i_k 0 \cdots 0}$ in this theorem, then one can conclude that

$$c_{i_1 \cdots i_N} - \sum_{k=1}^{N} c_{0 \cdots 0 i_k 0 \cdots 0} > 0 \tag{2.31}$$

implies that the corresponding zero-one variable $z_{i_1, \ldots, i_N}$ and cost $c_{i_1 \cdots i_N}$ can be removed from the problem since a lower cost can be achieved with the use of the variables $z_{0 \cdots 0 i_k 0 \cdots 0} = 1$. (This does not mean that one should set $z_{0 \cdots 0 i_k 0 \cdots 0} = 1$ for $k = 1, \ldots, N$.)

In the special case in which all costs with exactly one non-zero index are zero, this test is equivalent to

$$c_{i_1 \cdots i_N} > 0. \tag{2.32}$$

5.1.2   Problem Decomposition. Decomposition of the multidimensional assignment problem into a sequence of disjoint problems can improve the solution quality and speed of the algorithm, even on a serial machine. The following decomposition method, originally presented in the work of Poore, Rijavec, Barker, and Munger [Poore et al., 1993], uses graph theoretic methods.

Decomposition of the multidimensional assignment problem is accomplished by determining the connected components of the associated layered graph. To explain, let

$$\mathcal{Z} = \{z_{i_1 i_2 \cdots i_N} | z_{i_1 i_2 \cdots i_N} \text{ is not preassigned to zero}\} \tag{2.33}$$

denote the set of assignable variables. Define an undirected graph $\mathcal{G}(\mathcal{N}, \mathcal{A})$ where the set of nodes is

$$\mathcal{N} = \{z^k_{i_k} | i_k = 1, \ldots, M_k; \ k = 1, \ldots, N\} \tag{2.34}$$

and the set of arcs, by

$$\begin{aligned}
\mathcal{A} = \{(z^k_{j_k}, z^l_{j_l}) \mid & k \neq l, \ j_k \neq 0, \ j_l \neq 0 \text{ and there exists} \\
& z_{i_1 i_2 \cdots i_N} \in \mathcal{Z} \text{such that } j_k = i_k \text{ and } j_l = i_l\}.
\end{aligned} \tag{2.35}$$

The nodes corresponding to zero index have not been included in this graph, since two variables that have only the zero index in common can be assigned independently. Connected components of the graph are then easily found by constructing a spanning forest via a depth first search. Furthermore, this procedure can be used at each level in the relaxation, i.e., can be applied to each assignment problem (3.1) for $k = 3, \ldots, N$. Note that the decomposition algorithm depends only on the problem structure, i.e., on feasibility of the variables, and not on the cost function.

As an aside, this decomposition often yields small problems that are best and more efficiently handled by a branch and bound or an explicit enumeration procedure. (The reason is that there is a certain overhead associated with relaxation.) The remaining components are solved by relaxation. However, extensive decomposition can be a time sink and it is better to limit the number of components to, say, ten, unless one is using a parallel machine.

5.2   THE LAGRANGIAN RELAXATION ALGORITHM FOR THE ASSIGNMENT PROBLEM

This section presents a class of Lagrangian relaxation algorithms for the multidimensional assignment problem that have proved to be computationally efficient and accurate for the tracking purposes. The N dimensional assignment problem has $M_1 + \cdots + M_N$ individual constraints, which can be grouped into N constraint sets. Let $u^k = (u_0^k, u_1^k, \ldots, u_{M_k}^k)$ denote the $(M_k + 1)$ dimensional Lagrange multiplier vector associated with the kth constraint set, with $u_0^k = 0$ and $k = 1, \ldots, N$. The full set of multipliers is denoted by the vector $u = [u^1, \ldots, u^N]$. The multidimensional assignment problem is relaxed to an n-dimensional assignment problem by incorporating $N - n$ constraint sets into the objective function. There are several choices of n. The case $n = 0$ yields the linear programming dual; $n = 2$ yields a two dimensional assignment problem and is the one that has been highly successful in practice.

Although any constraint sets can be relaxed, sets $n + 1, \ldots, N$ are chosen for convenience. In the tracking problem using a sliding window, these are the correct ones from the point of view of the data structures that arise from the construction of tracks.

The *relaxed problem* for multiplier vector u is given by

$$L_n \left(u^{n+1}, \ldots, u^N \right) = \qquad (2.36)$$

$$\text{Minimize} \sum_{i_1=0}^{M_1} \cdots \sum_{i_N=0}^{M_N} \left(c_{i_1 \cdots i_N} + \sum_{k=n+1}^{N} u_{i_k}^k \right) z_{i_1 \cdots i_N} \qquad (2.37)$$

$$- \sum_{k=n+1}^{N} \sum_{i_k=0}^{M_k} u_{i_k}^k$$

$$\text{Subject To}: \sum_{i_2=0}^{M_2} \cdots \sum_{i_N=0}^{M_N} z_{i_1 \cdots i_N} = 1, \ i_1 = 1, \ldots, M_1, \qquad (2.38)$$

$$\sum_{i_1=0}^{M_1} \sum_{i_3=0}^{M_3} \cdots \sum_{i_N=0}^{M_N} z_{i_1 \cdots i_N} = 1, \ i_2 = 1, \ldots, M_2, \qquad (2.39)$$

$$\vdots$$

$$\sum_{i_1=0}^{M_1} \cdots \sum_{i_{n-1}=0}^{M_{n-1}} \sum_{i_{n+1}=0}^{M_{n+1}} \cdots \sum_{i_N=0}^{M_N} z_{i_1 \cdots i_N} = 1, \qquad (2.40)$$

$$i_n = 1, \ldots, M_n,$$

$$z_{i_1 \cdots i_N} \in \{0, 1\} \quad \text{for all} \quad \{i_1, \ldots, i_N\}. \qquad (2.41)$$

One can show that the above can be reduced to an n-dimensional assignment problem using the transformation

$$x_{i_1 i_2 \cdots i_n} = \sum_{i_{n+1}=0}^{M_{n+1}} \cdots \sum_{i_N=0}^{M_N} z_{i_1 \cdots i_n i_{n+1} \cdots i_N} \quad \text{for all} \quad i_1, i_2, \ldots, i_n \qquad (2.42)$$

$$c_{i_1 i_2 \cdots i_n} = \text{Min} \left\{ c_{i_1 \cdots i_N} + \sum_{k=n+1}^{N} u_{i_k}^k \mid \text{for all} \quad i_{n+1}, \ldots, i_N \right\} \qquad (2.43)$$

$$c_{0 \cdots 0} = \sum_{i_{n+1}=0}^{M_{n+1}} \cdots \sum_{i_N=0}^{M_N} \text{Min} \left\{ 0, c_{0 \cdots 0 i_{n+1} \cdots i_N} + \sum_{k=n+1}^{N} u_{i_k}^k \right\} \qquad (2.44)$$

Thus, the Lagrangian relaxation algorithm may be summarized as follows.

1. Solve the problem

$$\text{Maximize } L_n \left(u^{n+1}, \ldots, u^N \right) \qquad (2.45)$$

2. Give an optimal or near optimal solution $(u^{n+1}, \ldots, u^N)$, solve the above assignment problem for this given multiplier vector. This produces an alignment of the first n indices. Let these be enumerated by $\{(i_1^j, \ldots, i_n^j)\}_{j=0}^{J}$ where $(i_1^0, \ldots, i_n^0) = (0, \ldots, 0)$. Then, the variable and cost coefficient

$$x_{j i_{n+1} \cdots i_N} = z_{i_1^j \cdots i_n^j i_{n+1} \cdots i_N} \qquad (2.46)$$

$$c_{j i_{n+1} \cdots i_N} = c_{i_1^j \cdots i_n^j i_{n+1} \cdots i_N} \qquad (2.47)$$

satisfy the following $N - n + 1$ dimensional assignment problem.

$$\text{Minimize} \quad \sum_{j=0}^{J} \sum_{i_{n+1}=0}^{M_{n+1}} \cdots \sum_{i_N=0}^{M_N} c_{ji_{n+1}\cdots i_N} z_{ji_{n+1}\cdots i_N} \tag{2.48}$$

$$\text{Subject To:} \quad \sum_{i_{n+1}=0}^{M_{n+1}} \cdots \sum_{i_N=0}^{M_N} z_{ji_{n+1}\cdots i_N} = 1, \; j = 1,\ldots, J, \tag{2.49}$$

$$\sum_{j=0}^{J} \sum_{i_{n+2}=0}^{M_{n+2}} \cdots \sum_{i_N=0}^{M_N} z_{ji_{n+1}\cdots i_N} = 1, \tag{2.50}$$

$$\text{for } i_{n+1} = 1, \ldots, M_{n+1},$$

$$\vdots$$

$$\sum_{j=0}^{J} \cdots \sum_{i_{N-1}=0}^{M_{N-1}} z_{ji_{n+1}\cdots i_N} = 1, \; i_N = 1, \ldots, M_N \tag{2.51}$$

$$z_{ji_{n+1}\cdots i_N} \in \{0,1\} \text{ for all } (j, i_{n+1}, \ldots, i_N) \tag{2.52}$$

Let the nonzero zero-one variables in a solution be denoted by $z_{ji_{n+1}^j \cdots i_N^j} = 1$. Then the solution of the original problem is $z_{i_1^j \cdots i_n^j i_{n+1}^j \cdots i_N^j} = 1$ with all remaining values of z being zero.

This algorithm essentially is summarized by saying the N-dimensional assignment problem is relaxed to an n-dimensional assignment problem by relaxing $N - n$ constraint sets. The problem of restoring feasibility is posed as a $N - n + 1$ dimensional problem.

Notice that the problem of maximizing $L_n(u^{n+1}, \ldots, u^N)$ is one of nonsmooth optimization. Bundle methods [Schramm and Zowe, 1992, Kiwiel, 1985] have proved to be particularly successful for this purpose.

5.2.1   A Class of Algorithms.
In some earlier work [Poore and Rijavec, 1991a, Poore and Rijavec, 1993, Barker et al., 1995, Poore and Rijavec, 1994], we relaxed an N–dimensional assignment problem to an $(N-1)$–dimensional one, which is NP-hard for $N > 2$, by relaxing one set of constraints. The corresponding dual function $L_n(u^{n+1}, \ldots, u^N)$ is piecewise linear and concave, but the evaluation of the function and subgradients, as needed by the nonsmooth maximization, requires an optimal solution of an NP-hard n–dimensional assignment problem when $n > 2$. Due to the real-time needs, one must resort to suboptimal solutions to the relaxed problem; however, suboptimal solutions only provide approximate function and subgradient values. To moderate this

difficulty, we [Poore and Rijavec, 1994] used a concave, piecewise affine merit function to provide guidance for the function values for the nonsmooth optimization phase. This approach still computed approximate subgradients from good quality feasible solutions obtained from multiple relaxation and recovery cycles executed at lower levels in a recursive fashion. (The number of cycles can be any fixed number greater than or equal to one or it can be chosen adaptively by allowing the non-smooth optimization solver to converge to within user-defined tolerances.) In spite of these approximations, the numerical performance of these prior algorithms has been quite good [Poore and Rijavec, 1994].

A variation on the N–to–$(N-1)$ relaxation algorithm using a one cycle is due to Deb, Pattipati, Yeddanapudi, and Bar-Shalom [Deb et al., 1997]. Similar approximate function and subgradient values are used at each level of the relaxation process. To moderate this difficulty, they modify the accelerated subgradient method of Shor [Shor, 1985] by further weighting the search direction in the direction of violated constraints and report improvement over the accelerated subgradient method. They do not, however, use problem decomposition and a merit function as in our previous work [Poore and Rijavec, 1994].

When relaxing an N–dimensional assignment problem to an n–dimensional one, the one case in which the above difficulties are resolved is for $n=2$ and it is this algorithm that is currently used in our tracking system.

5.3   THE LINEAR PROGRAMMING DUAL

The constraints themselves imply that the dual of the multidimensional assignment problem with the zero-one constraint $x_{i_1 \cdots i_N} \in \{0,1\}$ relaxed to $0 \leq x_{i_1 \cdots i_N} \leq 1$ is equivalent to the one with these variables being relaxed to $0 \leq x_{i_1 \cdots i_N}$. The dual problem associated with this latter relaxation is

$$\text{Maximize} \quad -\sum_{k=1}^{N} \sum_{i_k=1}^{M_k} u_{i_k}^{k} \qquad (2.53)$$

$$\text{Subject To}: \quad c_{i_1 \cdots i_N} + \sum_{k=1}^{N} u_{i_k}^{k} \geq 0 \text{ for all feasible } \{i_1, \ldots, i_N\}. \quad (2.54)$$

Due to the unimodularity of the two-dimensional assignment problem, the optimal value of this linear programming dual is the same as the value of the maximum of the Lagrangian $L_n(u^{n+1}, \ldots, u^N)$ of the previous section when $n=2$. Thus, the use of linear programming techniques to compute the multipliers u for use in $L_2(u^3, \ldots, u^N)$ should provide an effective alternative to the use of non-smooth optimization. Since the multidimensional assignment problems arise from the use of a moving window, one always has a good initial primal and dual solution. Thus, an interior point method, which can make use of these prior solutions, may be the best method.

5.4   LAGRANGIAN DECOMPOSITION AND OTHER TRANSFORMATIONS

The algorithms described above are based on the same theme of relaxing a certain number of constraint sets and recovering a solution with a multidimensional assignment problem. Both the relaxed and recovery problems are of lower dimension than the original. Furthermore, the multipliers introduced above are each associated with a report or observation. The multipliers can be associated with an entire sequence of observations or with a subset of such a sequence. Lagrangian decomposition and other transformations provide such an approach.

As an example of Lagrangian decomposition, one can introduce the zero-one variable $y_{i_1 \cdots i_N}$ in addition to the variable $z_{i_1 \cdots i_N}$, replace the objective function by

$$\text{Minimize} \quad \sum_{i_1=0}^{M_1} \cdots \sum_{i_N=0}^{M_N} c_{i_1 \cdots i_N} \left(\frac{1}{2} z_{i_1 \cdots i_N} + \frac{1}{2} y_{i_1 \cdots i_N} \right), \quad (2.55)$$

leave the first n constraint sets with the variable $z_{i_1 \cdots i_N}$ and replace this variable in the last $N - n$ constraints with $y_{i_1 \cdots i_N}$, and finally add the constraint

$$z_{i_1 \cdots i_N} = y_{i_1 \cdots i_N} \quad \text{for all} \quad \{i_1 \cdots i_N\}. \quad (2.56)$$

Whereas in the first class of methods each multiplier $u_{i_k}^k$ corresponds to a report $z_{i_k}^k$, the relaxation of the constraint $z_{i_1 \cdots i_N} = y_{i_1 \cdots i_N}$ introduces multipliers $u_{i_1 \cdots i_N}$ and one might think of this as a multiplier corresponding to each track in the problem. Generally, but not always, there are far more of these multipliers than in the first relaxation. On the other hand, if there is only moderate track contention, then the nonsmooth optimization problem should be easier.

In addition to the Lagrangian relaxation and decomposition methods above, one can also transform the variables to obtain equivalent problems and then use Lagrangian relaxation on these. For example, for a three dimensional problem one can introduce the variables

$$x_{i_1 i_2} = \sum_{i_3=0}^{M_3} z_{i_1 i_2 i_3} \quad \text{for all} \quad i_1, i_2, \quad (2.57)$$

$$y_{i_2 i_3} = \sum_{i_1=0}^{M_1} z_{i_1 i_2 i_3} \quad \text{for all} \quad i_2, i_3, \quad (2.58)$$

to rewrite the three dimensional assignment problem. The subsequent relaxation of these constraints using multipliers $u_{i_1 i_2}$ and $v_{i_2 i_3}$, respectively, yields a particularly simple problem that can be composed into two two-dimensional assignment problems. Such transformations and relaxations have yet to be developed for the tracking problem.

5.5   IMPROVEMENT METHODS

The relaxation methods presented above have been enormously successful in providing quality solutions to the assignment problem. Improvement techniques are fundamentally important. Based on the relaxation and a branch and bound framework [Poore and Yan, 1999], significant improvements in the quality of the solution have been achieved for tracking problems. On difficult problems in tracking, the improvement can be significant. While straight relaxation may produce solutions to within 3% of optimality, the improvement techniques can produce solutions to within 0.5% of optimal, at least on one class of very difficult tracking problems.

Given the ever increasing need for more accurate solutions, further improvement methods such as local search methods should be developed.

6.   FUTURE DIRECTIONS

Having presented an overview of the problem formulation and some of the highly successful algorithms that have been used, we summarize here some of the open issues that should be addressed in the future.

6.1   PROBLEM FORMULATION

As discussed in the presentation, several alternate formulations such as more general set packings and coverings of the data should be pursued. The current formulation is sufficiently general to include other important cases, such as multisensor resolution problems and unresolved closely spaced objects in which an observation is necessarily assigned to more than one target. The assigning of a report to more than one target can be accomplished within the context of the multidimensional assignment problems by using inequality constraints. The formulation, algorithms, and testing have yet to be systematically developed. While allowing an observation to be assigned to more than one target is easy to model mathematically, allowing a target to be assigned to more than one observation is difficulty and may indeed introduce nonlinearity into the objective function due to a loss of the independence assumptions discussed in Section 3.

6.2 FRAMES OF DATA

The use of the multidimensional assignment problems to solve the central data association problem rests on the assumption that each target is seen at most once in each frame of data, i.e., the frame is a "proper" frame. For a mechanically rotating radar, this is reasonable easy to approximate as a sweep of the surveillance region. For electronically scanning sensors, which can switch from searching mode to tracking mode, this partitioning problem is less obvious. Although we have developed one such solution, the formulation and solution are not yet optimal.

6.3 SLIDING WINDOWS

The batch approach to the data association problem of partitioning the observations into tracks and false alarms is to add a frame of data (observations) to the existing frames and formulate and solve the corresponding multidimensional assignment problem. This dimension increases by the number of frames added. To avoid the intractability of this approach, we developed a moving window approach wherein one resolves the data association problem over the window of a limited number of frames of data in 1992 [Poore et al., 1992] and revised it in 1996 [Poore and Drummond, 1997] to use different length windows for track continuation (maintenance) and initiation. This allows one to improve the efficiency of the multiframe processing and maintain the longer window needed for track initiation. Indeed, numerical experiments todate show it to be far more efficient than a single pane window. One can easily imagine using different depths in the window for continuing tracks depending on the complexity of the problem. The efficiency in practice has yet to be determined.

A fundamental question is that of determining the dimension of the assignment problem that is most appropriate for a particular tracking problem. The goal would be to develop a method that adapts the dimension to the difficulty of the problem or to the need in the surveillance problem.

6.4 ALGORITHMS

Several Lagrangian relaxation methods were outlined in the previous section. The method that has been most successful has been the relaxation to a two dimensional assignment problem, maximization of the resulting relaxed problem with respect to the multipliers, and then restoring feasibility to the original problem by formulating this recovery problem as a multidimensional assignment problem of one dimension lower than the original. The process is then repeated until one reaches a two dimensional assignment problem which is solved optimally.

Such an algorithm can generally produce solutions that are accurate to within 3% of optimal on very difficult problem and optimal for easy problems. The use of an improvement algorithm based on branch and bound can considerably improve the performance to within 0.5% of optimal, at least on some classes of difficult tracking problems [Poore and Yan, 1999]. Other techniques such as local search should equally improve the solution quality.

Speed enhancements using the decomposition and clustering discussed in the previous section and as presented in 1992 [Poore et al., 1993] can improve the speed of the assignment solvers by an order of magnitude on large scale and difficult problems. Further work on distributed and parallel computing should allow one to enhance both the solution quality and speed.

Another direction of work is that of the computation of K-near optimal solutions just as K-best solutions for two dimensional assignment problems [Poore and Yan, 1999]. The difference is that these K-near optimal solutions give important information about the reliability of the computed tracks.

Finally, we have said little about other approaches to the assignment problem. Certainly, other approaches such as GRASP [Murphey et al., 1998a, Murphey et al., 1998b, Robertson, 2000] have also been successful. Relaxation has worked extremely well on the current problems.

References

[Bar-Shalom and Li, 1995] Bar-Shalom, Y. and Li, X. R., editors (1995). *Multitarget-Multisensor Tracking: Principles and Techniques*. OPAMP Tech. Books, Los Angeles.

[Barker et al., 1995] Barker, T. N., Persichetti, J. A., Poore, Jr., A. B., and Rijavec, N. (1995). Method and system for tracking multiple regional objects. US Patent Number 5,406,289.

[Bertsekas, 1998] Bertsekas, D. P. (1998). *Network Optimization*. Athena Scientific, Belmont, MA.

[Blackman and Popoli, 1999] Blackman, S. and Popoli, R. (1999). *Design and Analysis of Modern Tracking Systems*. Artech House, Norwood, MA.

[Chummun et al., 1999] Chummun, M., Kirubarajan, T., Pattipati, K., and Bar-Shalom, Y. (1999). Efficient multidimensional data association for multisensor-multitarget tracking using clustering and assignment algorithms. In *Proc. 2nd International Conference on Information Fusion*, Silicon Valley, CA.

[Deb et al., 1993] Deb, S., Pattipati, K. R., and Bar-Shalom, Y. (1993). A multisensor-multitarget data association algorithm for heterogeneous systems. *IEEE Transactions on Aerospace and Electronic Systems*, 29:560–568.

[Deb et al., 1997] Deb, S., Pattipati, K. R., Bar-Shalom, Y., and Yeddanapudi, M. (1997). A generalized s-dimensional assignment algorithm for multi-sensor multitarget state estimation. *IEEE Transactions on Aerospace and Electronic Systems*, 33:523–538.

[Drummond, 1999] Drummond, O. (1999). Target tracking. In *Wiley Encyclopedia of Electrical and Electronics Engineering, Vol. 21*, pages 377–391. Wiley.

[Garvey and Johnson, 1979] Garvey, M. and Johnson, D. (1979). *Computers and Intractability: A guide to the theory of NP-Completeness*. W.H. Freeman & Co., CA.

[Jonker and Volgenant, 1987] Jonker, R. and Volgenant, A. (1987). A shortest augmenting path algorithm for dense and sparse linear assignment problems. *Computing*, 38:325–340.

[Kirubarajan et al., 1997] Kirubarajan, T., Bar-Shalom, Y., and Pattipati, K. R. (1997). Multiassignment for tracking a large number of overlapping objects. In *Proceedings SPIE Conf. Signal & Data Proc. of small Targets, VOL. 3136*, San Diego, CA.

[Kirubarajan et al., 1998] Kirubarajan, T., Wang, H., Bar-Shalom, Y., and Pattipati, K. (1998). Efficient multisensor fusion using multidimensional assignment for multitarget tracking. In *Proceedings SPIE Conf. Signal Processing, Sensor Fusion and Target Recognition*, Orlando, FL.

[Kiwiel, 1985] Kiwiel, K. C. (1985). *Methods of descent for nondifferentiable optimization*. Lecture Notes in Mathematics 1133. Springer-Verlag, Berlin.

[Morefield, 1977] Morefield, C. L. (1977). Application of 0-1 integer programming to multitarget tracking problems. *IEEE Transactions on Automatic Control*, 22(3):302–312.

[Murphey et al., 1998a] Murphey, R., Pardalos, P., and Pitsoulis, L. (1998a). A GRASP for the multi-target multi-sensor tracking problem. In *Networks, Discrete Mathematics and Theoretical Computer Science Series*, volume 40, pages 277–302. American Mathematical Society.

[Murphey et al., 1998b] Murphey, R., Pardalos, P., and Pitsoulis, L. (1998b). A parallel GRASP for the data association multidimensional assignment problem. In *Parallel Processing of Discrete Problems, IMA Volumes in Mathematics and its Applications*, volume 106, pages 159–180. Springer-Verlag.

[Nemhauser and Wolsey, 1988] Nemhauser, G. L. and Wolsey, L. A. (1988). *Integer and Combinatorial Optimization*. John Wiley and Sons, New York.

[Poore, 1994] Poore, A. B. (1994). Multidimensional assignment formulation of data association problems arising from multitarget tracking and multisensor data fusion. *Computational Optimization and Applications*, 3:27–57.

[Poore and A.J. Robertson, 1997] Poore, A. B. and A.J. Robertson, I. (1997). A new class of lagrangian relaxation based algorithms for a class of multidimensional assignment problems. *Computational Optimization and Applications*, 8(2):129–150.

[Poore and Drummond, 1997] Poore, A. B. and Drummond, O. E. (1997). Track initiation and maintenance using multidimensional assignment problems. In Pardalos, P. M., Hearn, D., and Hager, W., editors, *Network Optimization*, volume 450 of *Lecture Notes in Economics and Mathematical Systems*, pages 407–422. Springer-Verlag.

[Poore and Rijavec, 1991a] Poore, A. B. and Rijavec, N. (1991a). Multidimensional assignment problems, and lagrangian relaxation. In *Proceedings of the SDI Panels on Tracking*, volume Issue No. 2, pages 3–51 to 3–74, Institute for Defense Analyses.

[Poore and Rijavec, 1991b] Poore, A. B. and Rijavec, N. (1991b). Multitarget tracking and multidimensional assignment problems. In *Proceedings of the 1991 SPIE Conference on Signal and Data Processing of Small Targets 1991*, volume 1481, pages 345–356.

[Poore and Rijavec, 1993] Poore, A. B. and Rijavec, N. (1993). A Lagrangian relaxation algorithm for multi-dimensional assignment problems arising from multitarget tracking. *SIAM Journal on Optimization*, 3(3):545–563.

[Poore and Rijavec, 1994] Poore, A. B. and Rijavec, N. (1994). A numerical study of some data association problems arising in multitarget tracking. In Hager, W. W., Hearn, D. W., and Pardalos, P. M., editors, *Large Scale Optimization: State of the Art*, pages 339–361, Boston, MA. Kluwer Academic Publishers B. V.

[Poore et al., 1992] Poore, A. B., Rijavec, N., and Barker, T. (1992). Data association for track initiation and extension using multiscan windows. In *Signal and Data Processing of Small Targets 1992, Proc. SPIE 1698, 1992*, volume 1698, pages 432–441.

[Poore et al., 1993] Poore, A. B., Rijavec, N., Barker, T., and Munger, M. (1993). Data association problems posed as multidimensional assignment problems: numerical simulations. In Drummond, O. E., editor, *Signal and Data Processing of Small Targets 1954*, pages 564–573. Proceedings of the SPIE.

[Poore and Yan, 1999] Poore, A. B. and Yan, X. (1999). Some algorithmic improvements in multi-frame most probable hypothesis tracking. In Drummond, O., editor, *Signal and Data Processing of Small Targets, SPIE*.

[Poore, 1996] Poore, Jr., A. B. (1996). Method and system for tracking multiple regional objects by multi-dimensional relaxation. US Patent Number 5,537,119.

[Poore, 1999] Poore, Jr., A. B. (issue on September 28, 1999). Method and system for tracking multiple regional objects by multi–dimensional relaxation, cip.US patent number 5,959,574.

[Popp et al., 1998] Popp, R., Pattipati, K., Bar-Shalom, Y., and Gassner, R. (1998). An adaptive m-best SD assignment algorithm and parallelization for multitarget tracking. In *Proceedings 1998 IEEE Aerospace Conference*, Snowmass, CO.

[Reid, 1996] Reid, D. B. (1996). An algorithm for tracking multiple targets. *IEEE Transactions on Automatic Control*, 24(6):843–854.

[Robertson, 2000] Robertson, A. (2000). A set of greedy randomized adaptive local search procedure (GRASP) implementations for the multidimensional assignment problem.

[Schramm and Zowe, 1992] Schramm, H. and Zowe, J. (1992). A version of the bundle idea for minimizing a nonsmooth function: Conceptual idea, convergence analysis, numerical results. *SIAM Journal on Optimization*, 2:121–152.

[Shea and Poore, 1998] Shea, P. J. and Poore, A. B. (1998). Computational experiences with hot starts for a moving window implementation of track maintenance. In *1998 SPIE Conference Proceedings 3373, Signal and Data Processing of Small Targets 1998*, Orlando, FL.

[Shor, 1985] Shor, N. Z. (1985). *Minimization Methods for Non-Differentiable Functions*. Springer-Verlag, New York.

[Sittler, 1964] Sittler, R. W. (1964). An optimal data association problem in surveillance theory. *IEEE Transactions on Military Electronics*, 8(2):125–139.

[Waltz and Llinas, 1990] Waltz, E. and Llinas, J. (1990). *Multisensor Data Fusion*. Artech House, Boston, MA.

Chapter 3

TARGET-BASED WEAPON TARGET ASSIGNMENT PROBLEMS

Robert A. Murphey

Air Force Research Laboratory, Munitions Directorate,
101 W. Eglin Blvd., Ste 330
Eglin AFB, FL 32542-6810
Murphey@eglin.af.mil

1. INTRODUCTION

In this chapter we consider a class of non-linear assignment problems collectively referred to as Target-based Weapon Target Assignment (WTA). The Target-based Weapon Target Assignment problem considers optimally assigning M weapons to N targets so that the total expected damage to the targets is maximized. We use the term target-based to distinguish these problems from those that are asset-based, that is problems where weapons are assigned to targets such that the value of a group of assets is maximized supposing that the targets themselves are missiles engaging the assets. The asset-based problem is most pertinent to strategic ballistic missile *defense* problems whereas the target-based problems apply to *offensive* conventional warfare types of problems. Previous surveys of WTA problems were made by Matlin [Matlin, 1970] and Eckler and Burr [Eckler and Burr, 1972] the first on offensive, target based problems and the second primarily about defensive, asset based problems.

If at some time t the numbers and locations of weapons and targets are known with certainty, then a single assignment may be made at time t such that all of the weapons are committed. This formulation is denoted *static* WTA. In its most general form, static WTA is known to be $\mathcal{NP}$-complete. A more difficult problem results when assignments may be made at any of several discrete points in time. These types of problems are known as *dynamic* WTA and are formulated in 2 ways. In the first type, the numbers and locations of targets are all known a priori. However, since the survival of a target with weapons assigned to it is probabilistic, the outcome of any time interval (i.e. which targets survive) is stochastic and so influences all future assignments. These problems

P.M. Pardalos and L.S. Pitsoulis (eds.), Nonlinear Assignment Problems, 39–53.
© *2000 Kluwer Academic Publishers.*

are known as shoot-look-shoot since some observation of the targets is required to condition the problem at the following interval. Shoot-look-shoot problems are readily adapted to a dynamic programming formulation. In the second type of dynamic WTA, we assume that the targets are not observable, hence the outcome of each interval is deterministic. However, we cannot assume that the numbers and locations of targets are known a priori. That is, at each interval, a subset of the N targets are known with certainty and the remainder are either not known or known only stochastically. This dynamic WTA formulation may be modeled as stochastic programs.

2.   STATIC ASSIGNMENT MODELS

Consider that there are M weapons and N independent targets. Define the decision variable x_{ij}, $i = 1, 2, \ldots M$, $j = 1, 2, \ldots, N$ as

$$x_{ij} = \begin{cases} 1: & \text{weapon } i \text{ assigned to target } j \\ 0: & \text{else} \end{cases} \qquad (3.1)$$

Given that weapon i engages target j, that is $x_{ij} = 1$, the outcome is random and assumes a Bernoulli distribution:

$$\begin{aligned} P(\text{target } j \text{ is destroyed by weapon } i) &= P_{ij} \\ P(\text{target } j \text{ is not destroyed by weapon } i) &= 1 - P_{ij} \end{aligned} \qquad (3.2)$$

Assume that each weapon engagement is independent of every other engagement. Then the outcomes of the engagements are independent Bernoulli distributed.

Let each target be assigned a positive real number V_j to indicate preference between targets. Assume that targets may be partitioned into classes and each class has a unique value to the decision maker. Let the set of all class values be contained in $\mathbb{V}$ such that $\mathbb{V}$ is of cardinality K. Our objective is to maximize the expected damage to the targets which is equivalent to minimizing the expected target value. With $q_{ij} = (1 - P_{ij})$, the resulting integer programming formulation is:

SWTA

$$\text{minimize} \sum_{j=1}^{N} V_j \prod_{i=1}^{M} q_{ij}^{x_{ij}}$$

subject to

$$\sum_{j=1}^{N} x_{ij} = 1, \quad i = 1, 2, \dots, M$$

$$V_j \in \mathbb{V} \in \mathbb{R}_+^K, \; j = 1, 2, \dots, N$$

$$x \in \mathbb{B}^{M \times N}$$

Note that $\mathbb{R}$ denotes the real number space, and $\mathbb{Z}$ and $\mathbb{B}$ denote the integer and binary subspaces respectively. The equality constraints of SWTA ensure that each weapon is assigned to exactly one target. Notice that nothing prevents all weapons from being assigned to a single target. This problem formulation was first suggested by Manne [Manne, 1958].

Although the nonlinear assignment problem developed in SWTA was shown to be *NP*-complete in [Lloyd and Witsenhausen, 1986], a useful lower bound may be developed and then applied in a branch and bound or other global integer optimization scheme. A lower bound for SWTA may be obtained by relaxing integrality of x:

RSWTA

$$\text{minimize} \sum_{j=1}^{N} V_j \prod_{i=1}^{M} q_{ij}^{x_{ij}}$$

subject to

$$\sum_{j=1}^{N} x_{ij} = 1, \qquad\qquad i = 1, 2, \dots, M$$

$$V_j \in \mathbb{V} \in \mathbb{R}_+^K, \qquad\qquad j = 1, 2, \dots, N$$

$$0 \le x_{ij} \le 1, \quad i = 1, 2, \dots, M, \; j = 1, 2, \dots, N$$

The objective of RSTWA is clearly convex while the constraints are linear resulting in a convex programming problem. A number of methods may be applied to solve RSWTA, however, for large M and N, dual methods should be much faster since the dual of RSWTA will only have M constraints. Indeed, the dual of RSWTA is simply

$$d(\lambda) = \text{minimize}_x \, L(x, \lambda)$$

where the Lagrangian is defined as:

$$L(x, \lambda) = \sum_{j=1}^{N} V_j \prod_{i=1}^{M} q_{ij}^{x_{ij}} + \sum_{i=1}^{M} \lambda_i \left(\sum_{j=1}^{N} x_{ij} - 1 \right).$$

Now it can be seen that $d(\lambda)$ has the separable form:

$$d(\lambda) = \sum_{j=1}^{N} \min_x \left(V_j \prod_{i=1}^{M} q_{ij}^{x_{ij}} + \sum_{i=1}^{M} \lambda_i x_{ij} \right) - \sum_{i=1}^{M} \lambda_i$$

and by Strong Duality, there is no duality gap.

2.1   UNIFORM WEAPONS

If the weapons are all assumed to be *uniform* that is, they are all the same, then the *uniform static WTA* problem may be written as:

SWTA – U

$$\text{minimize} \sum_{j=1}^{N} V_j q_j^{x_j}$$

subject to

$$\sum_{j=1}^{N} x_j = M,$$

$$V_j \in \mathbb{V} \in \mathbb{R}_+^K, \ j = 1, 2, \dots, N$$

$$x \in \mathbb{Z}^N$$

where now x_j is the number of weapons assigned to target j and the index i has been dropped from q since, given that the weapons are all the same, the probability of destroying target j depends only on the target.

denBroeder *et al* [denBroeder et al., 1959] presented an algorithm for solving SWTA optimally in $O(N + M \log(N))$ time termed the minimum marginal return (MMR). Essentially, MMR assigns weapons one at a time to the target which realizes the greatest decrease in value, where target value is defined to be $q_j^{x_j}$. denBroeder showed that this greedy strategy is optimal for the uniform weapon assumption. The algorithm is shown in Figure 3.1.

Hosein [Hosein, 1989], later showed that a variant of iterated 1-opt local search on any feasible assignment is also optimal for SWTA-U. The iterated 1-opt variant is easily implemented for this problem by removing one weapon from target j and replacing it on target k given that the move decreases the objective value.

Step 0. For each $j = 1, \ldots N$, let $x_j = 0$, and denote the probability of survival of target j by $S_j = V_j q_j^{x_j}$. Initialize weapon index $i = 1$.
While $i \leq M$ DO

<u>Step 1.</u> Find target k for which munition i has greatest effect:

$$k = \arg\max_j \{S_j(1 - q_j)\}.$$

<u>Step 2.</u> Add munition i to target k: $x_k = x_k + 1$ and revise probability of survival of target k: $S_k = S_k q_k$.
$i = i + 1$.

Figure 3.1   Minimum Marginal Return (MMR)

Note that if all targets are also uniform, q_j is replaced by q and V_j is dropped. It is easily shown that the optimal strategy in such cases is to simply spread the weapons evenly across all targets (see [Lloyd and Witsenhausen, 1986]).

2.2   RANGE RESTRICTED WEAPONS

Realistic WTA models dictate that not all weapons will be reachable by all targets. Consequently, the *range restricted uniform SWTA* problem is presented:

SWTA – URR

$$\text{minimize} \sum_{j=1}^{N} V_j q_j^{\sum_{i \in R_j} x_{ij}}$$

subject to

$$\sum_{j=1}^{N} x_{ij} = 1, \qquad i = 1, 2, \ldots, M$$

$$V_j \in \mathbb{V} \in \mathbb{R}_+^K, \qquad j = 1, 2, \ldots, N$$

$$x \in \mathbb{B}^N$$

where R_j is the set of weapon indices that are within reachable range of target j. By representing each weapon as a node and each target as a node in a bipartite graph partitioned on weapons and targets, Hosein [Hosein, 1989] showed that SWTA-URR was in-fact solvable as a network flow problem when an edge is added between weapon i and target j only if $i \in R_j$ for x_{ij}. A supply of one is provided at each weapon node and the target nodes are all sunk to a single sink node with demand M. The capacity of all weapon to target arcs is one while the cost of these arcs is always zero. The capacity of all arcs from the targets to the sink node are each M while the costs of these arcs are

$$F_{js}(x) = V_j \left[q_j^{\lfloor x \rfloor} + (x - \lfloor x \rfloor) \left(q_j^{\lceil x \rceil} - q_j^{\lfloor x \rfloor} \right) \right]$$

where flow x on arc $\{j, s\}$ is continuous between 0 and M. Since $F_{js}(x)$ is a convex combination, the resulting network flow problem is clearly convex. Furthermore, there will always exist an integer valued solution that is optimal for the convex minimum cost network flow problem that is also optimal for SWTA-URR.

3. THE DYNAMIC ASSIGNMENT MODEL

In static WTA problems the assignment is made at a single point in time. Now suppose that assignments may be made at any of T discrete intervals in time. There are 2 fundamental types of dynamic WTA problems in the literature. The most studied is the case when the numbers and locations of targets are all known a priori. However, since the survival of a target with weapons assigned to it is probabilistic, the outcome of any time interval (i.e. which targets survive) is stochastic and so influences all future assignments. These problems are known as *shoot-look-shoot* since some observation of the targets is required prior to making any future assignments. Shoot-look-shoot problems are readily adapted to a dynamic program yet prove very difficult to solve in the most general case. In the second type of dynamic WTA, we assume that the targets are not observable, hence the outcome of each interval is essentially deterministic. However, we no longer assume that the numbers and locations of targets are known a priori. That is, at each interval, a subset of the N targets are known with certainty and the remainder are either not known or known only stochastically. These dynamic WTA formulations may be modeled as stochastic programs.

3.1 SHOOT-LOOK-SHOOT PROBLEMS

In shoot-look-shoot problems, the assignment of M weapons to N targets is made over T discrete intervals in time. Therefore, the decision variable must be indexed by time as in $x_{ij}(t)$, $t = 1, 2, \ldots, T$. The assumption is that after an assignment, the outcome of each engagement can be observed without error. As a result, we discuss the evolution of the *target state* as a function of time and the previous assignment. At any time t, this state is simply

$$
U_j = \begin{cases} 1: & \text{target } j \text{ survives stage } t - 1 \\ 0: & \text{else} \end{cases}
$$

Since the survival of target j depends on the values of $x_{ij}(t-1)$ and q_{ij}, the target state is stochastic and evolves as:

$$
P(U_j = u) = u \prod_{i=1}^{M} q_{ij}^{x_{ij}(t-1)} + (1 - u) \left(1 - \prod_{i=1}^{M} q_{ij}^{x_{ij}(t-1)} \right)
$$

Similarly, the *weapon state* is defined as

$$
w_i = \begin{cases} 1: & \text{weapon } i \text{ not used in stage } t - 1 \\ 0: & \text{else} \end{cases}
$$

which evolves as:

$$
w_i = 1 - \sum_{j=1}^{N} x_{ij}(t-1)
$$

that is, weapon i is available for use in stage t only if it was not used in stage $t - 1$. Implied in this model is the fact that weapon i is available for use in stage t only if it was not used in *any previous* stage. This is based on an inductive argument that assumes that a solution of stage $t - 1$ considers what weapons were not used in stage $t - 2$ and so on.

Define the cost of a stage T assignment to be $F_1(u, w)$, the cost of a stage $T-1$ assignment to be $F_2(u, w)$, and so on. Now the dynamic shoot-look-shoot WTA problem (DWTA-SLS) is defined as the following recursive optimization problem:

DWTA – SLS

$$
\min F_1 = \min_{x} \sum_{u \in \mathbb{Z}^N} P(U = u) F_2^*(u, w)
$$

subject to

$$
x \in \mathbb{B}^{M \times N}
$$

$$
\sum_{j=1}^{N} x_{ij} = 1 - w_i, \quad i = 1, \ldots, M
$$

where the optimal solution to the stage $T - 1$ problem, denoted by $F_2^*(u, w)$, is obtained by solving DWTA-SLS with the stage $T - 1$ solution now defined by F_1. In a like manner, $F_3^*(u, w)$ is obtained and so on. Note however, that we need only use a 2-stage definition at any time. This formulation is clearly amenable to Dynamic Programming (DP) methods. Unfortunately, even for $T = 2$, the number of computations required for DP is unrealistically large, being that 2^M assignments are possible in stage 1, each for which there are N^{m_2} possible assignments in stage 2, where m_2 is the number of weapons left for stage 2.

Hosein [Hosein, 1989] studied DWTA-SLS in detail and found useful solutions for a simpler case when the probabilities of survival q_{ij} were no longer dependent on weapon or target but instead only dependent on the stage t. When this assumption is made our decision variable is now m_t, the number of weapons used in stage t. Hosein shows that these problems share an optimal strategy:

Optimal strategy of stage dependent probabilities. *For $t = 1, \ldots, T$, the m_t weapons used at stage t should be spread evenly over all the N targets.*

Of course we must still determine optimal values for $m_t, t = 1, \ldots T$. Hosein examines 3 special cases.

Case 1. *If $M \leq N$ then assign all weapons in the stage with the lowest $q(t)$.*
This is essentially the SWTA problem.

Case 2. *If $M \geq N$ and $q(t) \leq q(t + 1)$, $t = 1, \ldots, T - 1$, then $m_1^*/geqN$.*
Case 2 means that for $q(t)$ nondecreasing, the optimal first stage assignment is to use at least as many weapons as there are targets.

Case 3. *If $T > 1 + \frac{M-N}{2}$ for $M > N$ and $q(t) = q$ for all t, then $m_1^* = N$.*
This simply means that for large numbers of intervals and a constant q, the optimal strategy is to assign exactly N weapons, and by the fact that m_t weapons used at stage t should be spread evenly over all the N targets, we surmise that each target should get exactly one weapon in stage 1.

Finally, Hosein explored DWTA-SLS for very large numbers of targets by developing a limit on F_1 as $N \to \infty$.

3.2   STOCHASTIC DEMAND PROBLEMS

In the static WTA formulations, the decision maker knows at one instant in time the total number of targets N and their locations. Consider another WTA problem, where at some time t, only a subset of the total targets are known to the decision maker. Let the number of targets known at time t be denoted by $n(t)$ and $n(t) \leq N$ for any t. As time progresses, additional targets are discovered, hence $n(t)$ is nondecreasing with time. Since N is unknown, an assignment of weapons can only be made to the $n(t)$ weapons that are known or else reserved for targets that are expected in the future. These types of Dynamic

WTA problems essentially have a stochastic demand and were first introduced in [Murphey, 1999].

If the targets are ranked according to their value, as in the static WTA problems, then it is possible that at time t only low value targets have been found, in which case it is desirable to wait before making an assignment. If, after waiting time τ, all of the targets are discovered, that is $n(\tau) = N$, then a static WTA problem results. Assume however that there is a cost associated with waiting to make an assignment. This cost is due to the fact that each weapon has finite fuel, so once it is deployed, it has a finite time to detect and engage targets. Indeed, the target may pass out of the weapon's field of view altogether.

Define T to be the latest possible time to assign a weapon to a target. The *stochastic demand* problem (DWTA-SD) then attempts to find an assignment $x(t)$, which minimizes the value of targets over all $t \in [0, T]$. The cost function monotonically increases with time to account for the cost of waiting and, as in the Static WTA problems, is weighted by target values $V_j \in \mathbb{V}$. Assume that $\mathbb{V}$ does not change over time.

To make the problem easier to discuss, time will be discretized. A deterministic formulation for DWTA-SD is

DWTA-SD

$$\text{minimize} \quad \sum_{t=1}^{T} c(t) \sum_{j=1}^{n(i)} V_j \prod_{i=1}^{M} q_{ij}^{x_{ij}(t)}$$

subject to

$$\sum_{t=1}^{T} \sum_{j=1}^{n(t)} x_{ij}(t) = 1, \quad i = 1, 2, \ldots, M$$

$$V_j \in \mathbb{V} \in \mathbb{R}_+^K, \quad j = 1, 2, \ldots, n(T)$$

$$x(t) \in \mathbb{B}^{n(i)}, \quad t = 1, 2, \ldots, T$$

where $c(t)$ is a nondecreasing function that represents the cost of waiting. To model attrition, the number of munitions M may be a function of t. If so, $c(t)$ need only represent the cost of missed reacquisitions, which can be formulated as a linear scalar function of t. Clearly DWTA-SD is a very hard problem since at each time interval, a SWTA must be solved. Hence we consider DWTA-SD with a uniform weapon assumption:

DWTA-SDU

$$\text{minimize} \quad \sum_{t=1}^{T} c(t) \sum_{j=1}^{n(t)} V_j q_j^{x_j(t)}$$

subject to

$$\sum_{t=1}^{T} \sum_{j=1}^{n(t)} x_j(t) = M,$$

$$V_j \in V \in \mathbb{R}_+^K, \quad j = 1, 2, \ldots, n(T)$$

$$x(t) \in \mathbb{Z}^{n(t)}, \quad t = 1, 2, \ldots, T$$

On-line implementation of DWTA-SD or even DWTA-SDU is unrealistic since it determines the optimal decision after the final time T has passed, yet assignments for any time interval $t < T$ must be made during that interval. A realistic model must account for the fact that at interval I, $n(t)$ are stochastic for $t > I$. There are two approaches for a stochastic model; two-stage and multistage. We will only develop the two-stage model here since it is difficult to predict exactly *when* an unknown target might appear as would be required in a multistage model. It is fairly straightforward to model how many targets of each class remain undetected within a finite geographic region at any instant in time. As a result, the 2-stage model must be solved recursively at each time interval, each solution consisting of two subproblems: (1) a first stage problem which includes all targets detected during the current interval and all previous intervals, and (2) a second stage problem that includes all targets not yet detected. The first stage demand is deterministic while the second stage demand is stochastic and the trick is to balance assigning weapons to known targets against reserving weapons for possible, but not yet determined, future targets.

Let a random vector $\xi \in \Xi$ denote the number of targets in each class k, $k = 1, \ldots, K$ that have yet to be detected (e.g. targets in stage 2).

If we are given the probabilities of survival and the target values for each target class, then for any instance of random vector ξ, the second stage values of target value V_2, the probability of survival q_2 and the total number of yet to be detected (stage 2) targets n_2 are easily determined. The 2-stage stochastic programming formulation is as follows:

SP

$$Z_1(x) = \min_x f_1(x) + E_{\xi \in \Xi}[Z_2(x, \xi^j)] \qquad (3.3)$$

subject to

$$\sum_{i=1}^{n_1} x_i \leq M, \qquad (3.4)$$

$$x \leq b \qquad (3.5)$$

$$x \in \mathbb{Z}^{n_1} \qquad (3.6)$$

where $f_1(x)$ is the stage one cost function of the first stage assignment x and is integer-convex, that is, the continuous relaxation of $f_1(x)$ is convex. $E_{\xi \in \Xi}[Z_2(x, \xi^j)]$ is the expectation operator with respect to ξ. The constraint in (3.5) is used to limit the number of weapons assigned to any single target, preventing the assignment of large numbers of weapons to high value targets. The right hand side of (3.5) is determined by setting a threshold Φ_j on the reduced value of each target as in

$$q_i^{x_i} \leq \Phi_j$$

which can always be written in the form of (3.5) by applying a logarithm. $Z_2(x, \xi^j)$ is the solution to a second stage problem that clearly shows its dependence on the stochastic parameter $\xi \in \Xi$ and the first stage decision x as captured in the following program:

S2P

$$Z_2(x, \xi^j) = \min_y f_2^j(y) \qquad (3.7)$$

subject to

$$\sum_{i=1}^{n_1} x_i + \sum_{i=1}^{n_2(\xi^j)} y_i = M, \qquad (3.8)$$

$$y \in \mathbb{Z}^{n_2} \qquad (3.9)$$

where $f_2^j(y)$ is the stage two cost function of the number of weapons y assigned in the second stage. Since y depends on the outcome of ξ, $f_2^j(y)$ depends on ξ and furthermore is integer-convex. Specifically, $f_1(x)$ and $f_2^j(y)$ are the following:

$$f_1(x) = \sum_{i=1}^{n_1} V_1^i(q_1^i)^{x_i} \qquad (3.10)$$

$$f_2^j(y) = \sum_{i=1}^{n_2(\xi)} V_2^i(\xi)q_2^i(\xi)^{y_i} \qquad (3.11)$$

Due to the discrete support Ξ of ξ, SP may be replaced with the following so called *deterministic equivalent* program:

DESP

$$Z_1(x) = \min_x f_1(x) + \sum_{j=1}^{s} p^j Z_2(x, \xi^j) \qquad (3.12)$$

subject to

$$\sum_{i=1}^{n_1} x_i \leq M, \qquad (3.13)$$

$$x \leq b \qquad (3.14)$$

$$x \in \mathbb{Z}^{n_1} \qquad (3.15)$$

where s is the total number of scenarios (outcomes) for the random vector ξ.

Each stage of DESP/SP2 has an integer-convex objective and integer affine constraints. The classic approach to solving a stochastic programming problem with discrete random support is by the decomposition method. Decomposition methods decouple the stage 1 and stage 2 problems by first solving a variant of the stage 1 problem, often called the *current problem* (CP) with a scalar variable θ taking the place of the stage 2 solution so that

$$\theta \leq \sum_{j=1}^{s} p^j Z_2(x, \xi^j).$$

The stage 2 problem is solved for each $\xi^j, j = 1, \ldots s$. Information from the stage 2 problems is used to develop a supporting hyperplane constraint which bounds the first stage resource to θ (*optimality cut*). This constraint is added to CP which is solved once again. Iterations between the CP and stage 2 problems continue until the new solution is ϵ-close to the latest optimality cut.

Unfortunately, the integrality of the objective in DESP/SP2 no longer allows supporting hyperplanes for the stage 2 problem (see for example [Carœand Tind, 1998], [Laporte and Louveaux, 1993], [Shultz et al., 1998]). Nonetheless, a lower bound on the optimal solution can be obtained by applying a decomposition method on a relaxed version of DESP/SP2. The algorithm presented in Figure 3.2 finds a lower bound by solving a similar problem to ϵ-optimality. The problem solved replaces the integrality of y in S2P with a continuous version:

CS2P

$$Z_2(x, \xi^j) = \min_y f_2^j(y) \tag{3.16}$$

subject to

$$\sum_{i=1}^{n_1} x_i + \sum_{i=1}^{n_2(\xi^j)} y_i \leq M, \tag{3.17}$$

$$y \geq 0, y \in \mathbb{R}^{n_2} \tag{3.18}$$

Details of the method are in [Murphey, 1999] where it is shown that it terminates in a finite number of iterations with a solution that is ϵ-optimal for DESP/CS2P and consequently yields a lower bound solution for DESP/S2P. An algorithm for the current problem in step 1 of the lower bounding algorithm in Figure 3.2 is needed since non-Boolean integrality and non-linearity prevent us from using MMR as for SWTA-U. Whereas the SWTA constrains the sum of the decision variables to equality with M, the current problem has inequality constraints on the sum of the decisions as in equations (3.20) and (3.22). A natural approach for handling the constraints is to iteratively apply MMR to the current problem where the constraints (3.20) and (3.22) are replaced with the constraint

$$\sum_{j=1}^{N} x_j = m$$

where

$$m = \frac{\theta - d}{A}$$

and m is iterated from $0, 1, \ldots M$. Then for each value of m, the value of θ is uniquely determined and hence fixed in the objective.

References

[denBroeder et al., 1959] denBroeder, G., Ellison, R., and Emerling, L. (1959). On optimum target assignments. *Operations Research*, 7:322–326.

[Carœand Tind, 1998] Carœ, C. C. and Tind, J. T. (1998). L-shaped decomposition of two-stage stochastic programs with integer recourse. *Mathematical Programming*, 83:451–464.

```
Lower Bound Algorithm For DESP/SP2.
Step 0. Set θ = 0, L = 0, k = 1.
Step 1. Solve the current problem
```

Current problem

$$\bar{Z}_1(x) = \min_{x,\theta} f_1(x) + \theta \tag{3.19}$$

$$\text{subject to} \sum_{i=1}^{n_1} x_i \leq M, \tag{3.20}$$

$$x \leq b \tag{3.21}$$

$$A^k x + d^k \leq \theta \quad k = 1, \ldots, L \tag{3.22}$$

$$x \geq 0, \quad x \in \mathbb{Z}^{n_1} \tag{3.23}$$

If infeasible, STOP: problem is infeasible.
Else denote solution by (x^k, θ^k).

Step 2. For $j = 1, \ldots, s$ solve second stage primal:

$$Z_2(x^k, \xi^j) = \min_y f_2^j(y)$$

$$\text{subject to} \sum_{i=1}^{n_1} x_i^k + \sum_{i=1}^{n_2(\xi^j)} y_i = M,$$

$$y \geq 0, \quad y \in \mathbb{R}^{n_2}$$

and its dual:

$$\max_\lambda \min_{y \geq 0} f_2^j(y) + \lambda^j \left(\sum_{i=1}^{n_1} x_i^k + \sum_{i=1}^{n_2(\xi^j)} y_i - M \right)$$

Let the solutions to these problems be (y^{jk}, λ^{jk}).
Step 3. Define

$$A = \sum_{j=1}^s p^j \lambda^{jk}$$

$$d = \sum_{j=1}^s p^j \left(f_2^j(y^k) + \lambda^{jk}(e^T y^{jk} - M) \right)$$

where e is a vector of ones of comparable size to y^{jk}. If $\theta^k < Ae^T x^k + d + \epsilon$, for $\epsilon > 0$ where here e is a vector of ones of comparable size to x, then add cut to current problem: $A^k e^T x + d^k \leq \theta$. Let $k = k + 1$. Return to step 1.
Else STOP: (x^k, θ^k) is ϵ-optimal solution for DESP/CS2P.

Figure 3.2   Lower bound algorithm for DESP/SP2

[Eckler and Burr, 1972] Eckler, A. and Burr, S. (1972). *Mathematical Models of Target Coverage and Missile Allocation*. Military Operations Research

Society Press, Alexadria, VA.

[Hosein, 1989] Hosein, P. (1989). *A Class of Dynamic NonLinear Resource Allocation Problems*. PhD thesis, Massachusetts Institute of Technology, MA.

[Laporte and Louveaux, 1993] Laporte, G. and Louveaux, F. (1993). The integer l-shaped method for stochastic integer programs with complete recourse. *Operations Research Letters*, 13:133–142.

[Lloyd and Witsenhausen, 1986] Lloyd, S. and Witsenhausen, H. (1986). Weapons allocation is NP-complete. In *IEEE Summer Simulation Conference*.

[Manne, 1958] Manne, A. (1958). A target assignment problem. *Operations Research*, 6:346–351.

[Matlin, 1970] Matlin, S. (1970). A review of the literature on the missile allocation problem. *Operations Research*, 18:334–373.

[Murphey, 1999] Murphey, R. (1999). An approximate algorithm for a weapon target assignment stochastic program. In *Approximation and Complexity in Numerical Optimization: Continuous and Discrete Problems*. Kluwer Academic Publishers.

[Shultz et al., 1998] Shultz, R., Stougie, L., and van der Vlerk, M. (1998). Solving stochastic programs with integer recourse by enumeration: a framework using gröbner basis reductions. *Mathematical Programming*, 83:229–252.

Chapter 4

THE NONLINEAR ASSIGNMENT PROBLEM IN EXPERIMENTAL HIGH ENERGY PHYSICS

Jean-Francois Pusztaszeri

Logistics Division,
Sabre, Inc.
pusztaszeri@sabre.com

Abstract   This chapter describes a mathematical programming approach to solve the data association phase of the multiple-target tracking problem, and its implementation in the context of pattern recognition in High Energy Physics. The approach can be easily integrated within existing parameter estimation methods of dynamic systems commonly used in practice, and in particular within the framework of Kalman filtering. It also provides the only alternative to exhaustive search when optimality conditions of estimation methods are violated and when strict quality requirements are in place.

While it has been developed for a specific High Energy Physics experiment (ALEPH), the approach presented here is general enough to extend, with only little additional modeling effort, to other multiple-target tracking applications with similar operational requirements.

1.   MULTIPLE-TARGET TRACKING
1.1   OVERVIEW

Multiple-target tracking (or MTT) can be defined as the process of reconstructing trajectories of a generally unknown number of moving objects from a set of space-dependent or time-dependent signals that they trigger in a common detection region. This application is typically encountered in real-life surveillance systems typically involving large arrays of sensors, themselves interfaced with some extensive data-processing computational infrastructure. The actual track generation (or reconstruction) consists of four steps: the reconstruction of coordinate points (or observations) from raw detection signals, the association of the observations into tracks (commonly referred to as pattern recognition), the estimation of the track parameters, followed finally by consistency checks

P.M. Pardalos and L.S. Pitsoulis (eds.), Nonlinear Assignment Problems, 55–89.

on reconstructed trajectories. The first step in this sequence is handled by signal processing techniques. Data association requires methods drawn from control systems theory, artificial intelligence, and global optimization. Parameter e-valuation is done by means of static or dynamic statistical models, and the last step is generally based on application- dependent criteria. External constraints, such as real-time requirements, may apply to some (or all) of the phases.

The observations themselves may describe kinematic properties of the targets, such as position, velocity, or acceleration, or may consist of more general attributes (for example a signal magnitude), which help to determine target size or proximity. Kinematic and attribute observations are often available together, and may include a time stamp or occur periodically. The existence of at most one true observation per actual electronic sensor is generally assumed without loss of generality since observations from the same source may easily be combined in the (first) data reconstruction step. In most tracking environments, the presence of spurious observations created by random noise (or clutter) is assumed to be present. Excellent introductions to multiple-target tracking are provided by Bar-Shalom [Bar-Shalom and Fortmann, 1988] and by Blackman [Blackman, 1986].

1.2   DATA RECONSTRUCTION

This step consists of the transformation of raw signals provided by the detection system into a set of observations which serve as input to the data association problem described below. For most tracking systems, this is a stand-alone step which provides a parametric estimation of signal errors and performs a limited reduction of information such as merging observations originated from the same target, identifying observations which have been produced by more than one target, and applying a transformation of coordinates for simplified treatment. Tracking information may already be lost at this stage if, for example, the settings used to resolve sets of neighbor signals are too coarse. Statistical models are typically used at this stage to parameterize measurement errors, detector efficiency, systematic errors and process noise.

Reconstructed data, together with information on the underlying ballistic model and detector geometry, represent the input to the data association phase.

1.3   CALCULATING ASSOCIATION PROBABILITIES

This stage typically involves the calculation of "best" estimates for the state vector of a particular track hypothesis, at this stage defined as a subset of observations obtained from data reconstruction. The definition of "best" is a function of the complexity of the tracking environment. A least-square analysis, which may be adequate for some systems, will turn out to be inappropriate for others that require much finer statistical analysis tools.

For a multi-target system, this phase includes the construction of a likelihood function describing the state of the system and the correlation between hypotheses. The most general form of a likelihood function for an MTT system is notoriously complicated, and often includes models for the creation and deletion of targets, their track length, and their residual errors.

When performing multiple-target tracking in a Bayesian framework, the probabilities of feasible hypotheses are evaluated to obtain weighted state estimates and covariance (generally using a Kalman filter). In the simpler classical approach, the hypotheses with negligible probabilities are dropped, and only the most likely hypotheses are retained [Nagarajan et al., 1987].

1.4   DATA ASSOCIATION

This is the central, and in practice often the most time consuming, step of any multiple-target tracking procedure. It involves the generation of track hypotheses, the collection of which represents a partition of the set of observations. At either end of the spectrum defined by data association methods lie, respectively, purely sequential methods (greedy search over track-to-observation assignments), and so-called batch methods that process the entire observation sample in one step (global optimization). If the density of observations is large, the former tends to become extremely inaccurate while the latter are generally too compute-time intensive to be applied in practical settings, or if real-time requirements are in place. In the design of MTT systems, batch methods are typically used first, and requirements are progressively relaxed to satisfy compute-time requirements [Blackman, 1986].

The goal of data association is to find the optimal partitioning of the set of observations into true targets and noise. When performed over a dense set of observations, an MTT system should resolve all observations-to-track correlations, and this may represent a daunting task, but in sparse regions, data association problem is on the other hand trivial. A bad association is a less serious error than a wrongly identified or a lost target for some types of applications, so emphasis is generally placed on maintaining tracks for targets which have been identified, and possibly identifying new targets. For surveillance applications, track loss represents the worst possible error. In the context of High Energy Physics tracking, wrongly associated observations represent the most important source of errors.

1.5   PARAMETER CALCULATION

The parameterization of a track is entirely problem-dependent, since it is a function of the ballistic models used for specific applications. This step is essentially done if the estimators have been used directly in the likelihood function described earlier, but since the function is rarely used "as is" in practice, this step is considered to be independent from the calculation of association probabilities. When a recursive state estimate algorithm is used however, parameter calculation is generally merged with the data association step. In that case, the choice of tracking parameters depends on the availability of data to maintain meaningful tracking.

1.6   RECONSTRUCTION ANALYSIS

Once data association and parameter calculation is completed, a consistency check may be performed to evaluate the quality of the reconstruction. In High Energy Physics, this step is taken at analysis level, when physical hypotheses are matched against the results of the reconstruction. For the general class of surveillance tracking applications, and in particular those with real-time requirements, this step commonly takes the form of a performance feedback loop which is incorporated into the reconstruction algorithm.

2.   ESTIMATION OF TRACKING PARAMETERS

2.1   NOTATION

Under a restricted set of conditions, the data association and parameter estimation steps actually combine, as it will be seen below. This is shown by using the conventional notation . Given a time series $\vec{x}$, defined as the mapping $\vec{x} : N \times \omega \to \Re^n$, where N is the set of natural numbers and Ω is the set of outcomes from the probability space (Ω, ϵ, P), an n-dimensional generalized dynamic model for $\vec{x}(t)$ is a stochastic system of equations consisting of

- a vector difference equation, $\vec{x}(k + 1) = \phi(k)vecx(k) + \vec{u}(k), k = 0, 1, 2, \ldots$, where $\phi(k)$ is an arbitrary $n \times n$ operator (called the transport map), and $\vec{u}(k)$ is an $n-$ dimensional uncorrelated, normally distributed time series of mean zero, with covariance matrix $Q(k)$, known for all k. $\vec{u}(k)$ and $\vec{x}(k)$ are known as the process noise and state vector of the system respectively.

- A measurement equation, $\vec{z}(k) = H(k)\vec{x}(k) + \vec{w}(k)$, where H is a known $m \times n$ measurement operator, and $\vec{w}(k)$ is the Gaussian, uncorrelated measurement noise of mean zero and of covariance matrix $R(k)$ known for all k.

The time series $\vec{w}(k)$, $\vec{u}(k)$ and $\vec{x}(k)$ are generally assumed to be pairwise uncorrelated for all k. It is also assumed that the prior $\vec{x}(j)$ is uncorrelated with the process noise $\vec{u}(k)$, $j \leq k$. The *Best Linear Minimum Variance Estimator* (BLMVE) of $\vec{x}(k)$ with respect to

$$\vec{y}_j = [\vec{z}(0)\vec{z}(1) \ldots \vec{z}(j)]^T \qquad (4.1)$$

and with covariance matrix $P(k|j) = E((\hat{x}(k|j) - \vec{x}(k)\hat{x}(k|j) - \vec{x}(k)))$, is denoted by $\hat{x}(k|j)$. In the context of filtering methods, it is generally called as follows:

$$\hat{x}(k|j) = \begin{cases} \text{filtered estimate of } \vec{x}(k), j = k \\ \text{predicted estimate of } \vec{x}(k), j < k \\ \text{smoothed estimate of } \vec{x}(k), j > k \end{cases}$$

$\hat{x}(k|j)$ is also called the *conditional mean* of the state vector with respect to the set of measurements indexed by j.

2.2 OPTIMAL NONLINEAR PARAMETER ESTIMATION

When $\phi(k)$ and $H(k)$ are nonlinear, Bar-Shalom [Bar-Shalom and Fortmann, 1988] and others have shown that a general expression for the conditional mean is given by

$$\hat{x}(k|k) = \int \vec{x}(k)p\left[\vec{x}(k)|I^k\right]d\vec{x} \qquad (4.2)$$

where $p\left[\vec{x}(k)|I^k\right]$ is the conditional probability density function (PDF) of the state vector and I^k is the set of observations available at step k.

The conditional mean can, in principle, be computed by recursion over p,

$$p_{k+1} \propto p\left[\vec{z}(k+1)|\vec{x}(k+1)\right] \int p\left[\vec{x}(k+1)|\vec{x}(k)\right] p_k d\vec{x}_k \qquad (4.3)$$

but this expression is known to be analytically intractable and numerically expensive to implement algorithmically. $\hat{x}(k|k)$ is generally calculated by means of suboptimal methods.

It should be noted that if the dynamic system is linear, but involves non-Gaussian measurement and process noise distributions with known mean and variance, then there is no simple sufficient statistic with respect to the track parameters to model the state of the system (in an ideal case, $\hat{x}(k|k)$ is a sufficient statistic which models the entire past). Solving these systems to optimality requires the use of Equation 4.3 as well. The usual alternative in such cases is therefore to use the best linear estimator, assuming a Gaussian model and the existence of the first two moments, as described below.

2.3 OPTIMAL PARAMETER ESTIMATION FOR A LINEAR SYSTEM

If the state and measurement vectors are linear maps, *and* the measurement and process noise distributions are normally distributed, then the joint probability distribution function of the measurements $\vec{y}_k$ is written as follows:

$$p[\vec{y}_k] = \prod_{i=1}^{k} p[\vec{z}(i)|\vec{y}_{i-1}] \tag{4.4}$$

with $\vec{y}_0$ representing the initial state, or prior, of the system. If the individual PDF in Equation 4.4 are Gaussian, then

$$p[\vec{y}_k] = \left[\prod_{i=1}^{k} |2\pi S(i)|^{1/2}\right] exp\left[-1/2\sum_{i=1}^{k} \vec{v}^T(i)S^{-1}(i)\vec{v}(i)\right] \tag{4.5}$$

where $\vec{v}(k)$ is the residual vector of prediction (i.e., the difference between measured and estimated vectors at step k). The statistics $\epsilon_{\vec{v}}(i) = \vec{v}^T(i)S^{-1}(i)\vec{v}(i)$ are called the normalized measured residuals squared, and since they are the sums of squares of n_y independent Gaussian random variables with zero mean and a unit standard deviation, they follow a $\chi^2_{n_y}$ distribution. The modified log likelihood function derived from it,

$$\lambda(k) = \lambda(k-1) + \epsilon_{\vec{v}}(k) \tag{4.6}$$

follows a $\chi^2_{kn_y}$ distribution with a mean kn_y and variance $2kn_y$. This can be used to check the consistency of the state estimators. If all the assumptions stated above are met, the power of the χ^2 test may alone be used to determine the adequacy of the stochastic model based on the sample of observations. Also, if several observation sequences are to be tested, the sequence which minimizes Equation 4.6 (given the same number of degrees of freedom for all sequences) will best match the model. If an ordering of observations is available, observations can be retained or rejected purely on the basis of this test, and data association and parameter estimation are done in one step.

The residuals can be obtained from Kalman filtering estimation methods. The theorem of Kalman [Kalman and Bucy, 1961] provides a recursive (and linear) expression to calculate the BLMVE $\hat{x}(k|k)$ and its covariance matrix, and hence these residuals.

2.4 EXTENDED KALMAN FILTERING

One suboptimal approach for performing nonlinear recursive dynamic estimation is provided by the Extended Kalman Filter (EKF). Assuming additive, zero-mean, uncorrelated measurement and process noise models, no correlation between process noise, measurement noise, and initial state random vectors, and knowing the initial state of the system, the transport and measurement maps are expanded in Taylor series about the linear state estimate, and higher order terms are ignored. The resulting estimate is now viewed as an approximate conditional mean for the system (its covariance matrix represents likewise an approximate mean-square error estimate for the nonlinear system). Corrections to the covariance matrix of the process noise can be applied to reduce the impact of having neglected terms in the expansion.

Applying a nonlinear map to the approximate conditional mean gives rise to numerical instabilities. A *smoothing* step is generally added to the EKF procedure, and consists in correcting the updated state vector with a residual term, taking into account the entire measurement vector. The approximate conditional mean essentially becomes a maximum *a posteriori* estimate. The simplified form of this algorithm involves recomputing (for first order filters) the Jacobian of the map at the smoothed state vector, taking into account all observations in the sample, and recalculating the filtered estimate using this new Jacobian. This is to account for the fact that the Jacobian is initially calculated, not at the true (and unknown) state vector, but at its approximate conditional mean. See [Frühwirth, 1987] for a full description of the so-called recursively iterated EKF.

2.5 CONVENTIONAL DATA RECONSTRUCTION METHODS

For MTT applications with observations available "scan-by-scan" or by layers, the Kalman filter combined with a χ^2 test may be perfectly adequate if they are supplemented by a greedy search algorithm. At every step (or layer), only the best observation is retained (or no observation is selected), and the process is iterated over the layers. However, since the optimality conditions of the filter are rarely met in practice, this algorithm is easily misled into making the wrong track to observation assignment, and these errors propagate as the process goes on. Gating techniques can be used to reduce the choice of observations, corrections can be applied to the state estimation and to the covariance matrices of the filter [Blackman, 1986], and some limited branching over hypotheses with similar scores (track splitting) can be performed [Smith and Buechler, 1975]. However, these so-called *nearest-neighbor* search algorithm produce satisfactory results only under near-ideal conditions, in which targets are well separated and errors are well understood.

Half-way between nearest-neighbor procedures described above and batch processing methods are two mainstream MTT algorithms which rely on limited hypothesis branching (or deferred logic) in the presence of ambiguities in data associations. The *Multiple-hypothesis Tracking* (MHT) was developed by Reid [Reid, 1979] and involves developing the full tree of all subsets of observations, maintaining a score for each track hypothesis thus formed. In order to reduce the exponential growth of the tree, score thresholds are used to delete unrealistic hypotheses and merge similar hypotheses together. Normalizing probabilities for hypotheses with different numbers of degrees of freedom is achieved by allowing new tracks to be formed and existing tracks to be stopped if they cannot be further extended. The *Joint Probabilistic Data Association Filter* (JPDAF) is a special case of MHT developed by Bar-Shalom, Fortmann and Scheffe [Bar-Shalom et al., 1980]. The hypothesis tree is track-based, and is an extension of the conventional track splitting method. Unlike MHT, the algorithm relies on external track initiation and deletion mechanisms. Multiple hypotheses are combined before the next scan is processed, and as such, this method is best applied to environments which are known to contain a substantial number of false targets. Both methods require the exhaustive enumeration of all hypotheses at every scan, so a ranked list of the best hypotheses for each scan must be maintained. This is achieved by solving at most $n - 1$ two-dimensional assignment problems per scan, where n is the cardinality of the largest subproblem [Cox and Miller, 1995].

Adaptive methods are best suited for applications in which little is known about the trajectories of targets, for example in situations involving maneuvering targets, or when performing contour tracking or image synthesis. A wide variety of algorithms, based on neural networks and genetic algorithms, are available for this purpose. Much work has been done to apply adaptive methods to the High Energy Physics tracking problem. Denby [Denby et al., 1990], Peterson [Peterson, 1989], Stimpfl [Stimpfl-Abele and Garrido, 1991], among others, proposed a Hopfield-type neural network approach, while Gyulassy [Gyulassy and Harlander, 1991] used a generalization of the Radon transform to derive an elastic tracking algorithm relying on a lexicographic search for deformable "template" tracks. A hybrid algorithm based on simulated annealing and making use of a mean-field approximation on the state of indicator variables was developed by Peterson [Peterson and Anderson, 1987]. Practical implementation of these methods have often revealed performances equal to simpler nearest-neighbor search, which has hampered somewhat their more widespread use in experimental physics to this day. This situation may change with the next generation of particle colliders and detectors, and the far more complex instances of tracking problems associated with them.

Among batch processing methods, the earliest attempt to use integer programming to solve the general data association problem in MTT is attributed to Morefield [Morefield, 1977], who formulated a set packing model with each column of the packing representing a track hypothesis. The formulation requires mutual independence of the observations with respect to the track partition, in order to keep the likelihood of the partition separable. With a finite set of all possible partitions, and by assuming separability of the prior as well, the global optimization problem reduces to maximizing the likelihood over the discrete posterior distribution. The score of each partition can be obtained (say by Kalman filtering) and a linear objective function is constructed.

More recently, Poore [Poore and Rijavec, 1994a], [Poore and Rijavec, 1994b], [Poore, 1994] has proposed a generalized assignment model by assuming a noiseless and perfectly efficient detector environment. MTT data association is modeled as an n-dimensional linear assignment problems (where n is the number of scans). Simulated instances containing up to a hundred tracks were reported to be solved in near real-time conditions by means of a Lagrangian relaxation algorithm and a nested two-dimensional assignment formulation to solve the terminal nodes.

3. THE HIGH ENERGY PHYSICS TRACKING PROBLEM

3.1 APPLICATIONS OF MTT

The HEP tracking problem is divided into two main categories:

- Triggering applications: the number of product particles generated by a collision must be identified in real-time together with an "adequate" measure of track quality. For experiments with a very high trigger rate, the purpose of this step is to eliminate events which do not contain interesting physics before they are processed by the full reconstruction chain, thereby freeing the processing infrastructure for the analysis of worthwhile information. The power of discrimination needs to be as good as possible.

- Full reconstruction: events which satisfied trigger requirements need to be reconstructed with optimal precision, the quality of track parameters in an event determines the quality of the physics analysis based on it. Only once a full reconstruction of the tracks has been performed can the physics of the underlying collision/annihilation process be studied, and identification of particles be performed. As the reconstruction at this stage is generally done "off-line", real-time requirements are lifted to favor precision. The application which is presented here falls within this category.

3.2 LEP PHYSICS

The Large Electron-Positron (LEP) collider located at CERN near Geneva, Switzerland, is a storage ring, 27 km in circumference, in which beams of electrons and positrons moving in opposite orbits are made to collide at a rate of several collisions per second, at specific points along the ring. Most of the collisions are lost to elastic scattering (the two particles graze each other). This yields in practice only one collision every few seconds in which one may observe the annihilation of the electron-positron pair into a Z^0 vector gauge boson, one of the messengers of the weak force. This is defined as an *event*. The Z^0 itself decays into elementary particles, and the entire process is schematically written as follows:

$$e^+ e^- \rightarrow Z^0 \rightarrow q\bar{q}, \mu\bar{\mu}, e\bar{e}, \tau\bar{\tau}, \nu\bar{\nu} \qquad (4.7)$$

(the right-hand side represents different decay modes, and bars represent antiparticles). The first decay mode, $q\bar{q}$, represents the decay of Z^0 into hadrons (themselves subclassified into baryons and mesons) via the production of quarks and antiquarks. This decay mode has a branching ratio (or probability of occurrence) of about 70 %. All other decay channels in the expression above represent the decay of the Z^0 into leptons, which, unlike hadronic events, generally contain only a small number of particles emanating directly from the point of annihilation of the Z^0, which is also known as the *primary vertex*. Tracking chambers detect all charged particles which result from these interactions, and LEP-based experimental physics is entirely devoted to the study of the Z^0 through its decays. Results are used to test the theory of the standard model which provides a unifying framework for the strong , weak and electromagnetic fundamental forces. A general introduction to High Energy Physics is provided by Okun [Okun, 1985].

The *beam spot* is defined as a region of uncertainty about the expected point where the two beams collide. The *region of interaction* comprises the beam spot, together with a spherical neighborhood of 2cm in radius around it, corresponding to the decay length of the so-called B^0 decay. This region contains some of the most interesting physics produced by the decay of the Z^0 , and from the standpoint of tracking, represents the region where the points of origin of most tracks are located, may they be primary vertices, or *secondary vertices*, the points of origin of particles which do not come from the primary vertex. Elsewhere than in the region of interaction, the only other types of interactions which are directly related to the disintegration of the Z^0 are the so-called charged and neutral decays which are described in [Okun, 1985]. The total number of charged particles in an event is called the *multiplicity*. A comprehensive list of elementary particles and their properties is published and regularly updated in Physical Review [Group, 1994]. In what follows, charged particles will be referred to as targets.

3.3   DETECTOR LAYOUT

No detector could possibly fit within the region of interaction without affecting the behavior of the particles which are under study, or being destroyed by radiations produced by the beam. To circumvent this problem, it is the flight paths of charged particles, as they travel away from their origin, that are being reconstructed, and interpolated back towards the beam spot once all tracks have been identified. To obtain accurate estimates for track parameters is therefore essential to understand the physical processes taking place in the region of interaction.

The ALEPH detector, together with its three siblings, L3, Delphi and Opal, consists of concentric detection chambers shown in Figure 4.1. The design has been optimized for observing events which are produced at, or near, the origin of the coordinate system. The components of ALEPH are tracking chambers/detectors, respectively labeled in order of increasing radius VDET, ITC and TPC. The outer components of the detector, the calorimeters, are used to measure the energy deposited by charged particles, but fill a purpose which is independent of tracking altogether. A magnet generates an electromagnetic field throughout the entire chamber, with goal to deflect charged particles, and induce a curvature on the tracks so that the physics can be observed within the confines of the detector region. Because of the magnetic field, charged particles follow a helical flight path with the main helix axis parallel to the beam axis. Field fluctuations occasionally induce small local perturbations in the helices. Charged particles are detected in tracking chambers by means of the ionization

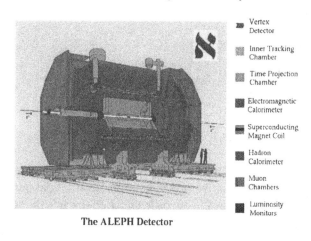

The ALEPH Detector

Figure 4.1 A cutout view of the ALEPH detector

they produce in gas or detector material. Ionization is also the main mechanism by which these particles lose energy during their time-of-flight.

The innermost ALEPH tracking device is a solid-state silicon strip *vertex detector* (VDET), made of two layers of overlapping silicon wafers with an inlay of orthogonal aluminum strips, as shown in Figure 4.2 (a and b). Radii of the two layers are 7cm and 11cm respectively. As a charged particle traverses a wafer, the ionization charge it produces in the silicon is picked up by the nearest pair of orthogonal strips The signals induced in the strips are used to identify the orthogonal pair, from which a three-dimensional point (knowing the position of the strips on the wafer) can be reconstructed. This device can be pictured as four copies of a one-dimensional detector embedded in a three-dimensional framework described by an orthogonal local coordinate system, the "z" direction used for observations which lie perpendicular to the beam axis, and $\rho\phi$ for observations which lie parallel to it. Approximately 5% of all tracks go through one or more wafer overlap(s). The silicon wafers are extremely thin (300μm). When traveling across the VDET wafer, charged

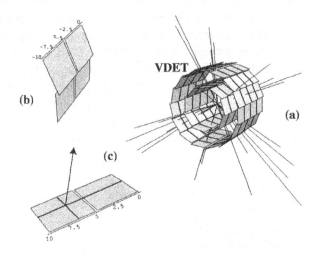

Figure 4.2   The Vertex Detector (VDET)

particles induce the ionization of the material (silicon), releasing a cloud of electrons and holes which, in the presence of an electric field normal to, and pointing out of, the wafers, drift in opposite direction towards the surface of the wafer. The energy of the carrier clouds are transferred to the physical strips at the surface, which trigger a signal via one of the 100 000 channels connected to the device. These are referred to as *hits*. Principles of detections, together with a detailed analysis of signal errors and data reconstruction, are given in [G. Batignani, 1992], [Comas et al., 1996] and [Pusztaszeri, 1996]. Efficiency and noise of the detector stand at 97% and 20 % respectively.

The shell enclosing the VDET is called the *Inner Tracking Chamber* (ITC). It is a cylindrical proportional multi-wire drift chamber with inner and outer radii of 14cm and 28cm respectively. The wires lie parallel to the beam axis, and the chamber is filled with a mixture of Ar (argon) and CO_2. A detection cell is defined by the volume enclosed by six field wires centered about an anode wire. The cylindrical volume is divided into eight layers of staggered cells. Separation between layers is about 2cm. The gas contained in the chamber is ionized by charged particles, producing electrons which, in the presence of an electric field generated by the field wires, drift towards the nearby anode wire and generate an avalanche.

At electronic level, the efficiency of the cells has been determined to be nearly 100 %. However, an ITC cell is a so-called single hit electronic device. At most one signal will be triggered by cell per event. Two or more charged particles traversing an ITC cell quasi-simultaneously will generate at most one signal, so the trace left by one particle may locally be shadowed by another. Errors of observations in the "z" direction (along the wire) is of the order of a few centimeters, which is enormous given the scale of the physics under study. Signal errors in the $\rho\phi$ direction are much better, in the order of a few microns. For tracking purpose therefore, the ITC really a two-dimensional detector, but the "z" information can be used for gating.

The outermost tracking region, the *Time Projection Chamber* or TPC, is a cylindrical drift volume with planar wire chambers at either end, and divided into sectors. Inner and outer radii of the TPC are 0.31cm and 1.80cm respectively for a length of 4.7m. The end-plate detectors are arranged in layers (or pad rows) with a separation of about 5cm between pad rows. The TPC is used to measure the charge (and the corresponding drift time) of a ionization cloud produced by a charged track traversing its volume and deposited in pads connected to a system of proportional wires located at either of its end-plates. Electron drift is achieved by the presence of an electric field along the beam axis. The raw data consists of pad hits containing the pad number, the arrival time and length of the pulse, and digitized pulse-heights. Spatial coordinates in z are reconstructed by means of the measured drift velocity, converted into a drift length and finally into a two-dimensional position. The $\rho\phi$ position is calculated differently according to the number of pads in the clusters, but in all cases is a function of the pulse-height of the pads.

The spatial resolution of the signals is known to within approximately $180\mu m$ in the $\rho\phi$ direction, and from 0.8 to 1.2 millimeters in the z direction. The errors are determined from the RMS of the width and lengths of the clusters. It should be noted that the electronic noise level in both the TPC and ITC is essentially zero. Occasionally, clouds of spurious observations may be observed as the detectors are hit by radiations and are about to trip, but these events are easily identified and are discarded by quality control procedures which are part of the data acquisition system. A view of the tracking chambers containing reconstructed hits *before* tracking takes place is shown in Figure 4.3.

3.4   JUSTIFYING A GLOBAL DATA ASSOCIATION ALGORITHM

The prime goal of the tracking procedure is to reconstruct a physics event which may in turn confirm of disprove theoretical hypotheses under study.

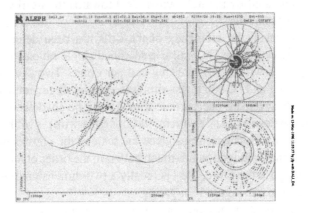

Figure 4.3   ALEPH hadronic event before reconstruction (3-D and axial views)

It is therefore critical that tracking should be free of any bias towards these hypotheses. Physical requirements dictate mainly that the number of targets must not be assumed a priori. From the Standard Model, the number of actual charged particles present in an hadronic event is known to follow a lognormal distribution with a mean of about twenty-one [D. Decamp, 1991]. However, the number of actual tracks at intermediate stages of the reconstruction procedure may be much higher than that, due to the presence of the following:

- So-called neutral decays, which involve a neutral (and therefore undetected) particle which has enough lifetime to reach the outer tracking region, and which decays into two charged particles of opposite charge. Upon reconstruction, the typical signature of this product is a "V" formed by the pair of tracks curling away from each other from a common point of origin. About one neutral decay per hadronic event can be observed on average.

- Charged decays, involving a charged particle with a "long" lifetime of up to 10^{-8} seconds, that decays into a charged particle with the same charge, and one or more neutral particles [Group, 1994]. This is represented, after reconstruction, by a helix which contains a kink at the point of decay of the incoming particle. Both sections of the helix often have nearly the same radius of curvature. Charged decays occur approximately once for every ten hadronic events. Insofar as track reconstruction and vertex identification is concerned, charged and neutral decays represent exactly the same problem, namely the identification of a substantial change of state of the track under observation.

- To the list of targets should be added the so-called "background track-s", defined as targets produced in interactions which are unrelated to the decay of the Z^0. Chief contributors to this set of background tracks are recoil electrons (interaction of an electron with material) observed on average once every ten to twenty events, electrons coming from either end plate with enough energy to work against the electric field (characterized by a very small helix parallel to, but far away from the beam pipe, and observed once every three to four events), additional loops of spiraling tracks (on average once per event), and very rare occurrences of nuclear interactions of charged particles with detector material. These interactions do not contribute to the physics understanding of processes under study by hadronic experiments, but they do generate correlated noise. Reconstructing tracks for these targets is therefore essentially for the purpose of noise removal.

- The process by which a charged particle is deflected when passing through a dense medium is known as *multiple scattering*. It is fully described in [Scott, 1963], and is based on the principle of Coulomb interaction between the particle and the nuclei of the medium. As the particle travels through the medium, it undergoes small changes of direction by interacting with these nuclei. The scattering distribution follows the model of Molière, and may be approximated by a normal distribution if the deflection angle is small. As the angle increases, however, the kinematic system tends to exhibit Rutherford scattering properties, and the tails of the corresponding deflection angle distribution deviates substantially from the Gaussian model. Tracks which undergo large-angle multiple scattering need to be treated as separate targets.

Additional tracking requirements are outlined below:

- During reconstruction, the end points of a given track are unknown, since charged particles may be created or may decay anywhere in the detector region. Particles may emanate directly from the primary vertex, or may be the products of secondary decays.

- Unlike most other MTT applications, the generation of targets is quasi-simultaneous in view of the very short lifetime (10^{-8} to 10^{-16} seconds) of particles generated by the Z^0 decay. Since the lifetime of the charged particles under consideration is much shorter than the refresh rate of the readout electronics of the detectors, all observations can be considered to have been generated simultaneously for an event. This also prohibits any possibility of ordering observations according to some time-stamp provided by the readout system. Defining a processing order for observations can only be made, therefore, with reference to the expected direction of flight of particles. But while most particles do travel away from their origin towards the outer detector regions, defining a radius-based scan sequence would be inappropriate, since it is very common to see high curvature (low momentum) tracks looping within a particular subdetector.

- By the same token, the set of observations is available at once, in total contrast with surveillance tracking applications in which data are ordered (and should be used) in a strict sequence by the tracking algorithm.

- Also, while surveillance tracking is devoted to maintaining the best and most recent estimate of target parameters, HEP tracking applications require entire tracks to be reconstructed before any physics analysis may take place.

- Unlike surveillance applications, tracking an hadronic event does not require a real-time treatment, which is lifted in favor of precision. Near-real time requirements are necessary, however, since analyses generally require examining a large sample of data, so a reasonable compute-time bound should be maintained on average. A practical limit to the all-inclusive average processing time (including data processing and association, together with track reconstruction) is that it should not exceed ten to fifteen seconds per event (and preferably be much less). Timing constraints arise mainly from anticipated computer performances (and indirectly from material budgets which need to be respected). This bound can be relaxed in special cases, where the analysis of a sample of events may require the best possible precision. As an example, one can cite the experimental confirmation of the existence of the top quark by the CDF and D0 collaborations at Fermilab near Chicago in 1995. The complete initial sample which led to the identification of the top quark signature consisted of only 27 (for CDF) and 17 (for D0) events respectively [F. Abe, 1995]. In HEP tracking applications, compute-time requirements are somewhat flexible.

- Formulation of a likelihood function that captures global data association is impractical in the HEP context, since multiple assignments and mutual exclusions of observations are very common there, and require the construction of conditional probabilities that can be very complex (see [Chang et al., 1994a], [Chang et al., 1994b], [Li and Bar-Shalom, 1991], [Li and Bar-Shalom, 1994] and [Trunk and Wilson, 1981]). The JPDAF and MHT algorithms, which are built on these conditional probabilities, may not therefore be applied so easily to the HEP tracking problem. It would be extremely cumbersome to compute the effective efficiency of a single-hit electronic device, such as the ITC, described above, by means of conditional probabilities. On the other hand, this can be formulated easily as a global constraint which is part of an integer programming formulation, as will be shown later.

3.5   SIMPLIFICATIONS

Alignment and other systematic errors have been neglected in the implementation described below: tracking chambers undergo regular maintenance, during which the detectors are removed and refitted. This is followed by a precise relative realignment of the chambers with respect to the reference frame of the experiment setup, in which each subdetector is treated as a rigid body. Due to the large relative difference in weight and size of the detectors, this process generally introduces systematic translational or axial position errors in the observations, which need to be taken into account during the data reconstruction phase. The calculation of systematic errors due to alignment is described in [Comas et al., 1996]. It is performed by first determining the internal positioning of the components of each subdetector, and then calculating the global relative positioning of the three devices. This, however, cannot be achieved by other means than explicitly reconstructing events (and therefore tracks), and measuring individually the residual *pull* of a track for each chamber. To achieve the greatest possible calibration precision (and therefore reduce systematic errors as much as possible), precision tracking is applied *a priori* to selected events with very low multiplicity and known topologies (for example, $\tau \bar{\tau}$ decays [Pusztaszeri, 1996]). It does not enter the model described below.

Nuclear interactions and energy loss are two physical processes that adversely affect the quality of tracking if neglected. The former is very rare and consists of the annihilation of the original particle while traversing some thick detector medium, and the generation of a new set of particles, the process bearing no relation to the original interaction under study. Energy loss is the process by which all particles lose energy when interacting with the detection medium, including gas. This is reflected by a change of curvature in the track of the particle. The curvature is itself inversely proportional to the momentum of the particle. In the ALEPH tracking environment, effects of energy loss were shown to be negligible away from detector walls [Comas et al., 1996], so this process has been ignored in the tracking model which assumes a constant track curvature. However, this assumption may not hold in the context of other experiments.

4.   COMBINATORIAL TRACK RECONSTRUCTION

4.1   MOTIVATION

As mentioned earlier, MHT and JPDAF do not lend themselves well to High Energy Physics tracking since it would be extremely difficult to derive a likelihood function that captures creation and deletion of targets, together with track length, as required by these methods. Also, even if a good approximation were attained to normalize hypotheses at every scan, the algorithms would return only the latest state estimates of the targets, which is only part of the information required to solve the HEP tracking problem (the latter requires optimal estimators of the state vector at any point on the track and especially near its point of origin). Finally, the scan sequence which is naturally available for surveillance applications is not by any means easily available in the HEP context.

HEP tracking involves solving a large number of similar instances of the same problem, so in the presence of common structures from one instance to the next, it is therefore natural to consider solving the data association problem with the help of adaptive methods, and in particular neural networks, the training sample consisting of a large number of simulated events. This approach may seem even more appealing since random perturbations in the deterministic trajectory model of particle motion are regularly observed, and sometimes even induce a change of state in the path of the particle.

However, fluctuations from the model are often not so large in practice, and when they are, deterministic corrections are often successful. What is left therefore is the possibility to account for the unknown prior distributions which enter the input to the tracking problem. However, the uncertainty in the input is generally dominated by data association errors which most often can be traced back to modeling flaws in the reconstruction algorithms. Also, no adaptive algorithm proposed up to now has taken into account correlations induced by competition of tracks for observations. Tracks are instead identified one by one, assuming at each step an unrealistically high noise rate to account for the presence of other targets (several authors report on MTT simulation runs involving noise rates ranging up to a thousand percent and applied to noise-free detectors). Also, it should be stressed that no clear compute-time bound, or convergence criteria, can generally be derived for these methods, putting their use for precision tracking into question. Finally, a risk exists of introducing a substantial bias in the reconstruction by using training samples consisting of simulated data, which themselves contain information on the physics under study.

Among the combinatorial models, the generalized assignment formulation proposed by Poore did not take into account detector noise and inefficiencies, a major drawback for the tracking problem under study here. Also, the set packing model of Morefield was formulated well before combinatorial methods could benefit from advances in linear programming and network flow algorithms. It seemed only worthwhile to revisit this model, given that it could be adapted easily to the existing tracking framework used in ALEPH.

4.2   TRACK RECONSTRUCTION IN THE ITC

A likelihood function used for general MTT problems has been derived by Blackman [Blackman, 1986]. If j represents the track index, it is in the form

$$Q = P_0(n, n_F) \prod_{j=1}^{n} P_{TL}(D_j) P_D(NU_j|D_j) \prod_{l=1}^{NU_j} P_{ER}(y_{jl}) \qquad (4.8)$$

where $P_0(n, nF)$, D_j, $PTL(D_j)$, $P_D(NU_j|D_j)$ and $PER(y_{jl})$ represent respectively the probability that n, n_F true targets and noise signals occur in the detection volume, the length of track in discrete units of detection, the probability that track has length D_j, the conditional probability of assigning NU_j signals to a track of length D_j and the probability of residual error y_{jl} for observation to be included in track . If this is viewed as the discrete posterior distribution of the many-target system, then the first factors represent a prior, which is assumed to be separable, while the product of residual errors represent the likelihood of the filter.

This expression is complete in the sense that it utilizes all information that could possibly be available in the general instance of a tracking problem, but it should not be applied to the HEP tracking case directly however. First, It can be made simpler by assuming that all targets of interest can be identified by explicit generation of hypotheses. While in Blackman's approach, a closed form solution for $P_0(n, n_F)$ is obtained by integration over an infinitesimal detection cell, we can simply take this probability to be unity. In our case, all targets identified in the outer region are of interest, since the background tracks described earlier, which have been kept up to this stage do contribute to assignment ambiguities, and therefore should be treated like any other track competing for observations within some dense region. Since actual electronic noise is negligible in the ITC, the expected number of noise signals (whether spurious or created by background tracks) is essentially zero, and does not contribute to the prior.

A serious difficulty lies in trying to compute the track length distribution $P_{TL}(Dj)$. As observed earlier, a number of physics postulates may be used to restrict the track parameter space, but these should generally not be used, on pain of introducing a bias in track properties. It would be conceivable to derive an empirical expression for the track length probability based on simulation studies (the length of a regular helix is after all only a function of the curvature and dip angle of the track), but doing so will limit the power of track reconstruction to merely exhibit the properties built into the simulation (Monte Carlo) event generator. Another way is to compute track length analytically, but this requires calculating the integral of the cross-section of every known physics decay channels obtained from the parton shower model of quantum chromodynamics [Comas et al., 1996], a feat beyond the scope of this work. The track length therefore may not enter into the prior formulation in our case simply because the corresponding computational problem is intractable.

The surviving contributions to the prior are therefore the effective efficiency of the detector with respect to a given target T_i , and the distribution of observations which belong to tracks not included in the cluster. Because of the ITC target "shadowing"effect discussed earlier, the probability of detection cannot be determined from the (nearly-perfect) efficiency of individual cells, as it is dependent on the proximity of other targets, which themselves are hypotheses under study. An interesting approximation to individual cell efficiency is obtained from considering instead the degree of correlation of any two pairs of tracks with respect to observation assignment. The average efficiency of cells which are covered by the gate of the two tracks is then 50 % (having assumed a perfect electronic efficiency). This in turn can be extended to the entire sample of tracks which enter into contention. The generation of correlated tracks can easily be performed by a clustering procedure, so a reasonable approximation for the average probability of detection is given by $P_D = P_{cell}/n$, where n is the fixed number of external targets used to generate the cluster under consideration, and P_{cell} is the probability of detection of an individual ITC cell. This formulation neglects the contribution of tracks outside the cluster, which should further contribute to reduce the cell efficiency, but as these tracks must have a substantially larger gating region, the approximation is a good one. By the same argument, the average contribution that observations which truly belong to these tracks contribute to the noise rate can be neglected.

Clustering has therefore the double advantage of breaking problems down into subproblems, the solutions of which are likely to be part of the overall global solution, and to provide a reasonable approximation for the prior of the objective function. This can be generalized to the statement that, if a sequential approach is relied upon, the effective efficiency and noise rate of a single-hit electronic detector may only be computed in the limiting case where only one layer is present. The identification of the core ambiguity of the track assignment problem, if it can be done, yields a reasonable estimator for the prior.

Under the assumptions stated above, the posterior distribution is then written

$$Q = \prod_{j=1}^{n} \left(\frac{P_{cell}}{n} \right)^{NU_j} \left(1 - \frac{P_{cell}}{n} \right)^{D_j - NU_j} \prod_{l=1}^{NU_j} \frac{e^{-d_{ij}^2/2}}{2\pi\sqrt{|S_{jl}|}} \qquad (4.9)$$

where NU_j and D_j are the number of observations and cell crossings assigned to track j respectively, while d_{jl} and S_{jl} are the norm and covariance matrix of the residual vector of that track after adding observation l, and are obtained from the EKF. Maximizing this expression is equivalent to maximizing its logarithm, which gives

$$L = \ln Q =$$

$$\sum_{j=1}^{n} \left[(D_j - NU_j) \ln \left(1 - \frac{P_{cell}}{n} \right) + \sum_{l=1}^{NU_j} \left(\ln \left(\frac{P_{cell}/n}{2\pi\sqrt{|S_{jl}|}} \right) - \frac{d_{jl}^2}{2} \right) \right],$$

$$n > P_{cell}$$

an expression which is separable in terms of individual track hypotheses. Letting $N_j^m = D_j - NU_j$, the individual score for a track hypothesis T_j is given by

$$c_j = N_j \ln \left(1 - \frac{P_{cell}}{n} \right) + \sum_{l=1}^{NU_j} \left(\ln \left(\frac{P_{cell}/n}{2\pi\sqrt{|S_{jl}|}} \right) - \frac{d_{jl}^2}{2} \right) \qquad (4.10)$$

Denoting by C the terms which do not depend on the quality of the fit, the score of a track is positive when

$$\sum_{l=1}^{NU_j} \ln \frac{1}{|S_{jl}|} > \sum_{l=1}^{NU_j} d_{jl}^2 + C \qquad (4.11)$$

corresponding to residual vectors of very small magnitudes, and with a very small difference between the products of variances and covariances of their components. Knowing the direction of the track with respect to the ITC cell (from outer tracking or from standalone hypothesis), the fit is performed independently for both the left and right observations.

It should be pointed out that, if the clustering procedure fails, or a more accurate parameterization of the prior is required, the objective function will be rendered nonlinear by the introduction of the effective efficiency which is now a function of the partition, and no longer one of the individual track hypotheses.

The individual score for a track hypothesis was derived from the posterior density of a partition of the set of observations into n tracks. The scores reflect the absence of noise in the model, but to require an actual partitioning of the set of data is placing too strong a requirement on the quality of the ballistic and noise model which enters the track residuals. Assigning a wrong observation to a track is a more serious error than dropping an observation from it, so to allow for occasional observations to be removed from consideration, it is preferable to use a set packing model. From the above, letting

$$x_j = \begin{cases} 1 & \text{if track } j \text{ belongs to a packing of } R \\ 0 & \text{otherwise} \end{cases}$$

the objective function to be maximized is

$$Z(\vec{x}) = \sum_{j=1}^{n} c_j x_j \qquad (4.12)$$

The packing constraints are defined in terms of the matrix a_{ij}, the columns of which are the incidence vectors of tracks with respect to ITC cells, i.e.,

$$a_{ij} = \begin{cases} 1 & \text{if observation } j \text{ is assigned to hypothesis } i \\ 0 & \text{otherwise} \end{cases}$$

and are thus defined as

$$\sum_j a_{ij} x_j \le 1, \forall i = \{1, \dots, |R|\} \qquad (4.13)$$

(any cell may be used at most once). A map of outer tracks to columns is used whenever necessary. If two columns are the same, only the column with the highest score is retained.

It should be noted that a pair of two-dimensional points is available from the track fit at every ITC cell, and both should be kept if they satisfy the gating conditions, since the decision of removing an observation can only be made once the hypothesis has been fully formed. This in effect doubles the number of hypotheses at every layer, and now seen as two independent hypotheses, the multiplexed track does not induce a violation of the packing constraints.

4.3   TRACK RECONSTRUCTION IN THE INNER VERTEX REGION

Unlike the other tracking detectors, the ALEPH Vertex Detector is a solid state device that provides *at most a pair* of three-dimensional observations for every track traversing its silicon wafers. This information is therefore not sufficient to define a track, and standalone tracking in the VDET is not possible. In order to benefit from observations provided by this detector, outer tracks obtained in the TPC and in the ITC must be extrapolated, and an assignment of observations to tracks must take place. As mentioned earlier, the inner shell of this detector is located only seven centimeters away from the beam spot region, and the resolution of observations there is in the order of a few microns. These two facts combined make the association of VDET observations to tracks very appealing, since doing so may substantially increase the resolution of the track as it is interpolated towards the vertex region.

Recalling that the VDET does not provide space-point observations like the other two chambers, a crucial step consists in finding the proper combination of observations for every track. Again, this does not represent a major difficulty if only one target is present, but as the VDET is located so close to the beam spot region, track separation in this region is altogether much smaller than anywhere else in the detection volume.

The VDET tracking model has been fully described in [Pusztaszeri et al., 1996]. Figure 4.4 shows a typical example of the association ambiguity taking place when two targets compete for the same observations, represented by two sets of orthogonal lines in the figure. The five standard-deviations elliptic cones of extrapolation of the outer tracks cover generally more than a unique combination of vertex detector observations in each layer. The ambiguity contained in this assignment provides the incentive for applying combinatorial optimization to improve the matching and overall tracking quality. Two partial tracks ex-

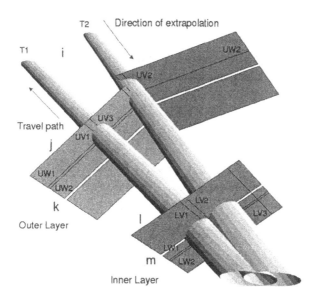

Figure 4.4   Five-dimensional assignment

trapolated from the outer region into the vertex detector, together with their five standard-deviation extrapolation errors. A possible assignment of observations is indicated by matching indissoluble observations are correlated in terms of track assignment.

The intersection of the cones with the wafer planes, together with the observations they cover, define the inputs to a global assignment. In this formulation, a given track may be assigned no observations (when it misses the detector altogether) and up to eight when its extrapolation footprint covers overlap regions in the two layers of the detector. The set of input tracks is indexed by i, and observations in the four layer-view combinations is indexed by j, k, l, m respectively.

A natural five-dimensional assignment formulation for this problem is derived by ensuring that each outer track is matched to *exactly* on VDET hit per layer and per view. This formulation is not imperiled by wafer overlap if care is taken to merge hits that lie in the same layer and same view, but on different wafers. Since noise and detector inefficiencies have been shown earlier to be non-negligible, a *noise* track is introduced for garbage collection in the assignment. To simulate inefficiencies, a set of *null* hits is defined for each layer and for each view.

Track assignment put aside, there is no correlation between orthogonal observations which lie on the same layer, so the observations form pairwise-disjoint sets with respect to layers and views. The amplitude of individual observations, however, referred to as their pulse-height is measured and allows to determine whether they may be used by one track (labeled from now on "single hits") or more tracks ("undecided hits").

The problem input consists therefore of a set of $n_i + 1$ outer tracks, $n_j + 1$ $\rho - \phi$ hits and $n_k + 1$ z hits on the outer layer, and $n_l + 1$ $\rho - \phi$ hits and $n_m + 1$ z hits on the inner layer:

$$i \in \{0, \dots, n_i\}$$
$$j \in \{0, \dots, n_{j,real,single}, \dots, n_{j,real}, \dots, n_j\}$$
$$k \in \{0, \dots, n_{k,real,single}, \dots, n_{k,real}, \dots, n_k\}$$
$$l \in \{0, \dots, n_{l,real,single}, \dots, n_{l,real}, \dots, n_l\}$$
$$m \in \{0, \dots, n_{m,real,single}, \dots, n_{m,real}, \dots, n_m\}$$

If UV_j, UW_k, LV_l and LW_m represent two orthogonal pairs of hits on the outer and lower layers of the vertex detector respectively, the assignment variable is written as

$$x_{ijklm} = \begin{cases} 1 & \text{if track } T_j \text{ is assigned to } \{UV_j, UW_k, LV_l, LW_m\}, \\ 0 & \text{otherwise}, \end{cases}$$

for all outer tracks T_i.

While it is possible to consider assignment problems of lower dimensions to solve this problem, doing so would be impractical. A sequential assignment across layer, but global at each layer, may so greatly constraint the gates of extrapolation that no feasible hit combination may be found once the first assignment is complete. Likewise, a global assignment across layers may benefit from the excellent precision of long extrapolated tracks, but might miss obvious hit combination at each layer that may have been formed by lower momentum tracks. Because of broad range of target parameters, the problem must be solved globally.

A linear approximation for the score c_{ijklm} of each five-dimensional assignment was derived in [Pusztaszeri et al., 1996]. The score c_{ijklm} of each individual five-dimensional combination is derived from the residuals obtained from Kalman filtering, and the objective function takes the form

$$Z(\vec{X}_{ijklm}) = \sum_{i,j,k,l,m=0}^{N} c_{ijklm} \cdot x_{ijklm} \qquad (4.14)$$

which is to be minimized.

With null hits present in the input set, real tracks must always be assigned a full hit combination, producing a first set of constraints

$$\sum_{j,k,l,m=0}^{N} x_{ijklm} = 1, \forall i \in \{1, \dots, n_i.\} \qquad (4.15)$$

A special constraint must exist for the noise track to ensure reasonable bounds on the number of noise hits, to match the noise level of the detector (captured by two parameters a and b):

$$a \leq \sum_{j,k,l,m=0}^{N} x_{0jklm} \leq b. \qquad (4.16)$$

Real hit constraints are index-symmetric for each layer and for each view, and are thus explicitly written for the $\rho - \phi$ view of the outer layer (indexed by j) only. Other constraints follow naturally by permutation of indices. For every real hit UV_j, the integer $M_j \geq 0$ indicates the number of its occurrences in a merged (fused) overlap hit. If that number is non-zero, the vector $G_j(M_j)$ contains the index map of the logical overlap hits (in general, one real hit may belong to more than one overlap pair). Single hits must be used exactly once, be it as part of the noise track, as part of an overlap hit combination, and if that is the case, subject to mutual exclusion between itself and the fused hit combination it is occurring in

$$\sum_{i,k,l,m=0}^{N} \left(x_{ijklm} + \sum_{m=1}^{M_j} x_{iG_j^m klm} \right) = 1, \forall j \in \{1, \dots, n_{j,real,single}\} \qquad (4.17)$$

Hits labeled "undecided" are subject to similar constraints. The probability of using these observations more than twice is negligible, given the difference in magnitude between true track separation and vertex detector resolution. This yields

$$1 \leq \sum_{i,k,l,m=0}^{N} \left(x_{ijklm} + \sum_{m=1}^{M_j} x_{iG_j^m klm} \right) \leq 2, \tag{4.18}$$

$$\forall j \in \{n_{j,real,single} + 1, \dots, n_{j,real}\}$$

Note that null hits are not subject to any constraints in this formulation. Yet, they could be part of a global constraint to ensure that the model matches the observed detector inefficiency, but this would require a substantial amount of parametric study, which was beyond the scope of this work.

5. IMPLEMENTATION

5.1 PATTERN RECOGNITION IN THE ITC

The model described earlier was implemented in the ALEPH reconstruction code. The implementation relied on the existence of a set of partial tracks reconstructed by means of nearest-neighbor search in the outer TPC tracking region for *some* well-separated targets. An attempt was made to decompose the original problem into subproblems using conventional gating and clustering methods (see [Blackman, 1986] for example). A five standard deviations gate in the position error of outer tracks was set, whenever the latter were available. For most high-multiplicity events, it was observed that the problems thus generated matched approximately the geometric decomposition of the event into separate antipodal jets, resulting in an average of two subproblems per event, each containing an approximately equal number of tracks. The average number of tracks per component was approximately ten. A distribution of the number of components generated after clustering had completed is shown in Figure 4.5. Over a sample of more than 30 000 hadronic events, a little less than twenty problems could not be decomposed, and most problems could be broken into three to five subproblems. This, however, does not give any indication of how much the complexity of the subproblems has been reduced. It is worth noting that the number of tracks in a cluster is roughly representative of the number of patterns (and therefore the number of columns) that enter the incidence matrix of the set packing instance.

In order to avoid resorting to some formal column generation scheme, an attempt was made to explicitly generate all columns of subproblems after completing a straightforward initial block diagonal decomposition based in part on outer tracking information. Exhaustive column generation proved to be possible for all but twenty events. The number of columns per component was observed in a range of 200 to 22 000, with a mean of about 2000. It took up to four minutes of CPU time on a RISC workstation to generate hypotheses for the largest subproblems.

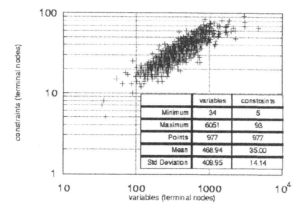

Figure 4.5   Component distribution after clustering

Set packing instances were solved using a commercial mixed-integer programming package (MINTO) interfaced with a linear programming solver (C-PLEX). Figure 4.6 shows the size distribution for the innermost terminal node of each decomposed problem. In most cases, these instances could be solved within one second of CPU time. A measure of the quality of reconstruction

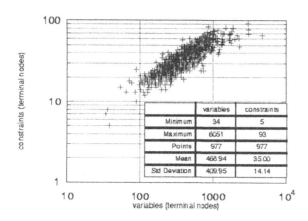

Figure 4.6   Size of terminal nodes

obtained by the global track finding algorithm was obtained by conducting a comparative statistical performance analysis with aim to calculate the errors in the *impact parameter* of a large sample of reconstructed tracks. The impact parameter of a track is defined as the distance of closest approach between that track and the primary vertex. In this process, the global algorithm was pitted against a conventional nearest-neighbor search algorithm implemented earlier in the ALEPH reconstruction code (JULIA). The distribution was obtained for tracks reconstructed by both the old and new ITC code, using a common subset of targets identified in the TPC. The results are shown in Figure 4.7. Tracks used in the sample were only required to contain a minimum of three assigned ITC observations, and the improvement in the position error was seen to be nearly systematic, with only 29 tracks, out of a sample of more than 25 000 retaining a better resolution after reconstruction by the old code. Performances of the old and new code were on par for what were presumably well separated targets, or high momentum tracks which did not enter into contention with one another for the assignment of observations. This set of "no improvement" tracks is shown on the $x = y$ line on the plot. Overall, a thirty percent gain in the mean of the distribution was attained, and standard deviation was slightly better. More than half of the instances under study were solved to optimality

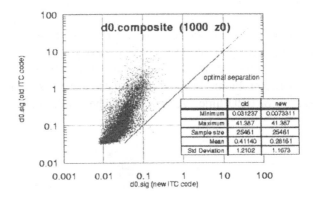

Figure 4.7   d0 improvements with new pattern recognition in the ITC. 1000 hadronic events representing approximately 25000 tracks identified by both the old and improved ITC code. Partial TPC tracks were used by both, if available

within a second of compute-time, and only one instance did not solve within the prescribed time allocation of 30 minutes. The resolution of the two-dimensional impact parameter was seen to have improved substantially after application of the combinatorial method.

5.2   TRACK RECONSTRUCTION IN THE INNER REGION

Tracks reconstructed earlier were extrapolated into the VDET region to provide the track set necessary for the five-dimensional assignment problem. A gate of five standard deviation was used for each track. Care was taken to reject outer tracks if their gate of extrapolation did not lie within the region of acceptance of the detector: tracks with a polar angle of more than 60 degrees were rejected. For an average event, approximately 30 tracks were eventually extrapolated towards the vertex region, and through the VDET.

A clustering heuristic was used to decompose the problems: a track was removed from the set of contending tracks (a "connected component") if its error of extrapolation was found to be much larger than its neighbors. With one or more tracks thus removed, the component would be checked for connectedness, and if possible, further decomposed. A sequential assignment of observations was performed over the set of nested components, with the components containing the smallest extrapolation errors fitted first, the observations assigned to them removed from the set, and the process iterating over the components containing tracks with larger errors.

A representative sample of three hundred hadronic events was reconstructed by both the Kalman filter-based nearest-neighbor algorithm (the old code that was eventually replaced) and by the new combinatorial tracker described in this chapter. The results were analyzed with the help of a graphical event display software. A search for differences in the graphic representation of each event was performed by using a convenient projection called a V-plot, showing three dimensional outer tracking information together with the hits of the vertex detector in a common plane. The projection, described in [Drevermann et al., 1995], consists of a map of cylindrical coordinates into the two-dimensional space in which is represented as centered parallel line segments scaled by a proper factor to connect the two values of defined for every . A "V"is formed by the two lines joining the end points of the segments at each side, and scaling ensures that the slopes of the lines are equal and proportional to the momentum of the particle. A "V" of irregular or asymmetric shape represents a track with a tracking error and/or one that does not point to the primary vertex, hence the convenience of using this projection when analyzing an event display output and looking for potential reconstruction errors.

For every event, a search for asymmetric V-plots was performed on both the "old" and "new" picture, and when found, a comparison was made between the two and the result recorded, this initial scan serving as a flag to further scrutinize the event. The search was restricted to a zoom area containing all tracks in the vicinity of the highest momentum track present in the event, and, because of the jet structure of many hadronic events, to a zoom area at the antipode of that track.

In average, approximately one event out of two did not contain meaningful information in the vicinity of the highest momentum track, and only contained a few isolated tracks. Classifying VDET pattern recognition performances of the new code as better, worse or equal with respect to the old code in terms of assignment errors, the study revealed that in nine cases, a wrong association in the old code was not corrected by the new code, performances being equally bad for both methods. Fifty two events, on the other hand, showed a clear improvement in the VDET pattern recognition (which could not be traced back to outer tracking). In the region restricted to the zoom area described above, about 12 events exhibited an error with the new pattern recognition program which was not present in the old reconstruction results. In all these cases, however, the error only involved a hit which was dropped, representing only a mild form of pattern recognition errors since tracks are already overdetermined. During the entire scan however, not a single instance of wrong hit assignment by the new code was observed.

The new code was observed to have the tendency to drop hits more easily than the old code, but to form complete cross-observation patterns at least on one layer. A common case was a perfect cross-observation assigned at the outer layer, with nothing at the innermost layer. The behavior of the old code was more greedy, as it was clear that additional changes had been brought to it to create as many complete cross-observation patterns as possible, but regardless of track vicinity. One should end by stating that many instances of small-scale contention often involving only two tracks were observed in this sample.

Reconstruction errors present in the old code solutions were more numerous and therefore easier to describe. From the 52 instances studied, most occurred when three or more tracks were entering contention for one or more hits, the momentum-based classification of track priority in the assignment queue being rendered useless. Two examples are given in Figure 4.8 and in Figure 4.9, showing on the left and right track fit results obtained with the old and new codes respectively. The outer tracking input, represented by the ellipses in the projection, has been altered slightly in the case of the first event, but not enough to modify the original assignment input (the gates are seen slightly reduced on the pictures). A better example yet is the second event (4.9), where, in spite of the fact that one of the tracks has moved towards the module gap, most of the tracks enter into contention for the same hits. In both cases, the improvement due to the combinatorial tracking algorithm is evident, from both the more regular V-plots and from a much better collimated set of tracks interpolated into the vertex region.

6.   CONCLUDING REMARKS

The method discussed in this chapter represents a successful attempt to improve the robustness of an optimal control technique, in this case the extended Kalman filter, in the context of a multiple-target tracking application in which optimality conditions of the filter could not be met. The usefulness of the approach was demonstrated by the systematic improvements in track resolution that were produced by the integer programming-based track finding algorithm when compared to a pure filter-based sequential nearest-neighbor search. A

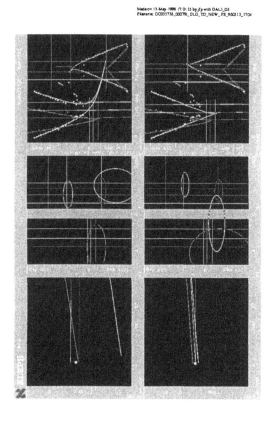

Figure 4.8   Comparative reconstruction performances in VDET (1)

practical implementation of the combinatorial framework was achieved with relative ease, since the set packing and generalized assignment models used in this context could be easily manipulated to fit the requirements imposed by the behaviors of targets and by the overall geometry and operating mode of the detection devices.

The approach is believed to extend to much larger problem sizes than presented here, if only because of the vast array of algorithms and approximation methods that is available to solve mathematical programming problems nowadays. For applications involving the offline reconstruction of tracks, column generation techniques could be applied to further improve tracking precision with compute time constraints in place.

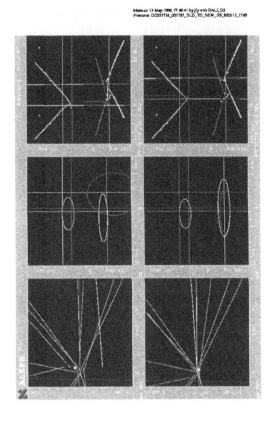

Figure 4.9   Comparative reconstruction performances in VDET (2)

GLOSSARY

ALEPH   Apparatus for LEP Physics (particle detector).

CERN   Centre Européen pour la Recherche Nucléaire, Geneva, Switzerland.

EKF   Extended Kalman Filter (parameter estimation method).

HEP   High Energy Physics.

JPDAF   Joint Probabilistic Data Association Filter (data association algorithm).

ITC   Inner Tracking Chamber (a tracking detector of ALEPH).

LEP   Large Electron-Positron Collider (one of the particle accelerators at CERN).

MHT   Multiple-Hypothesis Tracking (data association algorithm).

MTT   Multiple-Target Tracking.

TPC Time Projection Chamber (the largest tracking detector of ALEPH).
VDET Vertex Detector (the innermost tracking detector of ALEPH).
QCD Quantum Chromodynamics.

References

[Bar-Shalom and Fortmann, 1988] Bar-Shalom, Y. and Fortmann, T. E. (1988). *Tracking and Data Association*. Mathematics in Science and Engineering, Academic Press.

[Bar-Shalom et al., 1980] Bar-Shalom, Y., Fortmann, T. E., and Scheffe, M. (1980). Joint probabilistic data association filter for multiple targets in clutter. In *Proc. Conf. on Information Science and Systems*. Princeton, NJ.

[Blackman, 1986] Blackman, S. S. (1986). *Multiple-Target Tracking with Radar Applications*. Artech House.

[Chang et al., 1994a] Chang, K.-C., Mori, S., and Chong, C.-Y. (1994a). Evaluating a multiple-hypothesis multitarget tracking algorithm. *IEEE Transactions on Aerospace and Electronic Systems*, 30(1).

[Chang et al., 1994b] Chang, K.-C., Mori, S., and Chong, C.-Y. (1994b). Performance evaluation of track initiation in dense target environments. *IEEE Transactions on Aerospace and Electronic Systems*, 30(1).

[Comas et al., 1996] Comas, P., Knobloch, J., and Pusztaszeri, J.-F. (1996). Aleph event reconstruction manual. Technical Report 96-010, CERN. ALEPH Collaboration.

[Cox and Miller, 1995] Cox, I. J. and Miller, M. L. (1995). On finding ranked assignments with applications to multitarget tracking and motion correspondence. *IEEE Transactions on Aerospace and Electronic Systems*, 31(1).

[D. Decamp, 1991] D. Decamp, e. a. C. (1991). Measurement of the charged particle multiplicity distribution in hadronic z decays. *Physics Letters B*, 273:181–192.

[Denby et al., 1990] Denby, B., Lessner, E., and Lindsey, C. S. (1990). Tests of track segment and vertex finding with neural networks. In *Proceedings of the Conference on Computing in High Energy Physics - CHEP90, Santa Fe, U.S.A., April 8-13*. World Scientific.

[Drevermann et al., 1995] Drevermann, H., Kuhn, D., and Nilsson, B. S. (1995). Event display. In *Proceedings of the CERN School of Computing, CERN-ECP*, number 95-25.

[F. Abe, 1995] F. Abe, e. a. C. C. (1995). Observation of top quark production in ppbar collisions with the cdf detector at fermilab. Technical Report PUB-022E, Fermi National Laboratory, Batavia, U.S.A. CDF preprint.

[Frühwirth, 1987] Frühwirth, R. (1987). Applications of kalman filtering to track and vertex fitting. *Nuclear Instr. & Methods*, A262:444–450.

[G. Batignani, 1992] G. Batignani, e. a. (1992). Experience with the aleph silicon vertex detector. *Nuclear Instr. & Methods*, A315:121–124.

[Group, 1994] Group, P. D. (1994). Review of particle properties. *Physical Review D, Particles and Fields*, 50(1).

[Gyulassy and Harlander, 1991] Gyulassy, M. and Harlander, M. (1991). Elastic tracking and neural networks for complex pattern recognition. *Computer Physics Communications*, 66:31–46.

[Kalman and Bucy, 1961] Kalman, R. and Bucy, R. (1961). New results in linear filtering and prediction. *Journal of Basic Engineering*, 83D:360–368.

[Li and Bar-Shalom, 1991] Li, X. R. and Bar-Shalom, Y. (1991). Stability evaluation and track life of the pdaf for tracking in clutter. *IEEE Transactions on Automatic Control*, 36(5).

[Li and Bar-Shalom, 1994] Li, X. R. and Bar-Shalom, Y. (1994). Detection threshold selection for tracking performance optimization. *IEEE Transactions on Aerospace and Electronic Systems*, 30(3).

[Morefield, 1977] Morefield, C. L. (1977). Application of 0-1 integer programming to multitarget tracking problems. *IEEE Transactions on Automatic Control*, AC-22(3).

[Nagarajan et al., 1987] Nagarajan, V., Chidambara, M. R., and Sharma, R. N. (1987). Combinatorial problems in multitarget tracking - a comprehensive solution. *IEE Proceedings*, 134(1).

[Okun, 1985] Okun, L. B. (1985). *Particle Physics*. Harwood Academic Publishers.

[Peterson, 1989] Peterson, C. (1989). Track finding with neural networks. *Nuclear Instruments & Methods*, A279:537–545.

[Peterson and Anderson, 1987] Peterson, C. and Anderson, J. R. (1987). A mean field theory learning algorithm for neural networks. *Complex Systems Publications*, 66:31–46.

[Poore, 1994] Poore, A. B. (1994). Multidimensional assignment formulation of data association problems arising from multitarget tracking and multisensor data fusion. *Computational Optimization and Applications*, 3:27–57.

[Poore and Rijavec, 1994a] Poore, A. B. and Rijavec, N. (1994a). A lagrangian relaxation algorithm for multidimensional assignment problems arising from multitarget tracking. *SIAM Journal of Optimization*, 3(3):339–361.

[Poore and Rijavec, 1994b] Poore, A. B. and Rijavec, N. (1994b). A numerical study of some data association problems arising in multitarget tracking. In

Hager, W. W., Hearn, D. W., and Pardalos, P. M., editors, *Large Scale Optimization: State of the Art*, pages 339–361. Kluwer Academic Publishers B. V., Boston.

[Pusztaszeri, 1996] Pusztaszeri, J.-F. (1996). *Combinatorial Algorithms for Pattern Recognition in Composite Tracking Chambers*. PhD thesis, Swiss Federal Institute of Technology.

[Pusztaszeri et al., 1996] Pusztaszeri, J.-F., Rensing, P., and Liebling, T. (1996). Tracking elementary particles near their primary vertex: a combinatorial approach. *Journal of Global Optimization*, 9:41–64.

[Reid, 1979] Reid, D. B. (1979). An algorithm for tracking multiple targets. *IEEE Transactions on Automatic Control*, AC-24(6).

[Scott, 1963] Scott, W. T. (1963). The theory of small-angle multiple scattering of fast charged particles. *Reviews of Modern Physics*, 35(2).

[Smith and Buechler, 1975] Smith, P. and Buechler, G. (1975). A branching algorithm for discriminating and tracking multiple objects. *IEEE Transactions on Automatic Control*, AC-20:101–104.

[Stimpfl-Abele and Garrido, 1991] Stimpfl-Abele, G. and Garrido, L. (1991). Fast track finding with neural networks. *Computer Physics Communications*, 64:46–56.

[Trunk and Wilson, 1981] Trunk, G. V. and Wilson, J. D. (1981). Track initiation of occasionally unresolved radar targets. *IEEE Transactions on Aerospace and Electronic Systems*, AES-17(1).

Chapter 5

POLYHEDRAL METHODS FOR SOLVING THREE INDEX ASSIGNMENT PROBLEMS

Liqun Qi and Defeng Sun

School Of Mathematics
The University of New South Wales
*Sydney 2052, Australia**
L.Qi@unsw.edu.au, D.Sun@unsw.edu.au

Abstract   The *(axial) three index assignment problem*, also known as the *three-dimensional matching problem*, is the problem of assigning one item to one job at one point or interval of time in such a way as to minimize the total cost of the assignment. Until now the most efficient algorithms explored for solving this problem are based on polyhedral combinatorics. So far, four important facet classes $\mathcal{Q}, \mathcal{P}, \mathcal{B}$ and $\mathcal{C}$ have been characterized and $O(n^3)$ (linear-time) separation algorithms for five facet subclasses of $\mathcal{Q}, \mathcal{P}$ and $\mathcal{B}$ have been established. The complexity of these separation algorithms is best possible since the number of the variables of three index assignment problem of order n is n^3. In this paper, we review these progresses and raise some further questions on this topic.

1.   INTRODUCTION

Consider three disjoint n-sets, I, J, and K, and a weight c_{ijk} associated with each ordered triplet $(i, j, k) \in I \times J \times K$. The *(axial) three index assignment problem*, to be denoted AP3, also known as the *three-dimensional matching problem*, is to find a minimum-weight collection of n disjoint triplets $(i, j, k) \in I \times J \times K$. This problem is called *(axial)* to distinguish it from another three-index assignment problem, known as *planar*, in which one wants to find a minimum weight collection of n^2 triplets, forming n disjoint sets of n disjoint triplets.

* Supported by the Australian Research Council.

P.M. Pardalos and L.S. Pitsoulis (eds.), Nonlinear Assignment Problems, 91–107.

Mathematically, the AP3 can be stated as follows. A graph is *complete* if all of its nodes are pairwise adjacent. A maximal complete subgraph is called a *clique*. A graph is *k-partite* if its nodes can be partitioned into k subsets such that no two nodes in the same subset are joined by an edge. It is *complete k-partite*, if every node is adjacent to all other nodes except those in its own subsets. The complete k-partite graph with n_i nodes in its ith part (subset) is denoted $K_{n_1,n_2,...,n_k}$.

Consider now the complete tri-partite graph $K_{n,n,n}$ with node set $R = I \cup J \cup K, |I| = |J| = |K| = n$. $K_{n,n,n}$ has $3n$ nodes and n^3 cliques, all of which are triangles containing exactly one node from each of the three sets I, J and K. Let (i, j, k) denote the clique induced by the node set $\{i, j, k\}$. If a weight c_{ijk} is associated with each clique (i, j, k), then AP3 is the problem of finding a minimum-weight exact clique cover of the nodes of $K_{n,n,n}$, where an exact clique cover is a set of cliques that partitions the node R. It can be stated as $0 - 1$ programming problem:

$$
\left. \begin{array}{ll}
\min \quad \sum (c_{ijk}x_{ijk} : i \in I, j \in J, k \in K), \\[2mm]
\text{s.t.} \quad \sum (x_{ijk} : j \in J, k \in K) = 1, \quad \forall i \in I, \\[2mm]
\quad \sum (x_{ijk} : i \in I, k \in K) = 1, \quad \forall j \in J, \\[2mm]
\quad \sum (x_{ijk} : i \in I, j \in J) = 1, \quad \forall k \in K, \\[2mm]
\quad x_{ijk} \in \{0, 1\}, \qquad\qquad\qquad \forall i, j, k,
\end{array} \right\} \qquad (5.1)
$$

where I, J and K are three disjointed sets with $|I| = |J| = |K| = n$.
Let P_I denote the convex hull of feasible solutions to AP3, i.e.,

$$
P_I = \text{conv}\{x \in \{0, 1\}^{n^3}, x \in P\}, \qquad (5.2)
$$

where $P = \{x \in \Re^{n^3} : A_n x = e, x \geq 0\}$ and $e = (1, \cdots, 1)^T \in \Re^{3n}$.

Applications of AP3 mentioned in the literature include (Pierskalla [Pierskalla, 1967, Pierskalla, 1968]):

- In a rolling mill, schedule ingots through soaking pits (temperature stabilizing baths) to minimize idle time for the rolling mill (the next stage in the process).

- Find a minimum cost schedule of a set of capital investments, e.g., warehouses or plants, in different locations at different times.

- Assign military troops to locations over time to maximize a measure of capability.

- Launch a number of satellites in different directions at different altitudes to optimize coverage or minimize cost.

AP3 is known to be NP-complete [Garey and Johnson, 1979, Frieze, 1983]. For this type of problem, typically implicit enumeration methods are used. Among the algorithms and heuristics in the literature for this problem are those of Vlach [Vlach, 1967], Pierskalla [Pierskalla, 1967, Pierskalla, 1968] and Leue [Leue, 1972]; a primal-dual algorithm is described in Hansen and Kaufman [Hansen and Kaufman, 1973]. A branch-and-bound algorithm using Lagrangian dual and subgradient optimization was implemented in Fröhlich [Fröhlich, 1979] and discussed briefly in Burkard and Fröhlich [Burkard and Fröhlich, 1980]. Also see Burkard and Rudolf [Burkard and Rudolf, 1997]. The exact algorithms described in the literature shares a branching strategy based on fixing single variables to zero or one. Different from all the mentioned algorithms, in [Balas and Saltzman, 1991], Balas and Saltzman presented a branch-and-bound algorithm by using polyhedral combinatorics.

Balas and Saltzman [Balas and Saltzman, 1989] started to study the facial structure of P_I. They gave an $O(n^4)$ procedure to detect whether there is a clique facet of P_I, violated by a given noninteger point x. In [Balas and Qi, 1993], Balas and Qi gave an $O(n^3)$ procedure to do this. Since the number of variables of (5.1) is n^3, an $O(n^3)$ separation algorithm for a facet class of P_I is linear-time and its complexity is best possible. A clique facet defining inequality is a facet defining inequality with $\{0,1\}$ coefficients. The right hand side (RHS) of a clique facet defining inequality is 1. There are two distinct clique facet classes, which may be denoted by $\mathcal{Q}(1)$ and $\mathcal{P}(1)$ respectively. Both of them were identified by Balas and Saltzman [Balas and Saltzman, 1989]. In fact, Balas and Saltzman [Balas and Saltzman, 1989] also identified a general facet class $\mathcal{Q}$. The coefficients of a $\mathcal{Q}$ facet defining inequality are 0 and 1, while the RHS of a $\mathcal{Q}$ facet defining inequality can be an integer p satisfying $1 \leq p \leq n - 1$. We denote $\mathcal{Q}(p)$ for the facet subclass of $\mathcal{Q}$, with the RHS of its facet defining inequality as p. Beside giving linear-time separation algorithms for clique facet classes $\mathcal{Q}(1)$ and $\mathcal{P}(1)$, Balas and Qi [Balas and Qi, 1993] also gave a linear-time separation algorithm for the facet subclass $\mathcal{Q}(2)$. Qi, Balas and Gwan [Qi et al., 1994] identified a general facet class $\mathcal{P}$. The coefficients of a $\mathcal{P}$ facet defining inequality are also 0 and 1, while the RHS of a $\mathcal{P}$ facet defining inequality can be an integer p satisfying $1 \leq p \leq n - 3$. Again, let $\mathcal{P}(p)$ be the facet subclass of $\mathcal{P}$, with the RHS of its facet defining inequality as p. Then $\mathcal{P}(1)$ is the clique facet class mentioned before. Qi, Balas and Gwan [Qi et al., 1994] gave a linear-time separation algorithm for the facet subclass $\mathcal{P}(2)$. Gwan and Qi [Gwan and Qi, 1992] identified two general facet classes $\mathcal{B}$ and $\mathcal{C}$. The coefficients of a $\mathcal{C}$ facet defining inequality are still 0 and 1, while the coefficients of a $\mathcal{B}$ facet defining inequality are 0, 1 and 2. Let the facet subclasses of $\mathcal{B}$ and $\mathcal{C}$ be $\mathcal{B}(p)$ and $\mathcal{C}(p)$ respectively, where p is the RHS of the facet defining inequality. Then $p \geq 2$ is an even number for $\mathcal{B}(p)$ and $\mathcal{C}(1) = \mathcal{P}(1)$. Gwan and Qi [Gwan and Qi, 1992] gave a linear-time separation algorithm for the facet subclass $\mathcal{B}(2)$. It remains an open question if $\mathcal{C}(2)$ has a linear-time separation algorithm or not.

We review the four facet classes $\mathcal{Q}$, $\mathcal{P}$, $\mathcal{B}$ and $\mathcal{C}$ and their facial structures in Section 2. We describe linear separation algorithms for five facet subclasses $\mathcal{Q}(1)$, $\mathcal{P}(1)$, $\mathcal{Q}(2)$, $\mathcal{P}(2)$ and $\mathcal{B}(2)$ in Section 3. A polyhedral method, based on these linear separation algorithms, is described in Section 4. We raise some further questions in the last section.

Other papers on the three-index assignment problem include [Burkard et al., 1996], [Euler, 1987] and [Rudolf, 1991].

2.   FACIAL STRUCTURE

Let us denote the three index assignment problem of order n as AP3$_n$, the coefficient matrix of its equality constraint set in (5.1) as A_n, and I_n, J_n, and K_n the 3 associated index sets. The row and column index sets will be denoted by R_n and S_n respectively. Clearly, $|R_n| = |I_n| + |J_n| + |K_n| = 3n$ and $S_n = |I_n| \times |J_n| \times |K_n| = n^3$. When the order n of the three index assignment problem is clear in the text, we simply use A, R, S etc. instead of A_n, R_n or S_n.

Let a^s denote a column of A associated with $s \in S$. We may specify $s = \{i, j, k\}$ for $i \in, j \in J, k \in K$. For a set $Q \subseteq R$, we use Q_I, Q_J and Q_K in I, J and K, respectively. We also regard $s \in S$ as a three-element subset of R. So we may write $s \cup Q$ for $s \in S$ and $Q \subseteq R$. For $x \in \Re^{n^3}$ and $S' \subseteq S$, let

$$x(S') = \sum \{x_s : s \in S'\}. \tag{5.3}$$

In terms of $K_{n,n,n}$, A_n is the incidence matrix of nodes versus cliques: it has a row for every node and a column for every clique of $K_{n,n,n}$.

The *intersection graph* $G_A = (V, E)$ of a $0 - 1$ matrix A has a node s for every column a^s of A, and an edge (s, t) for every pair of columns a^s and a^t such that $a^s \cdot a^t \neq 0$. The intersection graph G_A of A is a *clique-intersection graph* of $K_{n,n,n}$, i.e., G_A has a node for every clique (triangle) of $K_{n,n,n}$, and an edge for every pair of triangles that share some node of $K_{n,n,n}$.

Let $\langle V \rangle$ be the subgraph induced by $V \subseteq S$, S^r be the support of row r of A, i.e., $S^r := \{s \in S : a_{rs} = 1\}$.

An important concept in polyhedral combinatorics is the Chvátal rank. Its definition can be found on pages 210 and 226 of [Nemhauser and Wolsey, 1988].

2.1   FACET CLASS Q

Let $Q \subseteq R, |Q| = 2p+1, 1 \leq p \leq n-1$, and $1 \leq |Q_L| \leq p$ for $L = I, J, K$.
Let

$$S(Q) = \{s \in S : |s \cup Q| \geq 2\}.$$

Then we have the following theorem.

Theorem 1 (Balas and Saltzman [Balas and Saltzman, 1989]) *The following
inequality*

$$x(S(Q)) \leq p \tag{5.4}$$

*defines a facet of P_I for $n \geq 3$. This facet is of Chvátal rank 1. The number of
distinct inequalities (5.4) is $O(2^{3n})$.*

Denote the facet class defined by inequalities (5.4) as $\mathcal{Q}(p)$ for $1 \leq p \leq n-1$,
and let

$$\mathcal{Q} = \bigcup_{p=1}^{n-1} \mathcal{Q}(p).$$

2.2   FACET CLASS P

Let $s, t \in S$ and assume that $s \cap t = \emptyset$. Let

$$C(s,t) := \{s\} \cup \{s' \in S : |s \cap s'| = 1, |s' \cap t| = 2\}.$$

Then, by Balas and Saltzman [Balas and Saltzman, 1989], $C(s,t)$ is a clique
of G_A and

$$x(C(s,t)) \leq 1 \tag{5.5}$$

defines a facet of P_I for all $n \geq 4$. This facet is of Chvátal rank 2. The number
of distinct inequalities (5.5) is $\frac{1}{4}n^3(n-1)^3$. Denote the facet class defined by
inequalities (5.5) as $\mathcal{P}(1)$.

For any $s, t \in S$ and $s \cap t = \emptyset$, $C(s,t)$ is a clique of G_A. For $p = 1$, $S(Q)$ is
also a clique of G_A. Note that the RHS of (5.5) is 1. Balas and Saltzman [Balas
and Saltzman, 1989] showed that $\mathcal{Q}(1)$ and $\mathcal{P}(1)$ include all clique facets of
G_A.

In [Qi et al., 1994], Qi, Balas and Gwan generalized the facet class $\mathcal{P}(1)$.
Suppose that $D, Q \subseteq R, D \cap Q = \emptyset, |D_L| + |Q_L| = p + 1$ and $1 \leq |Q_L| \leq r$
for $L = I, J, K, |Q| = 2r + 1, 1 \leq r \leq p \leq n - 3$. Let

$$C_1(D) := \{t \in S : t \in D\},$$

$$C_2(D,Q) := \{s \in S : |s \cap D| = 1, |s \cap Q| = 2\}$$

and

$$C(D,Q) := C_1(D) \cup C_2(D,Q).$$

Then

$$x(C(D,Q)) \leq p \qquad\qquad (5.6)$$

for $p = 1$ is the same as (5.5).

Theorem 2 (Qi, Balas and Gwan [Qi et al., 1994]) *For $n \geq 4$, (5.6) defines a facet of P_I. This facet is of Chvátal rank 2. The number of distinct inequalities (5.6) is*

$$\kappa(\mathcal{P}) \leq \sum_{p=1}^{n-3} \binom{n}{p+1}^3 2^{3p}.$$

Denote the facet class defined by inequalities (5.6) as $\mathcal{P}(p)$ for $1 \leq p \leq n-3$, and let

$$\mathcal{P} = \bigcup_{p=1}^{n-3} \mathcal{P}(p).$$

2.3   FACET CLASS B

Let $D \subset R$ have the same cardinalities in I, J and K, i.e., $|D_I| = |D_J| = |D_K| = r$, where $1 \leq r \leq n - 4$. Let $H \subseteq R$, $H \cap D = \emptyset$, $|H| = 2$, $|H_l| \leq 1$ for $L = I, J, K$. Let $Q = D \cup H$. If $|H_I| = |H_J| = 1$, then D is just like the body of a fat bull and H is like the head of the bull. Let

$$B(D) := \{t \in S : t \subseteq D\}$$

and

$$F(Q,D) := \{t \in S : |t \cap Q| \geq 2, 2 \geq |t \cap D| \geq 1\}.$$

Theorem 3 (Gwan and Qi [Gwan and Qi, 1992]) *Assume that $n \geq r+4$. Then the following equation*

$$2x(B(D)) + x(F(Q,D)) \leq 2r \qquad\qquad (5.7)$$

defines a facet of P_I. This facet has Chvátal rank 1.

Denote the facet class defined by inequalities (5.7) as $\mathcal{B}(2r)$ for $1 \leq r \leq n - 4$, and let

$$\mathcal{B} = \bigcup_{r=1}^{n-4} \mathcal{B}(2r).$$

2.4   FACET CLASS C

From Theorem 2 we know that (5.6) defines a facet of P_I for $n \geq 4$. One may think that D is a "handle" of a "comb" and Q is the sole "tooth" of this comb. Gwan and Qi [Gwan and Qi, 1992] extended the above case to more general cases, i.e., with several "teeth".

A subset of R is called a *uniform subset* of R if the cardinalities of the intersections of this set with I, J and K are the same.

Suppose that $D, T^1, \cdots, T^v \subseteq R$. Denote $T = \{T^1, \cdots, T^v\}$ and $T^0 = \cup_{\mu=1}^v$. Here, D is regarded as handle and $T^1, \cdots, T^v$ as the teeth of the comb. Several restrictions are needed on the handle and the teeth.

(a). Each tooth is an odd set, i.e., for $\mu = 1, \cdots, v$ and $L = I, J, K$, $|T^v| = 2r_\mu + 1, 1 \leq |(T^v)_L| \leq r_\mu$. Let $r = \sum_{\mu=1}^v r_\mu$.

(b). Any pair of teeth are disjoint, i.e., $T^v \cap T^\rho = \emptyset$ for all $\mu \neq \rho$.

(c). The whole index set $D \cup T^0$ is a uniform subset of R, and $\sigma = |(D \cup T^0)_I| = |(D \cup T^0)_J| = |(D \cup T^0)_K| \leq n - 2$.

(d). The cardinality of the handle satisfies $|D| = 3p - 2r + 2$ for some $p \geq r$.

(e). There is at most one link, i.e., one common index of a tooth and the handle, i.e., for $\mu = 1, \cdots, v, |D \cap T^\mu| \leq 1$.

(f). If a tooth T^μ is linked, assume l_μ is the link and $l_\mu \in L_\mu$ where L_μ is one of I, J and K. Then $T^\mu_{l_\mu} = \{l_\mu\}$, i.e., l_μ is the sole index in $T^\mu_{l_\mu}$.

In [Gwan and Qi, 1992], Gwan and Qi proved, under the above conditions, that at least one tooth is unlinked. Without loss of the generality, assume that the first ζ teeth are linked. Assume that the link for T^μ is l_μ and $l_\mu \in L_\mu$, where L_μ is in I or J or K. Let

$$C_1(D) := \{t \in S : t \subseteq D\},$$

$$C_2(D, T_\mu) := \{t \in S : |t \cap D| = 1, |t \cap T^\mu| = 2\}$$

for $\mu = 1, \cdots, v$, and

$$S(T_\mu, l_\mu, L_\mu) := \{t \in S : |t \cap T^\mu| \geq 2, l_\mu \in t\}$$

for $\mu = 1, \cdots, \zeta$. Let

$$C(D, T) := C_1(D) \cup (\cup_{\mu=1}^v C_2(D, T^\mu)) \cup (\cup_{\mu=1}^\zeta S(T_\mu, l_\mu, L_\mu)).$$

Theorem 4 (Gwan and Qi [Gwan and Qi, 1992]) *Under the above conditions,*

$$x(C(D, T)) \leq p, \tag{5.8}$$

defines a facet of P_I. This facet is of Chvátal rank 2.

Denote the facet class defined by inequalities (5.8) as $\mathcal{C}(p)$. Then $\mathcal{C}(1) = \mathcal{P}(1)$, but $\mathcal{C}(p) \neq \mathcal{P}(p)$ for $p \geq 2$. We denote the union of facet classes $\mathcal{C}(p)$ as $\mathcal{C}$.

3.   LINEAR-TIME SEPARATION ALGORITHMS FOR FACETS

In the previous section we have described several facet classes of the three index assignment polytope. In this section we review linear-time separation algorithms for five facet subclasses. A separation algorithm for a facet class of a combinatorial optimization problem is a procedure for finding a facet defining inequality in this facet class, violated by a given non-integer solution to the linear programming relaxation of this combinatorial optimization problem, or showing that no such inequality exists. A linear-time separation algorithm for the three index assignment problem is a separation algorithm with complexity $O(n^3)$, which is best possible.

3.1   A LINEAR-TIME SEPARATION ALGORITHM FOR $Q(1)$

For every $s \in S$, let

$$C(s) := \{t \in S : a^s \cdot a^t \geq 2\}.$$

Then, for $n \geq 3$, the inequality

$$x(C(s)) := \sum\{x_t : t \in C(s)\} \leq 1 \qquad (5.9)$$

defines a facet of P_I for every $s \in S$ (Theorem 1). Given a non-integer $x \in P$, Balas and Qi [Balas and Qi, 1993] presented the following procedure to detect whether any inequality induced by a clique of class $Q(1)$ is violated.

Algorithm 3.1. Suppose that x is a non-integer point in P. Let v be an integer greater than or equal to 4.

Step 1. Let $d_s = 0$ for all $s \in S$.

Step 2. Check x_t for all $t \in S$. If

$$x_t \geq \frac{1}{vn},$$

then set $d_s := d_s + x_t$ for all $s \in C(t)$. If $d_s > 1$, then stop: (5.9) is violated by x with this s. Otherwise, continue.

Step 3. For $s \in S$, if

$$d_s > \frac{v-3}{v},$$

then check whether the inequality (5.9) associated with s is violated and if so, then stop. Else continue.

Theorem 5 (Balas and Qi [Balas and Qi, 1993]) *Algorithm 3.1 determines in $O(n^3)$ steps whether a given non-integer $x \in P$ violates a facet defining inequality induced by a clique of class $Q(1)$. The value of v which minimizes the complexity of the algorithm is 6.*

3.2 A LINEAR-TIME SEPARATION ALGORITHM FOR $P(1)$

In this subsection, we describe Balas and Qi's $O(n^3)$ procedure to detect whether any inequality induced by a clique of class $\mathcal{P}(1)$ is violated.

Algorithm 3.2. Suppose that x is a non-integer point in P. Check x_s for $s \in S$. If

$$\frac{1}{4} < x_s < 1,$$

then check x_p for $p \in S$ satisfying

$$a^s \cdot a^p = 1.$$

If

$$x_p > \frac{1 - x_s}{3},$$

then for $t \in S$ such that $a^t \cdot a^s = 0$ and

$$a^t \cdot a^p = 2,$$

check where (5.5) is violated or not.

Theorem 6 (Balas and Qi [Balas and Qi, 1993]) *Algorithm 3.2 determines in $O(n^3)$ steps whether a given $x \in P$ violates a facet defining inequality induced by a clique of class $\mathcal{P}(1)$.*

3.3 A LINEAR-TIME SEPARATION ALGORITHM FOR $Q(2)$

Let $Q \subseteq R, |Q| = 2p+1, 1 \le p \le n-1$, and $1 \le |Q_L| \le p$ for $L = I, J, K$. Let

$$S(Q) = \{s \in S : |s \cup Q| \ge 2\}.$$

Then, by Theorem 1 $x(S(Q)) \le p$ defines a facet of P_I. A linear-time separation algorithm for $Q(1)$ has already been described. In this subsection, we describe a linear-time separation algorithm for $Q(2)$, due to Balas and Qi [Balas and Qi, 1993].

Denote the set of Q satisfying the above requirements with a fixed p by S^p. Suppose that $x \in P$. For $i \in I, j \in J, k \in K$, let

$$x(i, j, K) := \sum \{x_{ijk'} : k' \in K\}, \tag{5.10}$$

$$x(i, J, k) := \sum \{x_{ij'k} : j' \in J\}, \tag{5.11}$$

$$x(I, j, k) := \sum \{x_{i'jk} : i' \in I\}. \tag{5.12}$$

Before stating the separation algorithm for $Q(2)$ of Balas and Qi [Balas and Qi, 1993], we need some formulas to calculate $x(S(Q))$ for $Q \in S^2$. We consider two cases.

First, suppose that $s, t \in S$ satisfy

$$a^s \cdot a^t = 1.$$

Then

$$Q \equiv s \cup t \in S^2.$$

Assume that $s = (i_s, j_s, k_s)$ and $t = (i_t, j_t, k_t)$. It is not difficult to see that if $k_s = k_t$, then

$$\begin{aligned} x(S(Q)) &= x(C(s)) + x(C(t)) + x(i_s, j_t, K) \\ &\quad + x(i_t, j_s, K) - x_{i_s j_t k_s} - x_{i_t j_s k_t}; \end{aligned} \tag{5.13}$$

if $j_s = j_t$, then

$$\begin{aligned} x(S(Q)) &= x(C(s)) + x(C(t)) + x(i_s, J, k_t) \\ &\quad + x(i_t, J, k_s) - x_{i_s j_t k_t} - x_{i_t j_s k_s}; \end{aligned} \tag{5.14}$$

and if $i_s = i_t$, then

$$\begin{aligned} x(S(Q)) &= x(C(s)) + x(C(t)) + x(I, j_s, k_t) \\ &\quad + x(I, j_t, k_s) - x_{i_s j_s k_t} - x_{i_t j_t k_s}. \end{aligned} \tag{5.15}$$

Next, suppose that $s, t \in S$ satisfy

$$a^2 \cdot a^t = 2. \tag{5.16}$$

Again, assume that $s = (i_s, j_s, k_s)$ and $t = (i_t, j_t, k_t)$. Then define for the ordered pair (s, t),

$$x(s, t) := x(i_t, J, k_t) + x(i_t, j_t, K) - 2x_{i_t j_t k_t}, \tag{5.17}$$

if $i_s \neq i_t$;

$$x(s, t) := x(i_t, j_t, K) + x(I, j_t, k_t) - 2x_{i_t j_t k_t}, \tag{5.18}$$

if $j_s \neq j_t$;

$$x(s, t) := x(I, j_t, k_t) + x(i_t, J, k_t) - 2x_{i_t j_t k_t}, \tag{5.19}$$

if $k_s \neq k_t$. Notice that in (5.17) $j_t = j_s$ and $k_t = k_s$, in (5.18) $i_t = i_s$ and $k_t = k_s$, and in (5.19) $i_t = i_s$ and $j_t = j_s$. Thus, (5.17) can be written with

j_s and k_s substituted for j_t and k_t, and there are similar expressions for (5.18) and (5.19).

In the case that (5.16) holds, we may expand $s \cup t$ to a set $Q \in S^2$ by adding an appropriate element. For example, if $i_s \neq i_t$, we may add $j \neq j_s$ and let

$$Q_I = \{i_s, i_t\}, \quad Q_J = \{j_s, j\}, \quad Q_K = \{k_s\}, \tag{5.20}$$

or we may add $k \neq k_s$ and let

$$Q_I = \{i_s, i_t\}, \quad Q_J = \{j_s\}, \quad Q_K = \{k_s, k\}. \tag{5.21}$$

Notice that there are $2(n-1)$ ways to expand a given pair of s and t satisfying (5.16) to a set $Q \in S^2$. We may also calculate $x(S(Q))$ by expressions similar to (5.13)–(5.15). For example, in case (5.20), one can check

$$x(S(Q)) = x(C(s)) + x(s,t) + x(i_s, j, K) \\ + x(I, j, k_s) - 2x_{i_s j k_s} - 2x_{i_t j k_s}, \tag{5.22}$$

where $x(s,t)$ is given by (5.17). Note that if $i_s \neq i_t$, there are two possibilities, namely, (5.20) and (5.21). Therefore, there are six cases totally, i.e., two cases for $i_s \neq i_t$, two cases for $j_s \neq j_t$ and two cases for $k_s \neq k_t$. In each of these six cases, there is an expression of type (5.22) to calculate $x(S(Q))$. Later, we will refer to these as "expression of type (5.22)".

We are now ready to describe the linear-time separation algorithm for $Q(2)$ by Balas and Qi [Balas and Qi, 1993].

Algorithm 3.3. Suppose that x is a non-integer point in P. Suppose (5.9) holds for all $s \in S$.

Step 1. Let $d_s = 0$ for all $s \in S$.

Step 2. Check x_t for all $t \in s$. If $x_t \geq 1/(12n)$, then let $d_s := d_s + x_t$ for all $s \in S$ satisfying $a_s \cdot a_t \geq 2$.

Step 3. For all $s \in S$, if $d_s \geq 1/12$, then $d_s := x(C(s))$.

Step 4. For all $i \in I, j \in J, k \in K$, calculate $x(i, j, K), x(i, J, k)$ and $x(I, j, k)$ by (5.10), (5.11) and (5.12), respectively.

Step 5. For all $i \in I, j \in J, k \in K$, form the sets

$$L(i, J, K) := \{s \in S : d_s > 1/3, \, i_s = i\},$$

$$L(I, j, K) := \{s \in S : d_s > 1/3, \, j_s = j\}$$

and

$$L(I, J, k) := \{s \in S : d_s > 1/3, \, k_s = k\}.$$

Store them.

Step 6. For each $s = (i_s, j_s, k_s) \in S$ such that $d_s > 1/2$, check d_t for each $t \in L(s)$, where

$$L_s := (L(i_s, J, K) \cup L(I, j_s, K) \cup L(I, J, k_s))\backslash\{s\}.$$

If $a^t \cdot a^s = 1$ and $d_t > (2 - d_s)/3$, calculate $x(S(Q))$ by the appropriate expression (5.13), (5.14) or (5.15), with d_h substituted for $x(C(h)), h = s, t$, for every $Q \in S^2$ such that $s \cup t \in Q$.

If $x(S(Q)) > 2$ for some Q, stop: the corresponding inequality (5.4) with $p = 2$ is violated by x. Otherwise, continue.

Theorem 7 (Balas and Qi [Balas and Qi, 1993]) *Algorithm 3.3 determines in $O(n^3)$ steps whether a given non-integer $x \in P$ violates a facet defining inequality (5.4) with $p = 2$.*

3.4   A LINEAR-TIME SEPARATION ALGORITHM FOR $P(2)$

From Theorem 2, a facet defining inequality in $\mathcal{P}(2)$ has the form

$$x(C(D, Q)) \leq 2, \tag{5.23}$$

where $D \subseteq R, q \in S, D \cap q = \emptyset, |D_I| = |D_J| = |D_k| = 2$,

$$C(D, q) := C_1(D) \cup C_2(D, q),$$

$$C_1(D) := \{t \in S : t \in D\},$$

$$C_2(D, q) := \{s \in S : |s \cap D| = 1, |s \cap q| = 2\}.$$

Suppose that x is a given non-integer point in P and x violates (5.23), i.e.,

$$x(C(D, Q)) > 2. \tag{5.24}$$

Also, assume that x does not violate any clique facets of P_I, i.e., for any $s, t \in S$, $s \cap t = \emptyset$, the following inequalities hold:

$$x(S(s)) \leq 1, \tag{5.25}$$

$$x(C(s, t)) \leq 1. \tag{5.26}$$

For any index $l \in L$, where L is one of I, J and K, use $S(l, L)$ to denote the collection of $t \in S$ whose element in L is l.

In [Qi et al., 1994], Qi, Balas and Gwan gave the following separation algorithm for $\mathcal{P}(2)$.

Algorithm 3.4. Suppose that x is a non-integer point in P. Suppose (5.25) and (5.26) hold for all $s, t \in S$ such that $s \cap t = \emptyset$.

Step 1. Form $S_x := \{s \in S : 1/10 < x_s < 1\}$. For each index $l \in L$, where $L = I, J$ or K, form $S_x(l, L)$ as the set that indexes the 18 largest components of x among those indexed by $S(l, L)$.

Step 2. If there exists a pair $s, t \in S_x$ satisfying $s \cap t = \emptyset$ and $x_s > 1/8$, go to Step 4.

Step 3. For each $s \in S_x$, let

$$L_s := (\cup\{u \in S_x : u \cap s \neq \emptyset\})\backslash s.$$

If there exists a pair (s, t) such that $s \in S_x$, $t \in S$ and $t \subseteq L_s$, go to Step 4.

Step 4. For each pair (s, t) in Steps 2 and 3, let $G_1 = s \cup t$. If there exists a $u \in S_x(l, L)$ such that $l \in G_1$ and u is not belong to G_1, let $G_2 = G_1 \cup u$.

Step 5. For each possible triple s, t, u in Step 3, if $|G_2| = 8$, let $G_3 = G_2$ and go to Step 6. Otherwise, for each $l \in G_1$, if there exists a $v \in S_x(l, L)$ such that v is not belong to G and $|(G_2 \cup v)_{L'}| \leq 3$ for $L' = I, J, K$, let $G_3 = G_2 \cup v$. If $|G_3| = 9$, let $G = G_3$ and go to Step 7. Otherwise, go to Step 6.

Step 6. For each G_3 and each index $l \in L'\backslash(G_3)_{L'}$ satisfying $|(G_3)_{L'}| = 2$, add to G_3 and call it G now.

Step 7. For each G, let $D = s \cup t$ and $q = G\backslash D$. Check whether (5.23) is violated or not. If (5.23) is violated, stop.

Step 8. For each G and $q \subset G\backslash s$ satisfying $q \in s$ and $|q \cup t| = 1$, let $D = G\backslash q$ and check whether (5.23) is violated or not. If (5.23) is violated, stop.

If all pairs (s, t) in Step 2 have been considered, go to Step 3. After all pairs (s, t) in Step 3 have been considered, stop. Similarly, in Steps 4 and 5, each possible u and v should be considered to form G such that (5.23) is checked in Steps 7 and 8.

Theorem 8 (Qi, Balas and Gwan [Qi et al., 1994]) *Algorithm 3.4 determines in $O(n^3)$ steps whether a given non-integer $x \in P$ violates a facet defining inequality (5.23).*

3.5   A LINEAR-TIME SEPARATION ALGORITHM FOR $\mathcal{B}(2)$

Without loss of generality, we may assume that a facet defining inequality in $\mathcal{B}(2)$ has the form

$$2x_s + x(F(Q, s)) \leq 2, \tag{5.27}$$

where $s = (i_s, j_s, k_s) \in S, Q_I = \{i_s, i_q\}, Q_J = \{j_s, j_q\}, Q_K = \{k_s\}$, $i_s \neq i_q, j_s \neq j_q$, and

$$F(Q, s) := \{t \in S : |t \cap Q| \geq 2, 2 \geq |t \cap s| \geq 1\}.$$

For $i \in I, j \in J, k \in K$, $x(i, j, K), x(i, J, k)$ and $x(I, j, k)$ are defined by (5.10), (5.11) and (5.12), respectively. Since each of $x(i, j, K), x(i, J, k)$ and $x(I, j, k)$ sums up n components of x, we call each of them an *n-sum*. Also in each n-sum, only one index is summed up. So we call the other two indices *fixed indices* of that n-sum. We call i, j fixed indices of $x(i, j, K)$. Two n-sums are called *unrelated* if they do not have common fixed indices in between. Whereas two n-sums are called *related* if they have one common fixed indices in between.

There are seven n-sums associated with (5.27). They are: $A_1 = x(i_s, j_s, K)$, $A_2 = x(i_s, j_q, K)$, $A_3 = x(i_q, j_s, K)$, $A_4 = x(i_q, J, k_s)$, $A_5 = x(i_s, J, k_s)$, $A_6 = x(I, j_q, k_s)$, and $A_7 = x(I, j_s, k_s)$. Also

$$2x_s + x(F(Q, s)) = \sum_{i=1}^{7} A_i - x_s - 2x_{i_s j_q k_s} - 2x_{i_q j_s k_s} - 2x_{i_q j_q k_s}. \quad (5.28)$$

We call an n-sum a *big n-sum* if its value is greater than $1/4$.

Next, we describe a linear separation algorithm for $\mathcal{B}(2)$ due to Gwan and Qi [Gwan and Qi, 1992].

Algorithm 3.5. Suppose that x is a non-integer point in P.

Step 1. For all $i \in I, j \in J, k \in K$, calculate all the n-sums $x(i, j, K)$, $x(i, J, k)$ and $x(I, j, k)$.

Step 2. Check all pairs of big n-sums. For each pair, if the pair is an unrelated one, add another adequate index R to form Q and s; if the pair is a related one, add two other adequate indices in R to form Q and s. Consider all possible ways to form Q and s. Then use (5.28) to check whether (5.27) is violated in each other.

Theorem 9 (Gwan and Qi [Gwan and Qi, 1992]) *Algorithm 3.5 determines in $O(n^3)$ steps whether a given non-integer $x \in P$ violates a facet defining inequality (5.27).*

4.   A POLYHEDRAL METHOD

In [Balas and Saltzman, 1991], Balas and Saltzman described a branch-and-bound algorithm for solving the three index problem. In their paper, facets of class $\mathcal{Q}(1)$ were incorporated in a Lagrangian relaxation, which is solved by a modified subgradient procedure. It is the first algorithm to which polyhedral combinatorics theory is applied for solving AP3. A given non-integer solution to the linear relaxation problem of AP3, which is not violated by one of the facet defining inequalities in one class of facet may be violated by one of the facet defining inequalities in another facet class. So, algorithms, which use several linear-time separation algorithms one after another to see if any of facet defining inequalities is violated by the given non-integer solution, were introduced in [Qi et al., 1994, Gwan, 1993]. The following method uses the five known linear-time separation algorithms to generate cutting planes, namely facet defining inequalities in $Q(1)$, $P(1)$, $Q(2)$, $P(2)$ and $B(2)$, that are violated by the current solution. Otherwise, when no more violated inequalities can be found in the either of the five classes, we solve by branch and bound for the resulting problem, whose linear constraint set has been tightened by the cut generating procedure.

Let (**LP**) be the linear programming relaxation of the three index assignment problem, i.e., the problem obtained from (5.3) by replacing the condition $x_{ij} \in \{0,1\}$ with $x_{ij} \geq 0$.

Step 1. Solve (**LP**). Let $\bar{x}$ be the optimal solution found. If $\bar{x} \in \{0,1\}^{n^3}$, stop: $\bar{x}$ is optimal.

Step 2. Apply the linear-time separation algorithm for clique facets of the class $P(1)$. If a violated inequality is found, add it to the constraint set of (**LP**) and go to Step 1.

Steps 3,4,5,6 are analogous to Step 2, with facet class $P(1)$ replaced by the classes $Q(1), Q(2), P(2)$ and $B(2)$.

Step 7. Use a branch and bound algorithm to solve the current (**LP**).

The practical usefulness of a method of this type hinges on the question whether the computational effort spent on tightening the linear constraint set is less than the effort that it saves by making the branch and bound procedure more efficient on the tightened problem. Numerical results reported in [Gwan, 1993, Qi et al., 1994] show that the above procedure compares favorably with earlier procedures.

5.   OPEN QUESTIONS

In Table 5.1, we summarize results reviewed in the previous sections. We use "Class" to denote facet classes; "Coef" to denote coefficients of facet defining inequalities and "RHS" to denote their right hand sides; "Rank" to denote the Chvátal Rank; and "LT Alg" to denote the facet subclasses with known linear-time separation algorithms.

We have already mentioned an open question in the introduction: Is there a linear-time separation algorithm for $C(2)$? There are some other challenging open questions for this topic. The five facet classes with known linear-time separation algorithms have Chvátal rank 1 or 2. All of their defining inequalities have right hand sides 1 or 2 too. Are the right hand sides of the facet defining inequalities for all the linear-time separable facet classes 1 or 2? Do all facet classes defined by inequalities with right hand sides 2 have linear-time separation algorithms? Are there other facet classes such that the right hand sides of their defining inequalities are 2 and their Chvátal rank is 1 or 2? Are there other facet classes such that the right hand sides of their defining inequalities

Class	Coef	RHS	Rank	LT Alg
$\mathcal{Q}$	0,1	$1 \leq p \leq n-1$	1	$\mathcal{Q}(1), \mathcal{Q}(2)$
$\mathcal{P}$	0,1	$1 \leq p \leq n-3$	2	$\mathcal{P}(1), \mathcal{P}(2)$
$\mathcal{B}$	0,1,2	$2 \leq p = 2r \leq 2n-8$	1	$\mathcal{B}(2)$
$\mathcal{C}$	0,1	$p \geq 1$	2	$\mathcal{C}(1) = \mathcal{P}(1)$

Table 5.1   Summary of results for the 4 facet classes

are 2 and their Chvátal rank is more than 2? These are some challenging open questions for this topic.

References

[Balas and Qi, 1993] Balas,E. and Qi, L. (1993). Linear-time separation algorithms for the three-index assignment polytope. *Discrete Applied Mathematics*, 43:1–12.

[Balas and Saltzman, 1989] Balas, E. and Saltzman, M. (1989). Facets of the three-index assignment polytope. *Discrete Applied Mathematics*, 23:201–229.

[Balas and Saltzman, 1991] Balas, E. and Saltzman, M. (1991). An algorithm for the three-index assignment problem. *Oper. Res.*, 39:150–161.

[Burkard and Fröhlich, 1980] Burkard, R. and Fröhlich, K. (1980). Some remarks on the three-dimensional assignment problem. *Method of Oper. Res.*, 36:31–36.

[Burkard and Rudolf, 1997] Burkard, R. and Rudolf, R. (1997). Computational investigations on 3-dimensional axial assignment problems. *Belgian Journal of Operations Research, Statistics and Computer Science*.

[Euler, 1987] Euler, R. (1987). Odd cycles and a class of facets of the axial 3-index assignment polytope. *Applicationes Mathematicae (Zastosowania Matematyki)*, XIX:375–386.

[Frieze, 1983] Frieze, A. (1983). Complexity of a 3-dimensional assignment problem. *European J. Oper. Res.*, 13:161–164.

[Fröhlich, 1979] Fröhlich, K. (1979). *Dreidimenionale Zuordnungsprobleme*. Diplomarbeit, Mathematisches Institut, Universität zu Köln, Germany.

[Garey and Johnson, 1979] Garey, M. and Johnson, D. (1979). *Computers and Intractability: A Guide to the Theory of NP-Completeness*. Freeman, San Francisco, CA.

[Gwan, 1993] Gwan, G. (1993). *A Polyhedral Method for the Three Index Assignment Problem.* PhD thesis, School of Mathematics, University of New South Wales, Sydney 2051, Australia.

[Gwan and Qi, 1992] Gwan, G. and Qi, L. (1992). On facets of the three-index assignment polytope. *Australasian J. Combinatorics*, 6:67–87.

[Hansen and Kaufman, 1973] Hansen, P. and Kaufman, L. (1973). A primal-dual algorithm for the three-dimensional assignment problem. *Cahiers du CERO*, 15:327–336.

[Qi et al., 1994] Qi, L., Balas, E. and Gwan, G. (1994). A new facet class and a polyhedral method for the three-index assignment problem. In *Advances in Optimization and Approximation*, pages 256–274. Kluwer Academic Publishers, Dordrecht.

[Leue, 1972] Leue, O. (1972). Methoden zur lösung dreidimensionaler zuordnungsprobleme. *Angewandte Informatik*, pages 154–162.

[Nemhauser and Wolsey, 1988] Nemhauser, G. L. and Wolsey, L. A. (1988). *Integer and Combinatorial Optimization.* John Wiley & Sons, New York.

[Pierskalla, 1967] Pierskalla, W. (1967). The tri-substitution method for the three-dimensional assignment problem. *CORS J.*, 5:71–81.

[Pierskalla, 1968] Pierskalla, W. (1968). The multidimensional assignment problem. *Oper. Res.*, 16:422–431.

[Burkard et al., 1996] Burkard, R.E., Rüdiger, R., and Gerhard, W. (1996). Three-dimensional axial assignment problems with decomposable cost co-efficients. *Discrete Appl. Math.*, 65:123–139.

[Rudolf, 1991] Rudolf, R. (1991). *Dreidimensionale axiale Zuordnungsprobleme.* Technische Universität Graz, Austria. Master Thesis.

[Vlach, 1967] Vlach, M. (1967). Branch and bound method for the three index assignment problem. *Ekonomicko-Mathematický Obzor*, 3:181–191.

Chapter 6

POLYHEDRAL METHODS FOR THE QAP

Volker Kaibel

Fachbereich Mathematik, Sekr. 7-1
Technische Universität Berlin
Straße des 17. Juni 136
10623 Berlin, Germany
kaibel@math.TU-Berlin.DE

Abstract   For many combinatorial optimization problems investigations of associated polyhedra have led to enormous successes with respect to both theoretical insights
into the structures of the problems as well as to their algorithmic solvability. A-
mong these problems are quite prominent $\mathcal{NP}$-hard ones, like, e.g., the traveling
salesman problem, the stable set problem, or the maximum cut problem. In this
chapter we overview the polyhedral work that has been done on the quadratic
assignment problem (QAP). Our treatment includes a brief introduction to the
methods of polyhedral combinatorics in general, descriptions of the most important polyhedral results that are known about the QAP, explanations of the
techniques that are used to prove such results, and a discussion of the practical
results obtained by cutting plane algorithms that exploit the polyhedral knowledge. We close by some remarks on the perspectives of this kind of approach to
the QAP.

1.   INTRODUCTION

Polyhedral combinatorics is a branch of combinatorics and, in particular, of
combinatorial optimization that has become quite broad and successful since
it has sprouted in the 50's and 60's. Its scope is the treatment of combinatorial (optimization) problems by methods of (integer) linear programming.
Besides many beautiful results on several polynomially solvable combinatorial
optimization problems, methods of polyhedral combinatorics have also led to
enormous progress in the practical solvability of many $\mathcal{NP}$-hard optimization
problems. The most prominent example might be the traveling salesman problem, where today many instances with several thousand cities can be solved —
by extremely elaborate exploitation of polyhedral results on the problem.

P.M. Pardalos and L.S. Pitsoulis (eds.), Nonlinear Assignment Problems, 109–141.
© 2000 *Kluwer Academic Publishers.*

Among further examples one finds the maximum cut problem, the linear ordering problem, or the stable set problem. While these problems have been investigated by polyhedral methods extensively since the 70's, nearly no polyhedral results on the QAP were known until a few years ago. In this chapter, we give an overview on the (theoretical and practical) results on the QAP that have been obtained since then. In particular, we explain some crucial techniques that made possible to prove these results and that may be also useful for deriving further results in the future.

The objects of polyhedral combinatorics are combinatorial optimization problems with *linear* objective functions, while the QAP, e.g., in its original formulation by [Koopmans and Beckmann, 1957] as the task to find a permutation π that minimizes $\sum_{i=1}^{n} \sum_{k=1}^{n} a_{ik} b_{\pi(i)\pi(k)} + \sum_{i=1}^{n} c_{i\pi(i)}$ (with $A = (a_{ik}) \in \mathcal{Q}^{n \times n}$ the *flow-*, $B = (b_{jl}) \in \mathcal{Q}^{n \times n}$ the *distance-*, and $C = (c_{ij}) \in \mathcal{Q}^{n \times n}$ the *linear cost-matrix*), has a quadratic objective function. However, there are lots of different equivalent formulations of the problem with linear objective functions. The one that turns out to yield a suitable starting point for a polyhedral treatment is due to [Lawler, 1963]. He formulated the QAP by representing the permutations π by *permutation matrices* $X = (x_{ij}) \in \{0,1\}^{n \times n}$ in the following way:

$$
\text{(L)} \quad \min \quad \sum_{i,k=1}^{n} \sum_{j,l=1}^{n} q_{ijkl} y_{ijkl} + \sum_{i=1}^{n} \sum_{j=1}^{n} c_{ij} x_{ij}
$$

$$
\text{s.t.} \quad \sum_{j=1}^{n} x_{ij} = 1 \qquad (i \in \{1,\ldots,n\})
$$

$$
\sum_{i=1}^{n} x_{ij} = 1 \qquad (j \in \{1,\ldots,n\})
$$

$$
y_{ijkl} = x_{ij} x_{kl} \quad (i,j,k,l \in \{1,\ldots,n\})
$$

$$
x_{ij} \in \{0,1\} \qquad (i,j \in \{1,\ldots,n\})
$$

Of course, in case of a QAP instance of Koopmans & Beckmann type one takes $q_{ijkl} = a_{ik} b_{jl}$.

Several proposals have been made in the literature to replace the nonlinear constraints $y_{ijkl} = x_{ij} x_{kl}$ by some linear equations and inequalities and to derive lower bounding procedures from these different types of *linearizations*. [Adams and Johnson, 1994] (see also [Johnson, 1992]) gave the following formulation (notice that every solution to (L) satisfies $y_{ijkj} = 0$ for $i \neq k$,

$y_{ijil} = 0$ for $j \neq l$, and $y_{ijij} = x_{ij}$):

$$
\text{(AJ)} \quad \min \sum_{\substack{i,k=1 \\ i \neq k}}^{n} \sum_{\substack{j,l=1 \\ j \neq l}}^{n} q_{ijkl} y_{ijkl} + \sum_{i=1}^{n} \sum_{j=1}^{n} c_{ij} x_{ij}
$$

$$
\text{s.t.} \quad \sum_{j=1}^{n} x_{ij} = 1 \qquad\qquad (i \in \{1,\ldots,n\})
$$

$$
\sum_{i=1}^{n} x_{ij} = 1 \qquad\qquad (j \in \{1,\ldots,n\})
$$

$$
\sum_{\substack{l=1 \\ l \neq j}}^{n} y_{ijkl} = x_{ij} \qquad (i,j,k \in \{1,\ldots,n\}, i \neq k)
$$

$$
\sum_{\substack{k=1 \\ k \neq i}}^{n} y_{ijkl} = x_{ij} \qquad (i,j,l \in \{1,\ldots,n\}, j \neq l)
$$

$$
y_{ijkl} = y_{klij} \qquad (i,j,k,l \in \{1,\ldots,n\}, i < k, j \neq l)
$$

$$
y_{ijkl} \geq 0 \qquad (i,j,k,l \in \{1,\ldots,n\}, i \neq k, j \neq l)
$$

$$
x_{ij} \in \{0,1\} \qquad (i,j \in \{1,\ldots,n\})
$$

Additionally, they showed that this formulation yields a stronger *linear programming relaxation* (obtained from the formulation by relaxing the integrality constraints $x_{ij} \in \{0,1\}$ to $0 \leq x_{ij} \leq 1$) than most other linearizations available in the literature. In particular, the lower bounds obtained from solving the corresponding linear programs are at least as good as several well-known lower bounds, including the most popular one proposed by [Gilmore, 1962] and [Lawler, 1963].

[Resende et al., 1995] have done extensive computational studies with the lower bounds obtained from the linear programming relaxation of (AJ). It turned out that this yields rather good bounds in practice. The polyhedral approach to the QAP can be viewed as the attempt to sharpen these bounds by insights derived from investigations of the geometric structure of the feasible solutions to (L).

In order to provide the reader who is not yet acquainted with the subject of polyhedral combinatorics with the necessary background, we give a short introduction into this field in Section 2.. In Section 3. we define the central objects used to describe QAPs in terms of polyhedral combinatorics: the different versions of *QAP-polytopes*. A very useful technique for working with these polytopes is explained in Section 4.. Section 5. contains the most important results that are known on the QAP-polytopes, and Section 6. reports on practical results obtained by cutting plane algorithms that exploit these results. We close the chapter by a discussion of the future perspectives of the polyhedral approach to the QAP in Section 7..

2.   POLYHEDRAL COMBINATORICS

This short section can only touch the basic principles of polyhedral combinatorics with respect to those parts of the theory that we need for the later work on the QAP. There are many concepts and results besides the scope of this work. For an overview we refer to [Schrijver, 1995], for comprehensive treatments to [Schrijver, 1986] and [Nemhauser and Wolsey, 1988], and for a textbook [Wolsey, 1998].

Combinatorial Optimization and Linear Programming. The subjects of polyhedral combinatorics are, in general, (linear) combinatorial optimization problems of the following kind. Let U be a finite set, $c \in \mathcal{R}^U$ an *objective function vector*, and let a subset $\mathcal{F} \subseteq 2^U$ of the subsets of U be specified, the *feasible solutions*. The *objective function value* of a feasible solution $F \in \mathcal{F}$ is $c(F) = \sum_{u \in F} c_u$. The problem we are interested in is to find a feasible solution $F_{\text{opt}} \in \mathcal{F}$ with

$$c(F_{\text{opt}}) = \min\{c(F) : F \in \mathcal{F}\} . \qquad (6.1)$$

In the light of the formulation (L) (see Section 1.), it is clear that the QAP falls into this class.

Using the concept of the *incidence vector* χ^F of a feasible solution $F \in \mathcal{F}$ defined via

$$\chi_u^F = \begin{cases} 1 & \text{if } u \in F \\ 0 & \text{otherwise} \end{cases}$$

(i.e., χ^F is the characteristic vector of $F \subseteq U$), solving (6.1) is equivalent to finding $F_{\text{opt}} \in \mathcal{F}$ with

$$c^T \chi^{F_{\text{opt}}} = \min\{c^T \chi^F : F \in \mathcal{F}\} ,$$

i.e., equivalent to the task of minimizing a linear function over a finite set of vectors. This is equivalent (forgetting for a moment the minimizing element that we also want to find) to minimizing that linear function over the convex hull

$$\mathcal{P}_{\mathcal{F}} = \text{conv}\{\chi^F : F \in \mathcal{F}\}$$

$$= \left\{ \sum_{F \in \mathcal{F}} \lambda_F \chi^F : \sum_{F \in \mathcal{F}} \lambda_F = 1, \lambda_F \geq 0 \text{ for all } F \in \mathcal{F} \right\}$$

of these vectors, called the *associated polytope* to that combinatorial optimization problem.

The background for polyhedral combinatorics is a theorem found by [Minkowski, 1896] and [Weyl, 1935], who proved that a subset of a vector space $\mathcal{R}^n$ is the convex hull of a finite set of vectors if and only if it is the bounded solution space of a finite system of linear equations and inequalities. Hence, (writing every equation as two inequalities) a finite system $Ax \leq b$ of linear inequalities exists such that

$$\mathcal{P}_\mathcal{F} = \{x \in \mathcal{R}^U : Ax \leq b\}$$

holds.

Polyhedral combinatorics can be described as the discipline that tries to *find* such systems (called *linear descriptions*) for special polytopes arising from combinatorial optimization problems in the way described above, and to exploit these systems in order to derive structural insights into the problems they describe as well as algorithms to solve them. Once a linear system for a given problem is found, the theory of *linear programming* (for introductions see, e.g., [Chvátal, 1983, Padberg, 1995]) provides the tools for its exploitation in the above sense.

Linear programming deals with optimizing linear functions over (finite) systems of linear equations and inequalities. One of its central results is the *strong duality theorem*, stating that the *linear programs*

$$\begin{array}{rl} \min & c^T x \\ \text{s.t.} & Ax \leq b \end{array}$$

and

$$\begin{array}{rl} \max & b^T y \\ \text{s.t.} & A^T y = c \\ & y \geq 0 \end{array}$$

(called *dual* to each other) have the same optimal solution value (if both optimal values exist). If one succeeds in describing the polytope associated with a combinatorial optimization problem by a system of linear equations and inequalities, then the strong duality theorem of linear programming usually leads to a "short certificate" in the sense of [Edmonds, 1965a] and [Edmonds, 1965b], i.e., the guarantee that for an optimal solution to the problem there is a polynomially sized proof of its optimality. In other words, the corresponding decision problem (supposed to be contained in $\mathcal{NP}$) is contained in $\mathcal{NP} \cap \text{co-}\mathcal{NP}$.

In fact, in many cases, finding such a linear description even yields that the combinatorial optimization problem under investigation is solvable in polynomial time. This is due to the work of [Grötschel et al., 1981], [Karp and Papadimitriou, 1982] and [Padberg and Rao, 1980]. After [Khachiyan, 1979] had shown that the ellipsoid method can be used to solve linear programs in polynomial time, they extended this result by proving that a linear program

$$\begin{array}{rl} \min & c^T x \\ \text{s.t.} & Ax \leq b \end{array}$$

can basically be solved in time depending polynomially on the running time of a *separation procedure*. A separation procedure for a (maybe only implicitly given) linear system $Ax \leq b$ finds for every point $x^\star$ an inequality in that system that is violated by $x^\star$ if one exists and otherwise asserts that $x^\star$ satisfies the whole system, i.e., it solves the *separation problem* for the system $Ax \leq b$.

Thus, finding an arbitrarily large complete linear description for a polytope associated with some combinatorial optimization problem and proving that the separation problem for this system is polynomially solvable (in the input size of the original problem) immediately yields that the optimization problem is polynomially solvable, too. In fact, also the opposite direction is true (see [Grötschel et al., 1988]). For every polynomially solvable combinatorial optimization problem the separation problem belonging to a complete linear description of the associated polytope is solvable in polynomial time. These results are usually known as the *(polynomial) equivalence of optimization and separation*.

Proving that the underlying combinatorial optimization problem is polynomially solvable via showing that the associated linear programs are solvable in polynomial time is only one possibility that opens up if one succeeds in describing the respective polytope completely by a linear system. Another one might be derived directly from the strong duality theorem, if the dual linear program can also be interpreted as some combinatorial optimization problem. Such a combinatorial *min-max relation* often leads to combinatorial algorithms for the investigated problems.

Hence, for several reasons, there is a strong interest in finding linear systems that describe polytopes coming from combinatorial optimization problems. Several beautiful techniques have been developed for this purpose, e.g., the concepts of *total unimodularity, total dual integrality*, or the theory of *blocking-* and *anti-blocking polyhedra*.

However, we do not concentrate here on these parts of polyhedral combinatorics, since pursuing the goal of finding "useful" complete linear descriptions of the polytopes associated with the QAP (or any other $\mathcal{NP}$-hard combinatorial optimization problem) is not very promising. This is due to the above remark that finding such a description usually yields a "good characterization" of the underlying problem, i.e., it would lead to $\mathcal{NP} = \text{co-}\mathcal{NP}$ in our case, what is not to expect. In fact, [Karp and Papadimitriou, 1982] proved that, unless $\mathcal{NP} = \text{co-}\mathcal{NP}$ holds, no linear description of a polytope associated in the above way with an $\mathcal{NP}$-hard combinatorial optimization problem can exist with the property that for every inequality in the description its validity for the whole polytope can be proved in time polynomially depending on the size of the problem. Consequently, a "useful" description of the linear system describing a polytope coming from such a problem is unlikely to exist.

It follows from the previous discussion that we can only expect o find *partial* linear descriptions of the polytopes associated with the QAP. Notice that the "impossibility" of finding complete descriptions of polytopes coming from $\mathcal{NP}$-hard problems is only due to the inequalities. From the complexity point of view, there are no reasons against finding a *complete equation system* for such a polytope, i.e., a set of equations whose solution space is precisely the affine hull of the polytope.

What is a partial description good for? The aim of finding (as tight as possible) partial descriptions of polytopes coming from $\mathcal{NP}$-hard optimization problems is the computation of lower bounds (in case of a minimization problem) on the (yet unknown) optimal solution value of an instance. Lower bounds are very important tools in combinatorial optimization at all (regardless of polyhedral combinatorics). One reason for their importance is that they allow to give quality guarantees for feasible solutions that might have been obtained by heuristics, i.e., algorithms that do not necessarily give *optimal* solutions. Furthermore, they can be incorporated into implicitly enumerative algorithms like, e.g., branch-and-bound methods. There they can, applied to a problem defined on a subset of the solutions $\mathcal{F}$, give the guarantee that the overall optimal solution is not contained in that subset, and hence, one does not have to search this subset.

By solving the linear programs arising from a partial description, one clearly obtains lower bounds for the respective problem. However, usually also the partial descriptions contain exponentially many inequalities that cannot be handled by any linear programming solver in praxis. The way out is to develop *separation procedures* similar to the subroutines mentioned in the context of the ellipsoid method, which try to find for a given point $x^\star$ inequalities in the partial description that are violated by $x^\star$, i.e., which *cut off* $x^\star$. Such inequalities (or more precisely, the boundary hyperplanes of the halfspaces defined by them) are called *cutting planes*.

Now, one starts by solving a small linear program containing only some inequalities from the partial linear description. Often one takes an *integer programming formulation* here, i.e., a set of inequalities whose integer solutions are precisely the incidence vectors of the problem. After the linear program is solved, one checks, if, by chance, the resulting optimal solution vector $x^\star$ is an incidence vector of a feasible solution $F \in \mathcal{F}$. If this happens to be true, then, clearly, F must be an optimal solution. If the solution vector is not an incidence vector, then one calls the separation procedure in order to find inequalities that are violated by $x^\star$. If the procedure is successful, then the detected inequalities are added to the linear program, which is resolved, potentially giving a better bound. Iterating this process, one obtains a *cutting plane algorithm* for computing lower bounds for the problem. A branch-and-bound algorithm using this kind of lower bounding procedure is called a *branch-and-cut algorithm*.

If the pure cutting plane bounding procedure ends already with a bound that guarantees a known feasible solution to be optimal, one has obtained a very nice result, namely, a short, i.e., polynomially sized, *certificate* for the optimality of that solution. This is due to the fact that the final linear program of the cutting plane run provides a proof for optimality, in this case, and linear programming theory yields that one can remove equations and inequalities from every linear program such that the remaining linear program still has the same optimal solution value and its number of constraints does not exceed its number of variables. Hence, although the cutting plane algorithm may have run for a very long time, it might yield (if one is lucky) after all a short certificate either for the optimality or at least for a certain quality of the known solution (see, e.g., [Jünger et al., 1994] for provably good solutions for the example of the *traveling salesman problem*).

Due to efficiency reasons, one desires in particular to find non-redundant (partial) linear descriptions. For the equations in such a partial description this means that one wants to avoid that any among them is a linear combination of some others. For the inequalities analogous redundancies coming from linear combinations with nonnegative coefficients for inequalities (and arbitrary ones for equations) should be excluded. The questions concerning redundancies in the (partial) linear descriptions are strongly related to the geometry of the polytope associated with the combinatorial optimization problem. They lead to asking for the *dimension* of the polytope or for the possibility of proving that a certain inequality is unavoidable in a complete description of the polytope (and hence also cannot be redundant in any partial linear description). This is the point, where polyhedral theory takes over.

Polyhedral Theory.   Used in an auxiliary way, only a few aspects of the theory of polytopes become apparent in the context of our polyhedral investigations on the QAP. However, the general theory of polytopes is a fascinating and strongly developing field. For entering this wonderful area of mathematics as well as for the proofs of the statements in this section, we recommend the classical book of [Grünbaum, 1967], as well as [Ziegler, 1995] and [Klee and Kleinschmidt, 1995].

Polytopes can (that was the central point in the previous section) equivalently be defined as the convex hulls of finite point sets in $\mathcal{R}^n$ or as the bounded solution spaces of finite systems of linear inequalities (and equations), i.e., as the bounded intersections of finitely many halfspaces (and hyperplanes) of $\mathcal{R}^n$.

Two concrete examples are the *n-hypercube*

$$C_n = \text{conv}\,\{(x_1, \ldots, x_n) \in \mathcal{R}^n : x_i \in \{0,1\}\}$$
$$= \{(x_1, \ldots, x_n) \in \mathcal{R}^n : 0 \le x_i \le 1\,(1 \le i \le n)\}$$

and the *standard-$(n-1)$-simplex*

$$\Delta_{n-1} = \text{conv}\,\{e_1, \ldots, e_n\}$$
$$= \left\{(x_1, \ldots, x_n) \in \mathcal{R}^n \,\Big|\, \sum_{i=1}^{n} x_i = 1, x_i \ge 0\,(1 \le i \le n)\right\}$$

(see Figure 6.1).

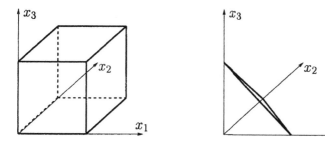

Figure 6.1 The 3-hypercube and the standard-2-simplex

Two polytopes $\mathcal{P} \subseteq \mathcal{R}^n$ and $\mathcal{Q} \subseteq \mathcal{R}^m$ are *affinely isomorphic* if there is an affine map $\phi : \mathcal{R}^n \longrightarrow \mathcal{R}^m$ (not necessarily an affine transformation) that induces a bijection between the points of $\mathcal{P}$ and $\mathcal{Q}$, or, equivalently, between the vertices of $\mathcal{P}$ and $\mathcal{Q}$. Whenever we say that two polytopes are *isomorphic* it is meant that they are affinely isomorphic.

If $\mathcal{P} \subseteq \mathcal{R}^n$ is a polytope, then its *dimension* $\dim(\mathcal{P})$ is defined as the dimension of its affine hull (i.e., the linear dimension of the linear subspace belonging to aff($\mathcal{P}$)). If $\dim(\mathcal{P}) = n$ holds, then $\mathcal{P}$ is called *full-dimensional*. We call the difference between the dimension of the vector space where the polytope is defined in and the dimension of the polytope the *dimensional gap*. The dimensional gap equals the minimal number of equations in a *complete equation system* for $\mathcal{P}$, i.e., an equation system having the affine hull of $\mathcal{P}$ as its solution space.

If a halfspace

$$\mathcal{H} = \{x \in \mathcal{R}^n : a^T x \le \alpha\}$$

of $\mathcal{R}^n$ with the *boundary hyperplane*

$$\mathcal{B} = \{x \in \mathcal{R}^n : a^T x = \alpha\}$$

contains the polytope $\mathcal{P}$, then $\mathcal{P} \cap \mathcal{B}$ is called a *face* of the polytope $\mathcal{P}$. Both the halfspace $\mathcal{H}$ as well as the inequality $a^T x \leq \alpha$ are said to *define* that face of $\mathcal{P}$. Adding any equation that is valid for the whole polytope to an inequality that defines a face of it yields another inequality defining the same face.

By the characterization of polytopes as the bounded intersections of finitely many halfspaces and hyperplanes, every face of a polytope is a polytope, again. Notice that both the polytope $\mathcal{P}$ itself as well as the empty set are faces of $\mathcal{P}$. The other faces of $\mathcal{P}$ are called *proper faces*. If $\mathcal{F}_1$ is a face of the polytope $\mathcal{P}$, and $\mathcal{F}_2$ is a face of $\mathcal{F}_1$, then $\mathcal{F}_2$ is also a face of $\mathcal{P}$.

Faces of $\mathcal{P}$ that have dimension 0, 1, $\dim(\mathcal{P}) - 2$, or $\dim(\mathcal{P}) - 1$ are called *vertices*, *edges*, *ridges*, and *facets* of $\mathcal{P}$, respectively. Facets of full-dimensional polytopes have the convenient property to be described by unique (up to multiplications with positive scalars) inequalities. Like every edge is the convex hull of two uniquely determined vertices, every ridge is the intersection of two uniquely determined facets. The following results are very basic for polyhedral theory.

(i) Every polytope is the convex hull of its vertices. If a polytope is the convex hull of some finitely many points, then its vertices must be among them (in particular, a polytope has finitely many vertices).

(ii) The vertices of a face $\mathcal{F}$ of a polytope $\mathcal{P}$ are precisely the vertices of $\mathcal{P}$ that are contained in $\mathcal{F}$ (in particular, every polytope has only finitely many faces).

(iii) Every polytope is the intersection of all halfspaces that define facets of it and all hyperplanes it is contained in. If a polytope is the intersection of some halfspaces (and hyperplanes), then all halfspaces that define facets of it must be among them.

(iv) Every face of a polytope $\mathcal{P}$ is the intersection of some facets of $\mathcal{P}$.

Point (iii) shows in particular that in any intersection of pairwise distinct facet defining halfspaces of a polytope there is no redundant halfspace. In other words, every (partial) linear description of a polytope $\mathcal{P}$ that contains (besides the equations only inequalities that define (pairwise distinct) facets of $\mathcal{P}$ is non-redundant, as long as the equations in it are non-redundant. Hence, aiming to find non-redundant (partial) descriptions of polytopes, one should concentrate on facet defining inequalities. Clearly, keeping the equation system in the (partial) linear description non-redundant means that the corresponding matrix should have full row rank.

There is another important property of facets from the point of view of poly-hedral combinatorics. Once one has found a class of inequalities contributing to a partial description of the polytope under investigation, one might (even know-ing already that it is non-redundant with respect to that partial description) ask if it is possible to improve that new class by, e.g., "playing with the coefficients". Only the fact that the new inequalities define facets of the polytope can tell us at which point we do not have to try to strengthen our inequalities any further, but preferably continue with the search for completely different ones.

A very convenient fact on polytopes is that a linear function always attains its minimum (and, clearly, also its maximum) over a polytope in a vertex of it. This is simply due to the fact that the optimal solution points for a linear optimization problem defined on a polytope always constitute a face of the polytope. That is the reason, why in the previous section we just had to forget the minimizing element *for a while*, when we passed from minimizing a linear function over all incidence vectors to minimizing that linear function over their convex hull.

3. POLYTOPES ASSOCIATED WITH THE QAP

In our treatment of the QAP we will use a formulation of the problem in terms of certain graphs — mainly, because this provides us with some convenient ways to talk about the problem. Since in many application-driven instances (in particular, this holds for several of the instances in the QAPLIB, a commonly used set of test instances compiled by [Burkard et al., 1997]) the number of objects to be assigned might be smaller than the number of locations that are available, we will use a model that deals with n locations and m objects (with $m \leq n$). Not surprisingly, this turns out to be much more efficient than introducing $n - m$ "dummy-objects" that do not have any interaction with anybody.

From now on, let $\mathcal{M} = \{1, \ldots, m\}$ be the set of objects and $\mathcal{N} = \{1, \ldots, n\}$ the set of locations (with $m \leq n$). Let us denote by $\mathcal{X}^{m \times n} \in \{0, 1\}^{m \times n}$ the set of all 0/1-matrices of size $m \times n$ that have *precisely* one "1" in every row and *at most* one "1" in every column (corresponding to the fact that every object has to be assigned to precisely one location, and every location can receive at most one object). Then the QAP can be formulated (see (L) in Section 1.) as the task to find $X^\star = (x^\star_{ij}) \in \mathcal{X}^{m \times n}$ such that

$$\sum_{i,k=1}^{m} \sum_{j,l=1}^{n} q_{ijkl} x^\star_{ij} x^\star_{kl} + \sum_{i=1}^{m} \sum_{j=1}^{n} c_{ij} x^\star_{ij}$$

becomes minimal, where $q_{ijkl} \in \mathcal{R}$ are the *quadratic costs* (we may — and will — assume $q_{ijij} = 0$) and $c_{ij} \in \mathcal{R}$ are the *linear costs*.

Let $\mathcal{G}_{m,n} = (\mathcal{V}_{m,n}, \mathcal{E}_{m,n})$ be the graph with nodes $\mathcal{V}_{m,n} = \mathcal{M} \times \mathcal{N}$ and edges

$$\mathcal{E}_{m,n} = \left\{ \{(i,j), (k,l)\} \in \binom{\mathcal{V}_{m,n}}{2} : i \neq k, j \neq l \right\},$$

where we denote by $\binom{M}{k}$ the set of subsets of a set M that have cardinality k. We usually denote an edge $\{(i,j), (k,l)\}$ by $[i,j,k,l]$ (notice that this implies $[i, j, k, l] = [k, l, i, j]$). The sets $\text{row}_i = \{(i, j) : 1 \leq j \leq n\}$ and $\text{col}_j = \{(i, j) : 1 \leq i \leq m\}$ are called the *i-th row* and the *j-th column* of $\mathcal{V}_{m,n}$, respectively. Thus, the graph $\mathcal{G}_{m,n}$ has all possible edges but the ones that have either both end points in the same row or in the same column.

By construction, the set $\mathcal{X}^{m \times n}$ of feasible solutions to our formulation of the QAP is in one-to-one correspondence with the m-cliques in $\mathcal{G}_{m,n}$. Now we put weights $c \in \mathcal{R}^{\mathcal{V}_{m,n}}$ and $d \in \mathcal{R}^{\mathcal{E}_{m,n}}$ on the nodes and edges, respectively, by letting $c_{(i,j)} = c_{ij}$ for every node $(i, j) \in \mathcal{V}_{m,n}$ and $d_{[i,j,k,l]} = q_{ijkl} + q_{klij}$ for every edge $[i, j, k, l] \in \mathcal{E}_{m,n}$. Then the QAP with linear costs c_{ij} and quadratic costs q_{ijkl} is equivalent to finding a cheapest node- and edge-weighted m-clique in the graph $\mathcal{G}_{m,n}$ weighted by (c, d).

This formulates the QAP as a *(linear) combinatorial optimization problem* in the sense of Section 2. (actually, it is still the same formulation as (L) in Section 1.). For a subset $W \subseteq \mathcal{V}_{m,n}$ of nodes of $\mathcal{G}_{m,n}$ let $\mathcal{E}_{m,n}(W)$ be the set of all edges of $\mathcal{G}_{m,n}$ that have *both* end points in W. We denote by $x^W \in \{0,1\}^{\mathcal{V}_{m,n}}$ the characteristic vector of W (i.e., $x_v^W = 1$ if and only if $v \in W$) and by $y^W \in \{0,1\}^{\mathcal{E}_{m,n}}$ the characteristic vector of $\mathcal{E}_{m,n}(W)$ (i.e., $y_e^W = 1$ if and only if $e \in \mathcal{E}_{m,n}(W)$). We call

$$QAP_{m,n} = \text{conv} \left\{ (x^C, y^C) : C \subseteq \mathcal{V}_{m,n} \text{ } m\text{-clique in } \mathcal{G}_{m,n} \right\}$$
$$\subseteq \mathcal{R}^{\mathcal{V}_{m,n}} \times \mathcal{R}^{\mathcal{E}_{m,n}}$$

the *QAP-polytope* (for QAPs with m objects and n locations). Notice that we have $\dim(\mathcal{R}^{\mathcal{V}_{m,n}} \times \mathcal{R}^{\mathcal{E}_{m,n}}) = mn + \frac{mn(m-1)(n-1)}{2}$.

Now recall the Koopmans & Beckmann type instances. There quite often the flows (a_{ik}) or the distances (b_{jl}) are *symmetric* (meaning that $a_{ik} = a_{ki}$ holds for all pairs of objects i, k or that $b_{jl} = b_{lj}$ holds for all pairs of locations, respectively). In either case, this implies that

$$q_{ijkl} + q_{klij} = a_{ik}b_{jl} + a_{ki}b_{lj} = q_{ilkj} + q_{kjil}$$

holds for all quadratic costs, yielding

$$d_{[i,j,k,l]} = d_{[i,l,k,j]}$$

for all pairs $[i, j, k, l]$, $[i, l, k, j]$ of edges in our graph formulation. Since furthermore no clique in $\mathcal{G}_{m,n}$ can contain both edges $[i, j, k, l]$ and $[i, l, k, j]$, this suggests that in case of a *symmetric instance* (i.e., $q_{ijkl} + q_{klij} = q_{ilkj} + q_{kjil}$ holds for all quadratic costs) we can identify each pair $y_{[i,j,k,l]}$, $y_{[i,l,k,j]}$ of variables, which will reduce the number of variables by roughly 50%.

In order to formalize this, we define a hypergraph $\hat{\mathcal{G}}_{m,n} = (\mathcal{V}_{m,n}, \hat{\mathcal{E}}_{m,n})$ with the same set of nodes as our original graph $\mathcal{G}_{m,n}$ has and with hyperedges

$$\hat{\mathcal{E}}_{m,n} = \left\{ \{(i, j), (k, l), (i, l), (k, j)\} \in \binom{\mathcal{V}_{m,n}}{4} : i \neq k, j \neq l \right\} .$$

A hyperedge $\{(i,j),(k,l),(i,l),(k,j)\} \in \hat{\mathcal{E}}_{m,n}$ is denoted by $\langle i,j,k,l\rangle$. Thus, any hyperedge is the union

$$\langle i,j,k,l\rangle = [i,j,k,l] \cup [i,l,k,j]$$

of two edges from $\mathcal{G}_{m,n}$. We call two edges whose union gives a hyperedge *mates* of each other. Notice that we have

$$\langle i,j,k,l\rangle = \langle k,l,i,j\rangle = \langle i,l,k,j\rangle = \langle k,j,i,l\rangle \ .$$

For an edge $e \in \mathcal{E}_{m,n}$ we denote the hyperedge belonging to e (and to its mate) by $\mathrm{hyp}(e)$. For any subset $W \subseteq \mathcal{V}_{m,n}$ define $\hat{\mathcal{E}}_{m,n}(W) = \{\mathrm{hyp}(e) : e \in \mathcal{E}_{m,n}(W)\}$. Thus, if $C \subseteq \mathcal{V}_{m,n}$ is a clique of $\mathcal{G}_{m,n}$ (which we will also call a clique of $\hat{\mathcal{G}}_{m,n}$) then $\hat{\mathcal{E}}_{m,n}(C)$ consists of all hyperedges in $\hat{\mathcal{G}}_{m,n}$ that contain two nodes of C (since C is a clique, these two nodes then must precisely be the intersection of that hyperedge with C, and they must be "diagonal" in the "rectangle" formed by the end nodes of e and its mate). Finally, we denote for every subset $F \subseteq \hat{\mathcal{E}}_{m,n}$ of hyperedges the characteristic vector of F by z^F.

Now, we define the geometric structure that is especially suitable for symmetric QAPs, the *symmetric QAP-polytope*:

$$\mathcal{SQAP}_{m,n} = \mathrm{conv}\left\{(x^C, z^C) : C \subseteq \mathcal{V}_{m,n} \ m\text{-clique in } \hat{\mathcal{G}}_{m,n}\right\}$$

$$\subseteq \mathcal{R}^{\mathcal{V}_{m,n}} \times \mathcal{R}^{\hat{\mathcal{E}}_{m,n}}$$

We have $\dim(\mathcal{R}^{\mathcal{V}_{m,n}} \times \mathcal{R}^{\hat{\mathcal{E}}_{m,n}}) = mn + \frac{mn(m-1)(n-1)}{4}$.

Next we will give integer programming formulations for both the general and the symmetric model. Recall that for a vector $v \in \mathcal{R}^I$ (I some finite set) and some subset $J \subseteq I$ we denote $v(J) = \sum_{j \in J} v_j$. We will need some further notations. For $i, k \in \mathcal{M}$ and $j \in \mathcal{N}$ let $((i,j) : \mathrm{row}_k) = \{[i,j,k,l] : l \in \mathcal{N} \setminus \{j\}\}$ be the set of all edges connecting (i,j) with the k-th row, and let $\Delta^{(i,j)}_{(k,j)} = \{\langle i,j,k,l\rangle : l \in \mathcal{N} \setminus \{j\}\}$ be the set of all hyperedges containing both nodes (i,j) and (k,j). The proofs of the following theorems can be found in [Kaibel, 1998] (for the symmetric version) as well as in [Kaibel, 1997] (for both versions). Figure 6.2 illustrates equations (6.4) and (6.9). We usually draw hyperedges just by drawing the two corresponding edges. Notice that in our drawings a solid line or a solid disc will always indicate a coefficient 1, while a dashed line or a gray disc stand for a coefficient -1.

Theorem 1 *Let $1 \leq m \leq n$.*

(i) A vector $(x,y) \in \mathcal{R}^{\mathcal{V}_{m,n}} \times \mathcal{R}^{\mathcal{E}_{m,n}}$ is a vertex of $QAP_{m,n}$, i.e., the characteristic vector of an m-clique of $\mathcal{G}_{m,n}$, if and only if it satisfies the

following conditions:

$$x(\text{row}_i) = 1 \qquad (i \in \mathcal{M}) \qquad\qquad (6.2)$$

$$x(\text{col}_j) \leq 1 \qquad (j \in \mathcal{N}) \qquad\qquad (6.3)$$

$$-x_{(i,j)} + y\left((i,j) : \text{row}_k\right) = 0 \qquad (i,k \in \mathcal{M}, i \neq k, j \in \mathcal{N}) \quad (6.4)$$

$$y_e \geq 0 \qquad (e \in \mathcal{E}_{m,n}) \qquad\qquad (6.5)$$

$$x_v \in \{0,1\} \qquad (v \in \mathcal{V}_{m,n}) \qquad\qquad (6.6)$$

(ii) *A vector* $(x,z) \in \mathcal{R}^{\mathcal{V}_{m,n}} \times \mathcal{R}^{\hat{\mathcal{E}}_{m,n}}$ *is a vertex of* $\mathcal{SQAP}_{m,n}$*, i.e., the characteristic vector of an* m*-clique of* $\hat{\mathcal{G}}_{m,n}$*, if and only if it satisfies the following conditions:*

$$x(\text{row}_i) = 1 \qquad (i \in \mathcal{M}) \qquad\qquad (6.7)$$

$$x(\text{col}_j) \leq 1 \qquad (j \in \mathcal{N}) \qquad\qquad (6.8)$$

$$-x_{(i,j)} - x_{(k,j)} + z\left(\Delta_{(k,j)}^{(i,j)}\right) = 0 \qquad (i,k \in \mathcal{M}, i < k, j \in \mathcal{N}) \qquad\qquad (6.9)$$

$$z_h \geq 0 \qquad (h \in \hat{\mathcal{E}}_{m,n}) \qquad\qquad (6.10)$$

$$x_v \in \{0,1\} \qquad (v \in \mathcal{V}_{m,n}) \qquad\qquad (6.11)$$

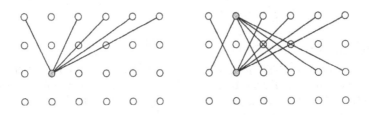

Figure 6.2   Equations (6.4) and (6.9)

Once one starts to investigate $\mathcal{QAP}_{m,n}$ and $\mathcal{SQAP}_{m,n}$ closer, it turns out very soon that the structures of the polytopes are quite different for the cases $m \leq n - 2$ and $m \in \{n - 1, n\}$. In fact, the case $m \leq n - 2$ is much nicer to handle than the other two. Here, we just want to give a result that shows that actually the case $m = n - 1$ reduces to the case $m = n$, thus leaving us with only one "less convenient case". Again the proof of the theorem can be found in [Kaibel, 1997] and [Kaibel, 1998].

Theorem 2 *For* $n \geq 2$*,* $\mathcal{QAP}_{n-1,n}$ *is affinely isomorphic to* $\mathcal{QAP}_{n,n}$*, and* $\mathcal{SQAP}_{n-1,n}$ *is affinely isomorphic to* $\mathcal{SQAP}_{n,n}$*.*

The affine maps from $\mathcal{R}^{\mathcal{V}_{n,n}} \times \mathcal{R}^{\mathcal{E}_{n,n}}$ to $\mathcal{R}^{\mathcal{V}_{n-1,n}} \times \mathcal{R}^{\mathcal{E}_{n-1,n}}$ and from $\mathcal{R}^{\mathcal{V}_{n,n}} \times \mathcal{R}^{\hat{\mathcal{E}}_{n,n}}$ to $\mathcal{R}^{\mathcal{V}_{n-1,n}} \times \mathcal{R}^{\hat{\mathcal{E}}_{n-1,n}}$ giving the isomorphisms in the theorem are quite simple. They are just the canonical projections that "forget" all coordinates belonging to nodes in row_n or to (hyper)edges that intersect row_n.

We conclude this section by some results on $\mathcal{QAP}_{m,n}$ and $\mathcal{SQAP}_{m,n}$ which can be found in a more detailed version (in particular including proofs) in [Kaibel, 1997], [Jünger and Kaibel, 1997b], [Jünger and Kaibel, 1996], and [Kaibel, 1998].

It is rather obvious that for $m' \leq m \leq n$ the polytopes $\mathcal{QAP}_{m',n}$ and $\mathcal{SQAP}_{m',n}$ arise as "canonical projections" of the polytopes $\mathcal{QAP}_{m,n}$ and $\mathcal{SQAP}_{m,n}$, respectively. This implies that whenever an inequality is valid (facet-defining) for $\mathcal{QAP}_{m,n}$ or $\mathcal{SQAP}_{m,n}$ and it has all its nonzero coefficients within the first m' rows, then the corresponding inequality is valid (facet-defining) for $\mathcal{QAP}_{m',n}$ or $\mathcal{SQAP}_{m',n}$, respectively.

But also $\mathcal{QAP}_{m,n}$ and $\mathcal{SQAP}_{m,n}$ have a relationship; the latter is an affine image (not under an isomorphism) of the first. In particular, if an inequality is valid for $\mathcal{QAP}_{m,n}$ and it is *symmetric* in the sense that each two edges $[i, j, k, l]$, $[i, l, k, j]$ have the same coefficient, then the inequality induces in the obvious way a valid inequality for $\mathcal{SQAP}_{m,n}$. Let us call a face of $\mathcal{QAP}_{m,n}$ *symmetric* if it *can* be defined by a symmetric inequality. Then it turns out that a symmetric face F of $\mathcal{QAP}_{m,n}$ (which induces a face o $\mathcal{SQAP}_{m,n}$, as indicated above) actually induces a facet of $\mathcal{SQAP}_{m,n}$ if and only if all faces of $\mathcal{QAP}_{m,n}$ that properly contain F are *not* symmetric. More generally, the face lattice of $\mathcal{SQAP}_{m,n}$ arises from the face lattice of $\mathcal{QAP}_{m,n}$ by deleting all faces that are not symmetric.

These various connections between the different polytopes we have to consider give several possibilities to carry over results between them. However, there are also interesting connections to polytopes "outside the QAP world". In particular, it turns out that $\mathcal{QAP}_{m,n}$ can be viewed as a certain face of a cut polytope. This rises hope that one might take profit from the rich knowledge on cut polytopes (see, e.g., [Deza and Laurent, 1997]) for the investigations of the QAP-polytopes. In fact, this turns out to be true, as we will see in Section 5..

Another connection to different kinds of polytopes is the following. It turns out that several well-known polytopes like the *traveling salesman polytope* or the *linear ordering polytope* can be obtained as quite simple projections (just "forgetting coordinates") of $\mathcal{QAP}_{n,n}$. This phenomenon corresponds to the possibility to obtain problems like the traveling salesman problem or the linear ordering problem as "immediate special cases" of the QAP (which explains to some extent the astonishing resistance of the QAP against all attacks to practically solve it).

4.   THE STAR-TRANSFORMATION

Having defined all these polytopes as convex hulls of certain characteristic vectors, according to the principles of polyhedral combinatorics explained in Section 2. the next (and crucial) step now is to find (partial) linear descriptions of the polytopes. That means, we aim for finding systems of linear equations whose solution spaces are the affine hulls of the polytopes as well as linear inequalities that are valid for the polytopes and, preferably, even define facets of them. While showing validity of equations and inequalities will always be done by ad-hoc arguments in our case, the proofs that the equation systems that we propose indeed completely describe the affine hulls or that some inequalities indeed define facets (i.e., faces of largest possible dimension) require some more elaborate techniques.

In both cases (equation systems and inequalities) we have to solve tasks of the following type: given $X \subseteq \{0,1\}^s$ and a system $Ax = b$ of linear equations that are valid for X (with $A \in \mathcal{Q}^{r \times s}$ having full row rank and $b \in \mathcal{Q}^r$) prove that $\mathrm{aff}(X) = \{x \in \mathcal{Q}^s : Ax = b\}$ holds.

In order to prove this, we proceed as follows. First, we identify a subset $\mathcal{B} \subseteq \{1, \ldots, s\}$ of variable indices with $|\mathcal{B}| = r$ such that the columns of A that correspond to $\mathcal{B}$ (called a *basis* of A) form a non-singular matrix. Then, by a dimension argument, it suffices to show

$$\mathrm{aff}(\{e_i : i \in \mathcal{B}\} \cup X) = \mathcal{Q}^s \ .$$

Our way to prove this equation then is to construct all unit vectors $e_1, \ldots, e_s \in \mathcal{Q}^s$ as linear combinations of $\{e_i : i \in \mathcal{B}\} \cup X$.

The crucial task in the proofs thus is to construct suitable linear combinations of certain vertices, i.e., of characteristic vectors of certain feasible solutions. If one starts to play around with vertices of the polytopes we have introduced in Section 3., one soon will find that this is quite inconvenient, because it turns out to be very hard to obtain vectors with a small and well-structured *support* (i.e., set of indices of nonzero components).

However, this is not a problem that is inherent to the geometry of the polytopes or even to the QAP itself. It is only a matter of the actual coordinate representation that we have chosen for modeling the problem. And indeed, it turns out that one can find different coordinate representations of the QAP-polytopes (in different, lower dimensional, ambient spaces) that make it much more comfortable to build linear combinations from the vertices.

The key is the observation that the last column (or any single column) in the (hyper-)graph model contains redundant information on the feasible solutions: the intersection of an m-clique $C \subseteq \mathcal{V}_{m,n}$ with the first $n-1$ columns already determines the intersection of C with the last column. In case of $m = n$ even the last column *and* the last row (or any single column *and* any single row) are redundant in this sense. This suggests to remove the last column (and the last row in case of $m = n$) from the model. If we remove the last column an m-clique $C \subseteq \mathcal{V}_{m,n}$ may become either an $(m - 1)$- or an m-clique in $\mathcal{V}_{m,n-1}$ (depending on whether C intersects the last column or not). If $m = n$ holds and we remove both the last column and the last row, then an n-Clique $C \subseteq \mathcal{V}_{n,n}$ becomes an $(n - 1)$- or an $(n - 2)$-clique in $\mathcal{V}_{n-1,n-1}$ (depending on whether (n, n) is contained in C or not).

For the QAP-polytopes, removing nodes and (hyper-)edges corresponds to orthogonal projections that "forget" the coordinates belonging to the nodes and (hyper-) edges that are removed. Since we just want to find different coordinate representations of the polytopes (with their geometry unchanged), we must ensure that these projections actually give affine isomorphisms of the polytopes.

For $m \leq n^\star$ define the polytopes

$$\mathcal{QAP}^\star_{m,n^\star} =$$

$$\text{conv}\big\{ (x^C, y^C) : C \subseteq \mathcal{V}_{m,n^\star} \ m\text{- or } (m - 1)\text{-clique in } \mathcal{G}_{m,n^\star} \big\}$$

and

$$\mathcal{SQAP}^\star_{m,n^\star} =$$

$$\text{conv}\big\{ (x^C, z^C) : C \subseteq \mathcal{V}_{m,n^\star} \ m\text{- or } (m - 1)\text{-clique in } \hat{\mathcal{G}}_{m,n^\star} \big\} \ ,$$

which are objects in $\mathcal{R}^{\mathcal{V}_{m,n^\star}} \times \mathcal{R}^{\mathcal{E}_{m,n^\star}}$ and $\mathcal{R}^{\mathcal{V}_{m,n^\star}} \times \mathcal{R}^{\hat{\mathcal{E}}_{m,n^\star}}$, respectively. The following result was shown (for the different versions of QAP-polytopes) in [Jünger and Kaibel, 1997b], [Jünger and Kaibel, 1996], and [Kaibel, 1997].

Theorem 3 *Let* $3 \leq m \leq n$.

(i) *The polytopes* $\mathcal{QAP}_{m,n}$ *and* $\mathcal{SQAP}_{m,n}$ *are affinely isomorphic to the polytopes* $\mathcal{QAP}^\star_{m,n-1}$ *and* $\mathcal{SQAP}^\star_{m,n-1}$, *respectively.*

(i) *The polytopes* $\mathcal{QAP}_{n,n}$ *and* $\mathcal{SQAP}_{n,n}$ *are affinely isomorphic to the polytopes* $\mathcal{QAP}^\star_{n-1,n-1}$ *and* $\mathcal{SQAP}^\star_{n-1,n-1}$, *respectively.*

The change of coordinate representation described in Theorem 3 is called the *star-transformation*. What have we gained by it? As desired, we have gained the possibility to obtain easily such simple vectors as the ones shown in Figure 4. as linear combinations of vertices. And in fact, these types of vectors are the basic ingredients for the proofs of all results coming up in Section 5..

Another advantage of applying the star-transformation to the QAP-polytopes is that the "dimensional gaps" between their dimensions and the dimensions of the spaces they are located in has become much smaller. This leads to much smaller systems of equations describing the affine hulls. In fact, it turns out that the affine hulls of $\mathcal{QAP}^\star_{m,n^\star}$ and $\mathcal{SQAP}^\star_{m,n^\star}$ can be described by very simple equation systems, which in particular have very convenient bases (see [Jünger and Kaibel, 1997b], [Jünger and Kaibel, 1996], and [Kaibel, 1997]).

Figure 6.3 Some very simple vectors that can easily be obtained as linear combinations of vertices of $\mathcal{QAP}^*_{m,n}$.

5.   FACIAL DESCRIPTIONS OF QAP-POLYTOPES

Let us first consider the affine hulls of $\mathcal{QAP}_{m,n}$ and $\mathcal{SQAP}_{m,n}$, i.e., the (linear) equations holding for these polytopes. In case of $m \leq n - 2$, it turns out that the equations in the integer programming formulations of Theorem 1 do already suffice.

Theorem 4 *Let* $1 \leq m \leq n - 2$.

(i) *The affine hull* $\mathrm{aff}(\mathcal{QAP}_{m,n})$ *is the set of solutions of the equations* (6.2) *and* (6.4). *If one removes for every pair* $i, k \in M$ *one of the equations in* (6.4) *then one obtains an irredundant system of equations describing* $\mathrm{aff}(\mathcal{QAP}_{m,n})$. *In particular, we have:*

$$\dim(\mathcal{QAP}_{m,n}) = \dim(\mathcal{R}^{\mathcal{V}_{m,n}} \times \mathcal{R}^{\mathcal{E}_{m,n}}) - \left(m^2 n - mn - \frac{1}{2}m^2 + \frac{3}{2}m \right)$$

(ii) *The affine hull* $\mathrm{aff}(\mathcal{SQAP}_{m,n})$ *is the set of solutions of the equations* (6.7) *and* (6.9). *Here, in the symmetric case, the equations already form an irredundant system. In particular, we obtain:*

$$\dim(\mathcal{SQAP}_{m,n}) = \dim(\mathcal{R}^{\mathcal{V}_{m,n}} \times \mathcal{R}^{\hat{\mathcal{E}}_{m,n}}) - \left(\frac{1}{2}m^2 n - \frac{1}{2}mn + m \right)$$

The proofs of both parts of the theorem can be found in [Kaibel, 1997].

Now let us turn to the case $m = n$. Here, one immediately finds that the inequalities (6.3) and (6.8) ("$x(\mathrm{col}_j) \leq 1$") actually must be satisfied with equality by the vertices of the polytopes $\mathcal{QAP}_{n,n}$ and $\mathcal{SQAP}_{n,n}$. Moreover, of course also the "column versions" of the equations (6.4) and (6.9) are satisfied by all vertices of the respective polytopes. In fact, it turns out that by these observations we have collected enough equations to describe the affine hulls of $\mathcal{QAP}_{n,n}$ and $\mathcal{SQAP}_{n,n}$.

Theorem 5 *Let* $n \geq 1$.

(i) *The affine hull* $\text{aff}(\mathcal{QAP}_{n,n})$ *is the set of solutions of the equations* (6.2), (6.4), *and*

$$x(\text{col}_j) = 1 \qquad (j \in \mathcal{N}) \tag{6.12}$$

$$-x_{(i,j)} + y\left((i,j) : \text{col}_l\right) = 0 \qquad (i,j,l \in \mathcal{N}, j \neq l) \ . \tag{6.13}$$

We have

$$\dim(\mathcal{QAP}_{n,n}) = \dim(\mathcal{R}^{\mathcal{V}_{n,n}} \times \mathcal{R}^{\mathcal{E}_{n,n}}) - \left(2n^3 - 5n^2 + 5n - 2\right) \ .$$

(ii) *The affine hull* $\text{aff}(\mathcal{SQAP}_{n,n})$ *is the set of solutions of the equations* (6.7), (6.9), *and*

$$x(\text{col}_j) = 1 \qquad (j \in \mathcal{N}) \tag{6.14}$$

$$-x_{(i,j)} - x_{(i,l)} + z\left(\Delta_{(i,l)}^{(i,j)}\right) = 0 \qquad (i,j,l \in \mathcal{N}, j < l) \ . \tag{6.15}$$

We have

$$\dim(\mathcal{SQAP}_{n,n}) = \dim(\mathcal{R}^{\mathcal{V}_{n,n}} \times \mathcal{R}^{\hat{\mathcal{E}}_{n,n}}) - \left(n^3 - 2n^2 + 2n - 1\right) \ .$$

The result of Part (i) was also proved by [Rijal, 1995] and [Padberg and Rijal, 1996]. The dimension of $\mathcal{QAP}_{n,n}$ (without an explicit system of equations) was already computed by [Barvinok, 1992] in his investigations of the connections between polytopes coming from combinatorial optimization problems and the representation theory of the symmetric group. For a proof in our notational setting see [Jünger and Kaibel, 1997b], where one may also find information which equations have to be removed in order to obtain an irredundant system. The proof of the symmetric version (Part (ii)) is in [Jünger and Kaibel, 1996]. Unlike with the case $m \leq n - 2$, the system given in the theorem is redundant also in the symmetric case. How to obtain an irredundant system can be found in [Kaibel, 1997]. Notice that [Padberg and Rijal, 1996] already conjectured Part (ii).

Now that we know everything about the equations that are valid for $\mathcal{QAP}_{m,n}$ and $\mathcal{SQAP}_{m,n}$, let us turn to inequalities. We start with the "trivial inequalities" (i.e., the bounds on the variables). If we state that some inequality is "implied" by some others then this will always mean that *all* solutions (not only the integral ones) to the other inequalities satisfy also the inequality under inspection.

Theorem 6 *Let* $3 \leq m \leq n$.

(i) *The inequalities*

$$y_e \geq 0 \qquad (e \in \mathcal{E}_{m,n})$$

define facets of $QAP_{m,n}$.

(ii) The inequalities

$$x_v \geq 0 \qquad\qquad (v \in \mathcal{V}_{m,n})$$
$$x_v \leq 1 \qquad\qquad (v \in \mathcal{V}_{m,n})$$
$$y_e \leq 1 \qquad\qquad (e \in \mathcal{E}_{m,n})$$

are implied by the equations (6.2), (6.4), and the nonnegativity constraints $y \geq 0$ on the edge variables.

(iii) The inequalities

$$x_v \geq 0 \qquad\qquad (v \in \mathcal{V}_{m,n})$$
$$z_h \geq 0 \qquad\qquad (h \in \hat{\mathcal{E}}_{m,n})$$

define facets of $SQAP_{m,n}$.

(iv) The inequalities

$$x_v \leq 1 \qquad\qquad (v \in \mathcal{V}_{m,n})$$
$$z_h \leq 1 \qquad\qquad (h \in \hat{\mathcal{E}}_{m,n})$$

are implied by the equations (6.7), (6.9), and the nonnegativity constraints $x \geq 0$ and $z \geq 0$ on the node and hyperedge variables.

The proof of the theorem can be found in [Kaibel, 1997]. The case $m = n$ is also in [Jünger and Kaibel, 1997b] and [Jünger and Kaibel, 1996]. Parts (i) and (ii) for the case $m = n$ have also been proved by [Rijal, 1995] and [Padberg and Rijal, 1996].

Let us now consider other inequalities for $m \leq n - 2$. Although in this case the "column versions" (6.13) and (6.15) do not hold for $QAP_{m,n}$ and $SQAP_{m,n}$, respectively, the corresponding inequalities

$$-x_{(i,j)} + y\left((i,j) : \text{col}_l\right) \leq 0 \qquad (i \in \mathcal{M}, j, l \in \mathcal{N}, j \neq l) \qquad (6.16)$$

and

$$-x_{(i,j)} - x_{(i,l)} + z\left(\Delta_{(i,l)}^{(i,j)}\right) \leq 0 \qquad (i \in \mathcal{M}, j, l \in \mathcal{N}, j < l) \qquad (6.17)$$

are valid.

There is one more interesting class of inequalities that in case of $m = n$ are valid as equations (and thus implied by the equations we have described above). For $j, l \in \mathcal{N}$ ($j \neq l$) let us denote by $(\text{col}_j : \text{col}_l)$ the set of all edges connecting column j with column l, and let $\langle \mathrm{col}_j : \mathrm{col}_l \rangle = \{\mathrm{hyp}(e) : e \in (\mathrm{col}_j : \mathrm{col}_l)\}$ be the set of hyperedges connecting column j with column l. Then the inequalities

$$x\left(\mathrm{col}_j \cup \mathrm{col}_l\right) - y\left(\mathrm{col}_j : \mathrm{col}_l\right) \leq 1 \qquad (j,l \in \mathcal{N}, j < l) \qquad (6.18)$$

and

$$x\left(\mathrm{col}_j \cup \mathrm{col}_l\right) - z\left(\langle \mathrm{col}_j : \mathrm{col}_l \rangle\right) \leq 1 \qquad (j,l \in \mathcal{N}, j < l) \qquad (6.19)$$

are valid for $\mathcal{QAP}_{m,n}$ and $\mathcal{SQAP}_{m,n}$, respectively.

Theorem 7 *Let* $4 \leq m \leq n - 2$.

   (i) *The inequalities (6.16) and (6.18) define facets of* $\mathcal{QAP}_{m,n}$.

   (ii) *The inequalities (6.3) are implied by the inequalities (6.18) and the e-quations (6.2) and (6.4).*

   (iii) *The inequalities (6.17) and (6.19) define facets of* $\mathcal{SQAP}_{m,n}$.

   (iv) *The inequalities (6.8) are implied by the inequalities (6.19) and the e-quations (6.7) and (6.9).*

The proofs of all parts of the theorem are in [Kaibel, 1997].

The final class of inequalities that we will consider is the class of *box-inequalities*. This was the first large (i.e., exponentially large) class of facet-defining inequalities discovered for the QAP-polytopes (and it is still the only one that is known). A large part of its importance, however, is not due to this theoretical property, but due to the fact, that using some of these inequalities as cutting planes one can indeed significantly improve the lower bounds obtained by classical LP-based bounding procedures — in several cases even up to the possibility to compute optimal solutions by a pure cutting plane algorithm (see Section 6.).

The starting point is the following trivial observation: if $\gamma \in \mathcal{Z}$ is an integer number then $\gamma(\gamma - 1) \geq 0$ must hold. Now suppose that $S, V \subseteq \mathcal{V}_{m,n}$ are disjoint subsets of nodes and that $\beta \in \mathcal{Z}$ is an arbitrary integer. Let $(x,y) \in \mathcal{QAP}_{m,n}$ be a vertex of $\mathcal{QAP}_{m,n}$. By the above observation the quadratic inequality

$$\bigl(x(T) - x(S) - \beta\bigr)\bigl(x(T) - x(S) - (\beta - 1)\bigr) \geq 0 \qquad (6.20)$$

holds. But since (x,y) is the characteristic vector of the nodes and edges in some m-clique $C \subseteq \mathcal{V}_{m,n}$ of $\mathcal{G}_{m,n}$ we have $x(S)x(T) = y(S : T)$ and $x(R)x(R) = x(R) + 2y(R)$ for every $R \subseteq \mathcal{V}_{m,n}$ (with $y(R) = y(\mathcal{E}_{m,n}(R))$), which allows us to rewrite the quadratic inequality into the following linear *ST-inequality*:

$$-\beta x(S) + (\beta - 1)x(T) - y(S) - y(T) + y(S:T) \leq \frac{\beta(\beta - 1)}{2} \quad (6.21)$$

It is obvious from (6.20) that the vertices of the face of $\mathcal{QAP}_{m,n}$ defined by this ST-inequality are precisely the characteristic vectors of those m-cliques $C \subseteq \mathcal{V}_{m,n}$ which satisfy

$$|C \cap T| - |C \cap S| \in \{\beta, \beta - 1\} \ .$$

As mentioned in Section 3., $\mathcal{QAP}_{m,n}$ can be viewed as a certain face of the cut-polytope (associated with the complete graph on $mn + 1$ nodes). We just mention that the ST-inequalities correspond to certain *hypermetric inequalities* for the cut-polytope. For details on this connection we refer to [Jünger and Kaibel, 1997a] and [Kaibel, 1997], and for answers to nearly every question on the cut-polytope (like "what are hypermetric inequalities?") and related topics to [Deza and Laurent, 1997].

In the light of the remarks at the end of Section 3. the symmetric ones among the ST-inequalities are of special interest. Here is a possibility to choose S and T such that the resulting inequality is symmetric. Let $P_1, P_2 \subseteq \mathcal{M}$ and $Q_1, Q_2 \subseteq \mathcal{N}$ with $P_1 \cap P_2 = \emptyset$ and $Q_1 \cap Q_2 = \emptyset$ and take $S = (P_1 \times Q_1) \cup (P_2 \times Q_2)$ as well as $T = (P_1 \times Q_2) \cup (P_2 \times Q_1)$. An ST-inequality arising from sets S and T of this type is called a *4-box-inequality* (see Figure 6.4). In [Jünger and Kaibel, 1997a] the following result is proved.

Theorem 8 *An ST-inequality is symmetric if and only if it is a 4-box-inequality.*

 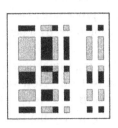 

Figure 6.4   The node sets of ST-inequalities in general, of 4-box inequalities, and of 4-box inequalities after suitable permutations of rows and columns. The set S is always indicated by the gray parts, the set T by the black ones

If one of the sets P_1, P_2, Q_1, or Q_2 is empty, then we call the corresponding 4-box-inequality a *2-box-inequality*. If P_1 or P_2 is empty and Q_1 or Q_2 is empty, the 4-box-inequality is a *1-box-inequality*. While theoretical investigations of the whole class of 4-box-inequalities seem to be too difficult, the 2-box-inequalities are studied extensively in [Kaibel, 1997]. In particular, the facet-defining ones among them are identified; it turns out that most of them define facets of $\mathcal{QAP}_{m,n}$. Rather than stating the result in its whole generality, we prefer to give a theorem that shows a large class of 1-box-inequalities that are facet-defining for $\mathcal{QAP}_{m,n}$ (and thus, for $\mathcal{SQAP}_{m,n}$). In particular, this class contains the inequalities that are used in Section 6. for computing lower bounds. The proof of the theorem is contained in [Jünger and Kaibel, 1997a].

Theorem 9 *Let* $7 \leq m \leq n$. *Let* $P \subseteq \mathcal{M}$ *and* $Q \subseteq \mathcal{N}$ *generate* $T = P \times Q \subseteq \mathcal{V}_{m,n}$, *and let* $\beta \in \mathcal{Z}$ *be an integer number such that*

- $\beta \geq 2$,

- $|P|, |Q| \geq \beta + 2$,

- $|P|, |Q| \leq n - 3$, *and*

- $|P| + |Q| \leq n + \beta - 5$

hold. Then the 1-box inequality

$$(\beta - 1)x(\mathcal{T}) - y(T) \leq \frac{\beta(\beta - 1)}{2}$$

defines a facet of $\mathcal{QAP}_{m,n}$ *and the corresponding 1-box-inequality*

$$(\beta - 1)x(T) - z(T) \leq \frac{\beta(\beta - 1)}{2}$$

defines a facet of $\mathcal{SQAP}_{m,n}$.

Before closing this section, we want to point out that the proofs of the results presented here heavily rely on the techniques described in Section 4. and on the various connections between the different polytopes arising from projections mentioned in Section 3.. While these projections are very helpful for avoiding to do work twice, the proofs that have to be done from the scratch are extremely simplified by the star-transformations. Even to prove the statements on the dimensions and the trivial facets of the polytopes (at least in case of $m = n$) are very tedious without using the star-transformation (see [Rijal, 1995]). For the more complicated situations, in particular with the box-inequalities, the proofs seem to be impossible without exploiting the star-transformation — at least for the author.

6.   CUTTING PLANE ALGORITHMS FOR QAPS

The insight into the geometry of the different QAP-polytopes described in the previous section can be used to compute lower bounds or even optimal solutions in the way explained in Section 2.. In [Jünger and Kaibel, 1997a], [Kaibel, 1998], and [Kaibel, 1997] experiments with cutting plane algorithms exploiting the results on the facial structures of the QAP-polytopes are described. Here, we just report on some of the experimental results in order to show that the polyhedral investigations of the QAP indeed help to improve the algorithmic solvability of the problem.

The cutting plane code that we have implemented is suited for symmetric instances and can handle both, the $m = n$ as well as the $m \leq n - 2$ case. In the first case, the initial LP consists of a complete equation system plus the nonnegativity constraints on the variables. In the second case, the initial LP contains, again, a complete system of equations, the nonnegativity constraints on the variables, and, additionally, the inequalities (6.17) and (6.19). Of course, one can transform every instance with $m \leq n - 2$ into an instance with $m = n$ by adding "dummy objects". In [Kaibel, 1998] it is shown that for such instances the bounds obtained from the initial LPs with and without dummy objects coincide. The bound from the initial LP in case of $m = n$ is (empirically) slightly weaker than the corresponding bound in the non-symmetric model (see [Jünger and Kaibel, 1996]), where the latter LP is equivalent to the linear programming relaxation of (AJ) (see Section 1.).

The algorithm first sets up the initial LP, and then solves it to obtain a lower bound on the optimal value of the QAP. If the LP-solution happens to be an integer vector (i.e., a 0/1-vector in this case), then, by Theorem 1, it is the characteristic vector of an optimal solution to the problem. Otherwise, the algorithm tries to separate the (fractional) LP-solution from the feasible solutions by searching for 4-box inequalities that are violated by it. If such inequalities are found (which is not guaranteed), then they are added to the LP, a new (hopefully better) lower bound is computed, and the process is iterated for a specified number of rounds.

Actually, our *separation algorithm* is quite simple. Since some initial experiments showed that 1-box inequalities with $\beta \in \{2, 3\}$ seem to be most valuable within the cutting plane algorithm, we have restricted to this type of inequalities in our experiments. Because there seems to be no obvious way to solve the separation problem (for this class of inequalities) fast and exactly (i.e., either to find violated inequalities or to affirm that no violated inequality exists within the class), we have implemented a rather simple heuristic for it. We just guess 1-box inequalities randomly and then try to increase their left-hand-sides (with respect to the current fractional solution) by changing the box in a 2-opt way. Actually, the experiments show that usually this primitive procedure detects quite a lot of violated inequalities. We then choose the most violated ones among them and add them to the LP. This way, about 0.2 to 0.4 times the number of initial constraints are added to the LP in every iteration of the cutting plane procedure. In order to control the size of the LP we also remove inequalities if they have been redundant (i.e., non-binding for the optimal LP-solution) for several cutting plane iterations in a row. For details on the algorithm we refer to [Kaibel, 1997], [Jünger and Kaibel, 1997a], and [Kaibel, 1998].

The LPs have been solved by the barrier method of the CPLEX 4.0 package. Using the (primal or dual) simplex algorithm did not pay off at all, which is due to the very high primal and dual degeneracy of the LPs. This is very much in accordance with the computational experiments performed by [Resende et al., 1995] with LPs that are equivalent to our initial LP in the non-symmetric $m = n$ case. All our experiments were carried out on a Silicon Graphics Power Challenge machine using the parallel version of the CPLEX barrier code on four processors.

Table 6.1 shows the results for all instances from the QAPLIB with $m = n \leq 20$ (they are all symmetric). For all these instances, optimal solutions are known (and published in the QAPLIB). The table gives the bounds obtained from the initial LP as well as the ones obtained by the cutting plane procedure. Notice that, since all objective functions are integral, we can round up every bound to the next integer number. The columns titled *qual* give the ratios of the respective bound and the optimal solution value. The running times for the cutting plane algorithm are specified in seconds. The column titled *iter* shows the number of cutting plane iterations, i.e., the number of LPs solved to obtain the bound.

The results show that the box-inequalities indeed are quite valuable for improving LP based lower bounds. For several instances the bounds even match the optimal solution values, and for most of the other instances, a large part of the gaps between the bounds obtained from the initial LPs and the optimal solution values is closed by adding box-inequalities.

While the quality of the bounds obtained from the cutting plane algorithm is quite good, for many instances the running times are rather large. We will address this point at the end of this section. But let us first turn to the experiments with instances where $m \leq n - 2$ holds. Table 6.2 and 6.3 report on the results obtained by the cutting plane code on the esc16- and esc32-instances (with $n = 16$ and $n = 32$, respectively) from the QAPLIB, where here the columns titled *box* contain the bounds computed by the cutting plane code. Since for these instances both the flow- as well as the distance-matrix are symmetric and integral, the optimal solution value must be an even integer number. Thus we can round up every bound to the next even integer.

For the esc16-instances, the column titled *opt* contains the optimal solution values and the column titled *speed up* contains the quotients of the running times with and without dummy objects. All these instances were solved to optimality for the first time by [Clausen and Perregaard, 1997], who used a Branch & Bound code running on a parallel machine with 16 i860 processors. The column titled *ClPer* shows the running times of their algorithm. The cutting plane code finds for all these instances (except for `esc16a`) the optimal solution value within (more or less) comparable running times.

For the `esc32`-instances the cutting plane code always produces the best known lower bounds. The column titled *upper* contains the values of the currently best known feasible solutions and the column titled *prev lb* contains the previously best known lower bounds. The instances `esc32e`, `esc32f`, and `esc32g` have been solved to optimality for the first time by a parallel Branch & Bound code of [Brüngger et al., 1996]. For the other instances (except for `esc32c`) the cutting plane algorithm improves the previously best known lower bounds, where the improvement for `esc32a` is the most significant one. The running times that have a "$\star$" in front are not measured exactly due to some problems with the queuing system of our machine.

While all these experiments show that the polyhedral investigations indeed pay off with respect to the goal of the computation of tight lower bounds, the running times of the cutting plane algorithm are (for most of the instances) quite large. In order to obtain a "practical" bounding procedure (that, in particular, might be incorporated into Branch & Bound frameworks) the algorithm has to be speeded up significantly. One approach into this direction is to implement more elaborate separation strategies. But the potential of this kind of improvements is limited, since already the initial LPs become really large for larger values of m and n. For instance, for `esc32a` the initial LP has 149600 variables and 22553 equations.

One way to reduce the sizes of the LPs is to exploit the fact that quite often the objective functions are rather sparse. For example, in case of a Koopmans & Beckmann instance the flow matrix might be sparse because there are lots of pairs of objects which do not have any flows between them. Actually, this is true for many instances in the QAPLIB. For example, `els19` has a (symmetric) flow matrix, where out of the 171 pairs of objects only 56 have a nonzero flow.

In our graph model, a pair $i, k \in \mathcal{M}$ of objects without any flow between i and k has the effect that all (hyper-)edges connecting row i and row k have objective function coefficient zero. This means that one might "project out" all variables corresponding to these (hyper-)edges and solve the problem over the corresponding projected polytope. For `els19` this reduces the number of variables from 29,602 down to 9,937.

Of course, (unlike with the projections used for the star-transformation) in general the projection will change the geometric properties of the polytope. Thus, one has to do theoretical investigations of the projected polytopes depending on the *flow graph*, i.e., the graph defined on the objects and having an edge for every pair of objects which does have some flow. This was suggested already by [Padberg and Rijal, 1996]. First results can be found in [Kaibel, 1997]. Again, lots of results presented in Section 5. can be carried over by the observation that an inequality that is valid (facet-defining) for the unprojected polytope immediately yields a valid (facet-defining) inequality for the projected polytope as long as the inequality has no nonzero coefficient on a (hyper-)edge that belongs to a pair of objects which do not have any flow between them. In particular, for every clique in the flow graph there are a lot of box-inequalities which are also valid (facet-defining) for the projected polytopes.

In [Elf, 1999] some computational experiments with a cutting plane algorithm working with the "sparse models" are performed. Table 6.4 shows results for the esc32 instances.

Comparing these results with the ones in Table 6.3 one finds that (at least for the esc32 instances) the running times of the cutting plane algorithm are reduced substantially by exploiting sparsity of the objective functions. While this might be paid by a weaker bound (esc32b, esc32c, esc32d, esc32h), it is also possible that the bound becomes better (esc32a). Notice that in the sparse model, it takes only about one minute to compute the optimal solution values of esc32e and esc32f.

7.   CONCLUSION

We close with three important aspects of the polyhedral work on the QAP that we have surveyed in this chapter.

- The techniques that have been developed for theoretical investigations of QAP-polytopes, like the projections between the different types of poly-topes and, especially important, the star-transformation, provide tools which make polyhedral studies on the QAP possible. In fact, they have led to first considerable insights into the polyhedral structure of the QAP, which before was one of the few problems among the classical combinatorial optimization problems about which we lacked any deeper polyhedral knowledge.

- The practical experiments with a cutting plane algorithm that exploits the polyhedral results show that this type of approach has a great potential for computing tight lower bounds and even optimal solutions. However, the tightness of the bounds is payed by considerable running times for larger instances.

- In order to overcome the relatively large running times one might exploit sparsity in the objective function. First theoretical and experimental studies have shown that at least special kinds of sparsity (coming from sparse flow structures on the objects) can be handled theoretically and can be used to improve the running times of cutting plane algorithms substantially.

These points suggest, in our opinion, the further lines of research in the area of polyhedral combinatorics of the QAP. The most promising possibility to really push the (exact) solvability of QAPs beyond the current limits by polyhedral methods is to extend the work on the sparse model. One direction here, of course, is the further investigation of the structures of the QAP-polytopes in the sparse model. In particular, one could search for facets of the polytopes that are not projections of facets of polytopes in the dense model. Another direction is to study models that do not only exploit sparsity of the flow structure of the objects, but also, simultaneously, sparsity of the distance structure of the locations. In several cases this would reduce the number of variables quite further.

One more aspect may make work on the polyhedral combinatorics of the QAP attractive: while investigating properties of the associated polytopes and developing better cutting plane algorithms for the QAP the LP-technology most probably will develop further in parallel. Thus, one might hope that the progress one achieves algorithmically is multiplied by a certain factor that (from the enormous improvements of LP-solvers in the recent years) can be estimated to be not too small.

References

[Adams and Johnson, 1994] Adams, W. P. and Johnson, T. A. (1994). Improved linear programming-based lower bounds for the quadratic assignment problem. In Pardalos, P. M. and Wolkowicz, H., editors, *Quadratic Assignment and Related Problems*, DIMACS Series in Discrete Mathematics and Theoretical Computer Science, pages 43–75.

[Barvinok, 1992] Barvinok, A. I. (1992). Combinatorial complexity of orbits in representations of the symmetric group. *Advances in Soviet Mathematics*, 9:161–182.

[Brüngger et al., 1996] Brüngger, A., Clausen, J., Marzetta, A., and Perregaard, M. (1996). Joining forces in solving large-scale quadratic assignment problems in parallel. Technical Report DIKU TR-23/96, University of Copenhagen, Copenhagen.

[Burkard et al., 1997] Burkard, R. E., Karisch, S. E., and Rendl, F. (1997). QAPLIB — A quadratic assignment problem library. *Journal of Global Optimization*, 10:391–403. http://serv1.imm.dtu.dk/~sk/qaplib/.

[Chvátal, 1983] Chvátal, V. (1983). *Linear Programming*. Freeman.

[Clausen and Perregaard, 1997] Clausen, J. and Perregaard, M. (1997). Solving large scale quadratic assignment problems in parallel. *Comput. Optim. Appl.*, 8(2):111–127.

[Deza and Laurent, 1997] Deza, M. M. and Laurent, M. (1997). *Geometry of Cuts and Metrics.* Springer Verlag.

[Edmonds, 1965a] Edmonds, J. (1965a). Maximum matching and a polyhedron with 0,1-vertices. *Journal of Research of the National Bureau of Standards—B, Mathematics and Mathematical Physics*, 69B:125–130.

[Edmonds, 1965b] Edmonds, J. (1965b). Paths, trees, and flowers. *Canadian Journal of Mathematics*, 17:449–467.

[Elf, 1999] Elf, M. (1999). LP-basierte Schranken für quadratische Zuordnungsprobleme mit dünner Zielfunktion. Master's thesis, Universität zu Köln.

[Gilmore, 1962] Gilmore, P. C. (1962). Optimal and suboptimal algorithms for the quadratic assignment problem. *SIAM Journal on Applied Mathematics*, 10:305–313.

[Grötschel et al., 1981] Grötschel, M., Lovász, L., and Schrijver, A. (1981). The ellipsoid method and its consequences in combinatorial optimization. *Combinatorica*, 1:169–197.

[Grötschel et al., 1988] Grötschel, M., Lovász, L., and Schrijver, A. (1988). *Geometric Algorithms and Combinatorial Optimization.* Springer-Verlag, Heidelberg.

[Grünbaum, 1967] Grünbaum, B. (1967). *Convex Polytopes.* Interscience, London. Revised edition: V. Klee and P. Kleinschmidt, Graduate Texts in Mathematics, Springer Verlag.

[Johnson, 1992] Johnson, T. A. (1992). *New Linear-Programming Based Solution Procedures for the Quadratic Assignment Problem.* PhD thesis, Graduate School of Clemson University.

[Jünger and Kaibel, 1996] Jünger, M. and Kaibel, V. (1996). On the SQAP-polytope. Technical Report No. 96.241, Institut für Informatik, Universität zu Köln. Submitted to: SIAM Journal on Optimization.

[Jünger and Kaibel, 1997a] Jünger, M. and Kaibel, V. (1997a). Box-inequalities for quadratic assignment polytopes. Technical Report 97.285, Angewandte Mathematik und Informatik, Universität zu Köln. Submitted to: Mathematical Programming.

[Jünger and Kaibel, 1997b] Jünger, M. and Kaibel, V. (1997b). The QAP-polytope and the star-transformation. Technical Report 97.284, Angewandte Mathematik und Informatik, Universität zu Köln. Submitted to: *Discrete Applied Mathematics*.

[Jünger et al., 1994] Jünger, M., Reinelt, G., and Thienel, S. (1994). Provably good solutions for the traveling salesman problem. *ZOR – Mathematical Methods of Operations Research*, 40:183—217.

[Kaibel, 1997] Kaibel, V. (1997). *Polyhedral Combinatorics of the Quadratic Assignment Problem*. PhD thesis, Universität zu Köln. www.math.TU-Berlin.de/~kaibel/diss.html.

[Kaibel, 1998] Kaibel, V. (1998). Polyhedral combinatorics of QAPs with less objects than locations. In Bixby, R. E., Boyd, E. A., and Rios-Mercado, R. Z., editors, *Proceedings of the 6th International IPCO Conference, Houston, Texas.*, volume 1412 of *Lecture Notes in Computer Science*, pages 409–422. Springer-Verlag.

[Karp and Papadimitriou, 1982] Karp, R. M. and Papadimitriou, C. H. (1982). On linear characterizations of combinatorial optimization problems. *SIAM Journal on Computing*, 11:620–632.

[Khachiyan, 1979] Khachiyan, L. G. (1979). A polynomial algorithm in linear programming. *Soviet. Math. Dokl.*, 20:191–194.

[Klee and Kleinschmidt, 1995] Klee, V. and Kleinschmidt, P. (1995). Convex polytopes and related complexes. In Graham, R. L., Grötschel, M., and Lovász, L., editors, *Handbook of Combinatorics*, volume 2, chapter 18, pages 875–918. Elsevier Science.

[Koopmans and Beckmann, 1957] Koopmans, T. C. and Beckmann, M. J. (1957). Assignment problems and the location of economic activities. *Econometrica*, 25:53–76.

[Lawler, 1963] Lawler, E. L. (1963). The quadratic assignment problem. *Management Science*, 9:586–599.

[Minkowski, 1896] Minkowski, H. (1896). *Geometrie der Zahlen*. Teubner Verlag, Leipzig. Reprinted by Chelsea, New York 1953 and Johnson, New York 1963.

[Nemhauser and Wolsey, 1988] Nemhauser, G. L. and Wolsey, L. A. (1988). *Integer and Combinatorial Optimization*. Wiley-Interscience Series in Discrete Mathematics and Optimization. John Wiley & Sons, Chichester New York.

[Padberg and Rijal, 1996] Padberg, M. and Rijal, M. P. (1996). *Location, Scheduling, Design and Integer Programming*. Kluwer Academic Publishers.

[Padberg, 1995] Padberg, M. W. (1995). *Linear Optimization and Extensions*. Springer-Verlag, Berlin Heidelberg.

[Padberg and Rao, 1980] Padberg, M. W. and Rao, M. R. (1980). The russian method and integer programming. Technical report, CBA Working Paper, New York University, New York.

[Resende et al., 1995] Resende, M. G. C., Ramakrishnan, K. G., and Drezner, Z. (1995). Computing lower bounds for the quadratic assignment problem with an interior point solver for linear programming. *Operations Research*, 43:781–791.

[Rijal, 1995] Rijal, M. P. (1995). *Scheduling, Design and Assignment Problems with Quadratic Costs*. PhD thesis, New York University.

[Schrijver, 1986] Schrijver, A. (1986). *Theory of Linear and Integer Programming*. Wiley-Interscience Series in Discrete Mathematics. John Wiley & Sons, Chichester New York.

[Schrijver, 1995] Schrijver, A. (1995). Polyhedral combinatorics. In Graham, R. L., Grötschel, M., and Lovász, L., editors, *Handbook of Combinatorics*, volume 2, chapter 30, pages 1649–1704. Elsevier Science.

[Weyl, 1935] Weyl, H. (1935). Elementare Theorie der konvexen Polyeder. *Comm. Math. Helv.*, 7:290–306.

[Wolsey, 1998] Wolsey, L. A. (1998). *Integer Programming*. John Wiley & Sons, Inc.

[Ziegler, 1995] Ziegler, G. M. (1995). *Lectures on Polytopes*, volume 152 of *Graduate Texts in Mathematics*. Springer-Verlag, New York. (Revised edition: 1998).

| name | initial LP | | cutting planes | | | | gap |
	bound	qual	bound	qual	iter	time	reduced
chr12a	9552	1.000	9552	1.000	1	16	1.000
chr12b	9742	1.000	9742	1.000	1	16	1.000
chr12c	11156	1.000	11156	1.000	1	21	1.000
had12	1619	0.980	1652	1.000	3	435	1.000
nug12	521	0.901	577	0.997	13	23981	0.971
rou12	222212	0.943	235278	0.999	18	26541	0.981
scr12	29558	0.941	31410	1.000	5	1326	1.000
tai12a	220019	0.980	224416	1.000	3	371	1.000
tai12b	30581825	0.775	39464925	1.000	4	761	1.000
had14	2660	0.976	2724	1.000	4	2781	1.000
chr15a	9371	0.947	9896	1.000	7	25036	1.000
chr15b	7895	0.988	7990	1.000	3	2838	1.000
chr15c	9504	1.000	9504	1.000	1	105	1.000
nug15	1031	0.896	1130	0.982	6	19906	0.827
rou15	322945	0.912	340470	0.961	7	25315	0.561
scr15	48817	0.955	51140	1.000	4	5083	1.000
tai15a	351290	0.905	366466	0.944	7	25449	0.411
tai15b	51528935	0.995	51765268	1.000	7	17909	1.000
esc16b	278	0.952	292	1.000	2	762	1.000
esc16c	118	0.738	160	1.000	4	4929	1.000
esc16h	704	0.707	996	1.000	4	4886	1.000
had16	3549	0.954	3717	0.999	8	23381	0.982
nug16a	1414	0.878	1567	0.973	8	19296	0.781
nug16b	1080	0.871	1209	0.974	5	16512	0.801
nug17	1491	0.861	1644	0.949	4	16007	0.633
tai17a	440095	0.895	454626	0.924	5	25606	0.281
chr18a	10739	0.968	10948	0.986	5	22335	0.580
chr18b	1534	1.000	1534	1.000	1	507	1.000
had18	5072	0.946	5300	0.989	5	23367	0.795
nug18	1650	0.855	1810	0.937	5	19390	0.569
els19	16502857	0.959	17074681	0.992	3	17440	0.806
chr20a	2170	0.990	2173	0.991	2	22488	0.121
chr20b	2287	0.995	2295	0.999	2	13645	0.710
chr20c	14007	0.990	14034	0.992	2	14794	0.196
had20	6559.4	0.948	6732	0.972	2	22783	0.475
lipa20a	3683	1.000	3683	1.000	1	1145	1.000
lipa20b	27076	1.000	27076	1.000	1	935	1.000
nug20	2165	0.842	2314	0.900	3	17845	0.367
rou20	639679	0.882	649748	0.896	3	13143	0.117
scr20	94558	0.859	96562	0.878	3	15122	0.130
tai20a	614850	0.874	625942	0.890	3	34135	0.125
tai20b	84501940	0.690	104534175	0.854	2	10143	0.528

Table 6.1   Results on instances with $m = n$ (dense model)

name	m	opt	init LP	box	iter	time	speed up	ClPer
esc16a	10	68	48	64	3	522	4.87	65
esc16d	14	16	4	16	2	269	2.74	492
esc16e	9	28	14	28	4	588	3.37	66
esc16g	8	26	14	26	3	58	14.62	7
esc16i	9	14	0	14	4	106	28.18	84
esc16j	7	8	2	8	2	25	32.96	14

Table 6.2   Results on the esc16 instances (dense model)

name	m	upper	prev lb	init LP	box	iter	time
esc32a	25	130	36	40	88	3	62988
esc32b	24	168	96	96	100	4	⋆60000
esc32c	19	642	506	382	506	8	⋆140000
esc32d	18	200	132	112	152	8	⋆80000
esc32e	9	2	2	0	2	2	576
esc32f	9	2	2	0	2	2	554
esc32g	7	6	6	0	6	2	277
esc32h	19	438	315	290	352	6	119974

Table 6.3   Results on the esc32 instances (dense model)

name	bound	iter	time
esc32a	92	3	8673
esc32b	96	4	13058
esc32c	394	15	18716
esc32d	120	12	7472
esc32e	2	2	74
esc32f	2	2	82
esc32g	6	4	228
esc32h	280	15	22716

Table 6.4   Results of cutting plane algorithm with "sparse model"

Chapter 7

SEMIDEFINITE PROGRAMMING APPROACHES TO THE QUADRATIC ASSIGNMENT PROBLEM

Henry Wolkowicz*

Department of Combinatorics and Optimization

Waterloo, Ontario N2L 3G1, Canada

hwolkowi@orion.math.uwaterloo.ca

Abstract   The Quadratic Assignment Problem, QAP, is arguably the hardest of the NP-hard problems. One of the main reasons is that it is very difficult to get good quality bounds for branch and bound algorithms. We show that many of the bounds that have appeared in the literature can be ranked and put into a unified Semidefinite Programming, SDP, framework. This is done using redundant quadratic constraints and Lagrangian relaxation. Thus, the final SDP relaxation ends up being the strongest.

1. INTRODUCTION

The Quadratic Assignment Problem, QAP, can be considered to be the hardest of the NP-hard problems. This is an area where dimension $n = 30$ is considered to be large scale, and more often than not, is too hard to solve to optimality. One of the main reasons is that it is very difficult to get good lower bounds for fathoming partial solutions in branch and bound algorithms. In this chapter we consider several different bounding strategies. We show that these can be put into a Semidefinite Programming, SDP, framework.

Bounds for QAP can be classified into four types: Gilmore-Lawler type; eigenvalue based; reformulation or linear programming type; and semidefinite programming based. A connection between the reformulation type bounds and the Gilmore-Lawler type bounds has been made using Lagrangian relaxation, see [Frieze and Yadegar, 1983, Adams and Johnson, 1994]. The connection

---

*Research partially supported by The Natural Sciences Engineering Research Council Canada.

P.M. Pardalos and L.S. Pitsoulis (eds.), Nonlinear Assignment Problems, 143–174.
© 2000 *Kluwer Academic Publishers.*

between Lagrangian and semidefinite relaxations is now well known, see e.g. [Shor, 1987, Poljak et al., 1995, Wolkowicz, 2000b]. In this chapter we unify many of the bounds in the literature using the Lagrangian relaxation approach. Our main theme is to show that with the correct choice of redundant constraints, we can illustrate the equivalence of many of the bounds with Lagrangian relaxations and therefore show in a transparent way how the bounds rank against each other.

The following (7.1) is the trace (Koopmans-Beckmann [Koopmans and Beckmann, 1957]) formulation of the QAP (see e.g. [Pardalos et al., 1994, Pardalos and Wolkowicz, 1994] for various formulations and many useful applications), where the variable X is a permutation matrix and e is the vector of ones. As a model for facility location problems, where there are n facilities (locations), the matrix B represents distances between locations, the matrix A represents flows between facilities, and the matrix C represents fixed costs. We use the fact that assignment problems can be modeled using permutation matrices, and permutation matrices are 0,1 matrices with row and column sums 1. This formulation illustrates the quadratic nature of the objective function.

$$
QAP \qquad
\begin{aligned}
\mu^* := \quad & \max && q(X) = && \left(\text{Trace}\,(AXB - 2C)X^T\right) \\
& \text{subject to} && Xe = e \\
& && X^Te = e \\
& && X_{ij} \in \{0,1\} \quad \forall i,j.
\end{aligned}
\tag{7.1}
$$

Rather than restricting the data to being the product of flows and distances, a more general formulation was given in [Lawler, 1963]; see also [Adams and Johnson, 1994].

1.1   PRELIMINARIES

1.1.1   Notation.

$\mathcal{M}_t$ the space of $t \times t$ *real matrices*

$\mathcal{S}_t$ the space of $t \times t$ *symmetric matrices*

$t(n)$ $\frac{n(n+1)}{2}$, the dimension of $\mathcal{S}_t$

$\langle A, B \rangle$ Trace $A^T B$, the trace inner product of two matrices, Section 1.1.2

$\mathcal{P}^t$ **or** $\mathcal{P}$ the cone of positive semidefinite matrices in $\mathcal{S}_t$

$M_1 \succeq M_2$ $M_1 - M_2$ is positive semidefinite

A^* the adjoint of the linear operator A, (7.2)

$A \circ B$ $(A_{ij}B_{ij})$, the Hadamard (elementwise) product of A and B

$A \otimes B$ the Kronecker product of A and B

$\text{vec}(X)$ the vector formed from the columns of the matrix X

$\text{Mat}(x)$ the matrix formed, columnwise, from the vector X

$\text{Diag}(v)$ the diagonal matrix formed from the vector v

$\text{diag}(M)$ the vector of the diagonal elements of the matrix M

E the matrix of ones

e the vector of ones

u the normalized vector of ones, $u = e/\|e\|$

V the orthogonal matrix to u, so $[u \mid V]$ is orthogonal

e_i the i-th unit vector

E_{ij} the matrix $E_{ij} := e_i e_j^T$

$\mathcal{R}(M)$ the range space of the matrix M

$\mathcal{N}(M)$ the null space of the matrix M

$\mathcal{E}$ $\{X : Xe = X^T e = e\}$, the set of matrices with row and column sums one

$\mathcal{Z}$ $\{X : X_{ij} \in \{0,1\}\}$, the set of (0,1)-matrices

$\mathcal{N}$ $\{X : X_{ij} \geq 0\}$, the set of nonnegative matrices

$\mathcal{O}$ $\mathcal{O} := \{X : XX^T = X^T X = I\}$, the set of orthogonal matrices

Π the set of permutation matrices, (7.3)

$\langle x, y \rangle_-$ $\min_{P \in \Pi} \langle x, Py \rangle$, the minimal scalar product of two vectors

$r(A)$ Ae, the vector of row sums of A

$s(A)$ $e^T Ae$, the sum of elements of A

Y_X the lifting of the matrix X, with $x = \text{vec}(X)$,

$$Y_X := \begin{bmatrix} x_0 & x^T \\ x & xx^T \end{bmatrix}, \quad x_0^2 = 1$$

$\mathcal{G}_J(Y)$ Gangster operator, an operator that "shoots" holes or zeros in the matrix Y, (7.44)

$\mathcal{P}G(Y)$ Gangster operator projected onto its range space, (7.47)

Arrow $(\cdot)$ the Arrow operator, (7.34)

$B^0\text{Diag}\,(\cdot)$ the Block Diag operator, (7.35)

$O^0\text{Diag}\,(\cdot)$ the Off Diag operator, (7.36)

arrow $(\cdot)$ the arrow operator, (7.38)

$b^0\text{diag}\,(\cdot)$ the block diag operator, (7.39)

$o^0\text{diag}\,(\cdot)$ the off diag operator, (7.40)

QAP the trace formulation of QAP, (7.1)

LAP the linear assignment problem (QAP with no quadratic term)

$QAP_{\mathcal{E}}$ an equivalent formulation of QAP, (7.27)

$QAP_{\mathcal{O}}$ an equivalent formulation of QAP, (7.28)

1.1.2   Background. We work with $n \times n$ real matrices and use the trace inner product $\langle A, B \rangle = \text{trace}\,A^T B$. We also work with several linear operators and their adjoints. Though linear operators in finite dimensions are equivalent to matrices, we find that using adjoints, rather than the transposes of the equivalent matrix representations, simplifies things in the long run. Note that for a linear operator $\mathcal{A}$, the adjoint operator, $\mathcal{A}^*$, satisfies

$$\langle \mathcal{A}x, y \rangle = \langle x, \mathcal{A}^*y \rangle, \quad \forall x, y. \tag{7.2}$$

The bounds we discuss involve relaxations of QAP. The constraints in QAP can be expressed in several ways, e.g. the permutation matrices satisfy

$$\Pi = \mathcal{O} \cap \mathcal{E} \cap \mathcal{N} = \mathcal{O} \cap \mathcal{N} = \mathcal{O} \cap \mathcal{Z} = \mathcal{E} \cap \mathcal{Z}. \tag{7.3}$$

Relaxations can be interpreted to mean that we ignore part of the definitions of permutation matrices. Usually we ignore the hard (or combinatorial) parts, e.g. $\mathcal{N}$ and/or $\mathcal{Z}$.

In [Li et al., 1994], the authors present a review of three categories of existing bounds: first is Gilmore-Lawler (GLB) related bounds [Gilmore, 1962, Lawler, 1963] and the authors' new lower bounds (see also [Resende et al., 1995]); second is eigenvalue related bounds [Finke et al., 1987, Rendl and Wolkowicz, 1992, Hadley et al., 1992, Hadley et al., 1990]; third is reformulation type bounds e.g. [Assad and Xu, 1985, Carraresi and Malucelli, 1992, Carraresi and Malucelli, 1994, Adams and Johnson, 1994]. In [Adams and Johnson, 1994], the authors present a new lower bound based on a mixed 0-1 linear formulation which is derived by constructing redundant quadratic inequalities and then defining additional continuous variables to replace all product terms. They show that this technique provides a strengthened version of the majority of lower bounding techniques. The major tool that they use is Lagrangian relaxation.

A seemingly independent category appears to be bounds based on semidefinite programming relaxations, e.g. [Zhao et al., 1998]. In this chapter we show how these SDP type bounds can fit into and unite the eigenvalue type bounds and, in fact, the other bounds as well. The main theme of this chapter is to show how the SDP bounds arise using Lagrangian relaxation and thus provide strengthened versions of the other bounds. Indeed, we follow a similar approach to [Adams and Johnson, 1994] in that we use many redundant quadratic constraints. However, the linearization that we do is different. Rather than defining additional continuous variables to replace product terms, we use the hidden (semidefinite) constraint that: *a quadratic function bounded below must be convex (positive semidefinite Hessian).* We see that the addition of the correct redundant constraints can be a very powerful tool in strengthening relaxations. In fact, in Section 3.4.1, we show that our SDP bound is always stronger than the one in [Adams and Johnson, 1994].

Due to the equivalence between Lagrangian and SDP relaxations, we often do not differentiate between the two in this chapter.

1.1.3   Outline.   This chapter is organized as follows. In Section 2. we present various eigenvalue bounds and their duality properties. This includes the hierarchical structure: the basic eigenvalue bound Section 2.1; the eigenvalue bounds using transformations (perturbations) Section 2.2; and the projected eigenvalue bound 2.3.

We then study the SDP relaxation for QAP in Section 3.. We derive the relaxation studied in [Zhao et al., 1998] using Lagrangian relaxation, i.e. the relaxation is the Lagrangian dual of the Lagrangian dual of the quadratic model of the QAP obtained after adding redundant quadratic constraints. We discuss the geometry of the relaxation in Section 3.2 including the so-called *gangster operator* that results in a simplified relaxation at the end. Concluding remarks are given in Section 4..

2.   EIGENVALUE TYPE BOUNDS

Linear bounds such as the Gilmore-Lawler bound deteriorate quickly as the dimension increases, e.g. [Finke et al., 1987, Hadley et al., 1990]. One of the earliest nonlinear bounds for QAP was based on eigenvalue techniques. We now look at several different eigenvalue bounds for QAP in increasing improvement, viewed using the SDP and Lagrangian relaxations. In Section 2.1 we look at the basic eigenvalue bound on the homogeneous QAP. Then Section 2.2 looks at improvements to this bound using transformations (perturbations) of the data. An improved bound is the projected bound in Section 2.3 which avoids one class of the perturbations. It is quite interesting to see how this bound can also be viewed using Lagrangian relaxation and adding redundant constraints. This view allows one to easily see that one class of the transformations (perturbations) are not helpful, i.e. the Lagrangian relaxation finds the best of these transformations automatically, see Section 2.4.

2.1   HOMOGENEOUS QAP

The first eigenvalue bounds for QAP are based on ignoring all but the orthogonality constraints, see e.g. [Finke et al., 1987, Hadley et al., 1990] and the survey article [Pardalos et al., 1994]. This was applied to the homogeneous QAP, i.e. the case where $C = 0$.

The bounds were based on a generalization of the eigenvalue problem. Let D be an $n \times n$ symmetric matrix. By abuse of notation, we define the quadratic function $q(x) = x^T D x$. Then the Rayleigh Principle yields the following formulation of the smallest eigenvalue.

$$\lambda_{\min}(D) = \min_{x^T x = 1} q(x) \quad (= x^T D x).$$

This result can be proved easily using Lagrange multipliers, i.e. the optimum x must be a stationary point of the Lagrangian $q(x) + \lambda \left(1 - x^T x\right)$. We can get an equivalent SDP problem using Lagrangian duality and relaxation. Note that

$$
\begin{aligned}
\lambda_{\min}(A) &= \min_{x^T x = 1} x^T A x \\
&= \min_x \max_\lambda x^T A x + \lambda \left(1 - x^T x\right) & (7.4) \\
&\geq \max_\lambda \min_x x^T A x + \lambda \left(1 - x^T x\right) & (7.5) \\
&= \max_{A - \lambda I \succeq 0} \min_x x^T A x + \lambda \left(1 - x^T x\right) & (7.6) \\
&= \max_{A - \lambda I \succeq 0} \min_x x^T (A - \lambda I) x + \lambda & (7.7) \\
&= \max_{A - \lambda I \succeq 0} \lambda = \lambda_{\min}(A). & (7.8)
\end{aligned}
$$

The second equality (7.4) follows from the hidden constraint on the inner maximization problem, i.e. if $x^T x \neq 1$ is chosen then the inner maximization is $+\infty$. If we add this hidden constraint $x^T x = 1$ to the minimization problem, then we recover the Rayleigh Principle. The next inequality (7.5) comes from interchanging min and max. The following equality (7.6) (and equivalently (7.7)) comes again from a hidden constraint, i.e. the quadratic function $x^T (A - \lambda I) x$ must be convex or the inner minimization is $-\infty$. This then yields the equivalence to the smallest eigenvalue problem (7.8) again. Thus we see the equivalence of this norm 1 problem with an SDP and with its Lagrangian dual. The trick to getting the equivalence was to use the hidden constraints.

Note that the above strong duality result still holds if the quadratic objective function $q(x)$ has a linear term. In this case the problem is called *the Trust Region Subproblem, TRS*. (See [Stern and Wolkowicz, 1995, Theorem 5.1] for the strong duality theorem.) However, strong duality can fail if there are two constraints, i.e. the so-called CDT problem [Celis et al., 1984]. Thus we see that going from one to two constraints, even if both constraints are convex, can result in a duality gap. Therefore, the following strong duality result in Theorem 1 below is very surprising.

We now relax the QAP to a quadratic problem over orthogonal constraints by ignoring both the nonnegativity and row and column sum constraints in (7.3), i.e. we consider the constraints

$$X^T X = I, \qquad X \in \mathcal{M}_n.$$

(The set of such X is sometimes known as the Stiefel manifold, e.g. [Edelman et al., 1999, Stiefel, 5 36].) Because of the similarity of the orthogonality constraint to the norm constraint $x^T x = 1$, the result of this section can be viewed as a matrix generalization of the strong duality result for the Rayleigh Principle given above. Thus we consider the homogeneous version of the QAP and its orthogonal relaxation

$$\text{QAP}_{\mathcal{O}} \qquad \mu^{\mathcal{O}} := \quad \begin{array}{ll} \min & \text{Trace } AXBX^T \\ \text{s.t.} & XX^T = I. \end{array} \qquad (7.9)$$

Though this is a nonconvex problem with many nonconvex constraints, this problem can be solved efficiently using Lagrange multipliers and eigenvalues, see e.g. [Hadley et al., 1992], or using the classical Hoffman-Wielandt inequality, e.g. [Bhatia, 1987]. The optimal value is the minimal scalar product of the eigenvalues of A and B. We include a simple proof for completeness using Lagrange multipliers. As was done for the ordinary eigenvalue problem above, we note that Lagrange multipliers can be used in two ways. First, one can use them in the necessary conditions (Karush-Kuhn-Tucker) for optimality, i.e. in the stationarity of the Lagrangian. This is how we apply them now. (The other use is in Lagrangian duality or Lagrangian relaxation where the Lagrangian is positive semidefinite. This is done below.) Also, the Lagrange multipliers here are symmetric matrices since the image of the constraint $X^T X - I$ is a symmetric matrix.

Proposition 1 *Suppose that the orthogonal diagonalizations of A, B are $A = V\Sigma V^T$ and $B = U\Lambda U^T$, respectively, where the eigenvalues in Σ are ordered nonincreasing, and the eigenvalues in Λ are ordered nondecreasing. Then the optimal value of $\text{QAP}_{\mathcal{O}}$ is $\mu^{\mathcal{O}} = \text{Trace } \Sigma\Lambda$, and the optimal solution is obtained using the orthogonal matrices that yield the diagonalizations, i.e. $X^* = VU^T$.*

Proof. The constraint $G(X) := XX^T - I$ maps $\mathcal{M}_n$ to $\mathcal{S}_n$. The Jacobian of the constraint at X acting on the direction h is $J(X)(h) = Xh^T + hX^T$. (This can be found by simply expanding and neglecting the second order term.) The adjoint of the Jacobian acting on $S \in \mathcal{S}_n$ is $J^*(X)(S) = 2SX$, since

$$\text{Trace } SJ(X)(h) = \text{Trace } h^T J^*(X)(S).$$

But $J^*(X)(S) = 0$ implies $S = 0$, i.e. J^* is one-one for all X orthogonal. Therefore J is onto, i.e. the standard constraint qualification holds at the optimum. It follows that the necessary conditions for optimality are that the gradient of the Lagrangian

$$L(X,S) = \text{Trace } AXBX^T - \text{Trace } S(XX^T - I), \quad (7.10)$$

is 0, i.e.

$$AXB - SXI = 0.$$

Therefore,

$$AXBX^T = S = S^T,$$

i.e. $AXBX^T$ is symmetric which means that A and XBX^T commute and so are mutually diagonalizable by the orthogonal matrix U. Therefore, we can assume that both A and B are diagonal and we choose X to be a product of permutations that gives the correct ordering of the eigenvalues.   ∎

The second use of Lagrange multipliers is in forming the Lagrangian dual. The Lagrangian dual of QAP$_\mathcal{O}$ is

$$\max_{S=S^T} \min_X \text{Trace } AXBX^T - \text{Trace } S(XX^T - I). \quad (7.11)$$

However, there can be a nonzero duality gap for the Lagrangian dual, see [Zhao et al., 1998, Anstreicher and Wolkowicz, 1999] and Example 1 below. The inner minimization in the dual problem (7.11) is an unconstrained quadratic minimization in the variables vec (X), with Hessian

$$B \otimes A - I \otimes S.$$

We apply the hidden semidefinite constraint again. This minimization is unbounded only if the Hessian is not positive semidefinite. In order to close the duality gap, we need a larger class of quadratic functions. Here is where our theme comes in, i.e. we find some redundant quadratic constraints to add. Note that in QAP$_\mathcal{O}$ the constraints $XX^T = I$ and $X^TX = I$ are equivalent. We add the redundant constraints $X^TX = I$ and arrive at

$$\text{QAP}_{\mathcal{OO}} \quad \mu^O := \min \text{ Trace } AXBX^T \quad (7.12)$$
$$\text{s.t.} \quad XX^T = I, \; X^TX = I. \quad (7.13)$$

Using symmetric matrices S and T to relax the constraints $XX^T = I$ and $X^TX = I$, respectively, we obtain a dual problem

$$\text{DQAP}_{OO} \quad \mu^O \geq \mu^D := \quad \max \quad \text{Trace } S + \text{Trace } T$$
$$\text{s.t.} \quad (I \otimes S) + (T \otimes I) \preceq (B \otimes A)$$
$$S = S^T, \ T = T^T.$$

We now prove the strong duality presented in [Anstreicher and Wolkowicz, 1999]. We include two proofs. The first proof is from [Anstreicher and Wolkowicz, 1999]; it uses the well known strong duality for LAP, the linear assignment problem; and, it uses the fact that we know the optimal value from Proposition 1. The second proof exploits the LAP duality results from the first proof; but, it illustrates where convexity and complementary slackness arise without using Proposition 1.

Theorem 1 *Strong duality holds for* QAP$_{OO}$ *and* DQAP$_{OO}$, *i.e.* $\mu^D = \mu^O$ *and both primal and dual are attained.*

Proof I. Let $A = V\Sigma V^T$, $B = U\Lambda U^T$, where V and U are orthonormal matrices whose columns are the eigenvectors of A and B, respectively, σ and λ are the corresponding vectors of eigenvalues, and $\Sigma = \text{diag}(\sigma)$, $\Lambda = \text{diag}(\lambda)$. Then for any S and T,

$$(B \otimes A) - (I \otimes S) - (T \otimes I) = (U \otimes V)\left[(\Lambda \otimes \Sigma) - (I \otimes \bar{S}) - (\bar{T} \otimes I)\right]$$
$$(U^T \otimes V^T),$$

where $\bar{S} = V^T SV$, $\bar{T} = U^T TU$. Since $U \otimes V$ is nonsingular, $\text{Trace } S = \text{Trace } \bar{S}$ and $\text{Trace } T = \text{Trace } \bar{T}$, the dual problem DQAP$_{OO}$ is equivalent to

$$\mu^D = \quad \max \quad \text{Trace } S + \text{Trace } T$$
$$\text{s.t.} \quad (\Lambda \otimes \Sigma) - (I \otimes S) - (T \otimes I) \succeq 0 \qquad (7.14)$$
$$S = S^T, \ T = T^T.$$

However, since Λ and Σ are diagonal matrices, (7.14) is equivalent to the ordinary linear program:

$$\text{LD} \quad \max \quad e^T s + e^T t$$
$$\text{s.t.} \quad \lambda_i \sigma_j - s_j - t_i \geq 0, \quad i, j = 1, \ldots, n.$$

But LD is the dual of the linear assignment problem:

$$\text{LP} \qquad \min \sum_{i,j} \lambda_i \sigma_j y_{ij}$$

$$\text{s.t.} \quad \sum_{j=1}^{n} y_{ij} = 1, \quad i = 1, \ldots, n$$

$$\sum_{i=1}^{n} y_{ij} = 1, \quad j = 1, \ldots, n$$

$$y_{ij} \geq 0, \quad i, j = 1, \ldots, n.$$

Assume without loss of generality that $\lambda_1 \leq \lambda_2 \leq \ldots \leq \lambda_n$, and $\sigma_1 \geq \sigma_2 \geq \ldots \geq \sigma_n$. Then LP can be interpreted as the problem of finding a permutation $\pi(\cdot)$ of $\{1, \ldots, n\}$ so that $\sum_{i=1}^{n} \lambda_i \sigma_{\pi(i)}$ is minimized. But the minimizing permutation is then $\pi(i) = i$, $i = 1, \ldots, n$, and from Proposition 1 the solution value μ^D is exactly μ^O.

Proof II. Using the above notation in Proof I, we diagonalize A and B. We can write (7.12) with diagonal matrices, i.e.

$$\text{QAP}_{OO} \qquad \mu^O := \quad \min \quad \text{Trace } V\Sigma V^T X U \Lambda U^T X^T$$

$$\text{s.t.} \quad XX^T = I, \ X^T X = I.$$

With

$$Y = V^T X U, \tag{7.15}$$

we get the equivalent problem

$$\text{QAP}_{OO} \qquad \mu^O := \quad \min \quad \text{Trace } \Sigma Y \Lambda Y^T$$

$$\text{s.t.} \quad YY^T = I, \ Y^T Y = I. \tag{7.16}$$

The Lagrangian for this problem is

$$L(Y, S, T) = \text{Trace } \Sigma Y \Lambda Y^T - \text{Trace } S(YY^T - I) - \text{Trace } (YTY^T - T).$$

Stationarity for the Lagrangian is

$$0 = \nabla L(Y, S, T) = \Sigma Y \Lambda - SYI - IYT.$$

As shown in Proof I, the dual program is equivalent to the ordinary linear program LD which is the dual of the LAP, LP above. Let Y be the optimal permutation of LP above and let S, T be the optimal solutions of LD above. Then the constraints of LD guarantee that the Hessian of the Lagrangian $L(Y, S, T)$

is positive semidefinite, i.e. the Lagrangian is convex in Y. In addition, complementary slackness between LD and LP is equivalent to the stationarity condition. Therefore, we have feasibility (and so complementary slackness), stationarity and convexity of the Lagrangian, i.e. these are the ingredients needed to guarantee optimality. Therefore Y is optimal for (7.16). After using the transformation (7.15), we get the optimal X for the original problem. ∎

Remark 1 *Though we have strong duality between the above dual pairs, it is not known what happens if a linear term ($C \neq 0$) exists. The second proof of the above theorem could be used to study this case, i.e. one needs to use the optimal solution found from the dual LD to obtain an optimal solution for the original problem. To prove optimality one needs to use the following necessary conditions for sufficiency to hold: primal feasibility; stationarity of the Lagrangian (equivalently complementary slackness between the dual and the dual of the dual); and convexity of the Lagrangian.*

In [Wolkowicz, 2000a], it is shown that a duality gap can occur, for this orthogonal relaxation, if $C \neq 0$. This becomes clear once one notices that the optimal value of the dual is independent of the signs of the individual elements of C. Whereas, in the pure linear case, the optimal value is found using the sum of the singular values of C, see e.g. [Wolkowicz, 2000a, Proposition 2.3]. However, one can close this duality gap, in the pure linear case, by using the objective function Trace YCX^T *and relaxing the constraints to* $XX^T + YY^T = I$, *see [Wolkowicz, 2000a]. Thus, instead of doubling the number of constraints, we double the number of variables.*

2.2   PERTURBATIONS

Though the eigenvalue bound may be better then the linear type bounds, it still deteriorates very quickly as the dimension grows. One approach to improve this bound is to perform perturbations (transformations) on A and B that do not change the objective value but reduce the influence of the quadratic part of the objective function.

Note that the quadratic part can be bounded using Proposition 1, while the linear part is solved independently as a linear assignment problem, LAP. We let QAP(A,B,C) denote the optimal objective function value of the QAP defined by matrices A, B, C, and we let LAP(C) denote the optimal value of the LAP defined by C. The following eigenvalue related bound was proposed in [Finke et al., 1987].

$$QAP(A, B, C) \geq \langle \lambda(A), \lambda(B) \rangle_- + LAP(C). \qquad (7.17)$$

To improve the bound in 7.17, transformations are applied to A, B and C that leave $q(X)$ unchanged over Π, but move a part of the quadratic over to the linear part. (The advantage for this is that the linear part is solved exactly.) Two types of transformations are known to have this property:

1. adding a constant to A or B either row or column-wise and appropriately modifying C;

2. changing the main diagonal of A or B and appropriately modifying C.

To be more specific, suppose $g, f, r, s \in \Re^n$. We define

$$
\begin{aligned}
A(g,r) &:= A + ge^T + eg^T + \operatorname{diag}(r) \\
B(f,s) &:= B + fe^T + ef^T + \operatorname{diag}(s) \\
C(g,f,r,s) &:= C + 2Aef^T + 2ge^T B - 2ngf^T - 2\sum_k g_k ef^T \\
&\quad + \operatorname{diag}(A)s^T + r\operatorname{diag}(B)^T - 2gs^T - 2rf^T - rs^T.
\end{aligned}
$$

Then it can easily be verified, see [Finke et al., 1987, Frieze and Yadegar, 1983], that

$$
\begin{aligned}
\text{Trace } (AXB^T + C)X^T &= \text{Trace } (A(g,r)XB^T(f,s) + C(g,f,r,s))X^T \\
&\quad \forall g, f, r, s \in \Re^n,\ \forall X \in \Pi.
\end{aligned}
$$

$$(7.18)$$

Relation (7.18) shows that we may choose any transformation $d := (g, f, r, s) \in \Re^{n \times 4}$ to derive bounds for QAP. There are several strategies for making reasonable choices for the transformations, [Finke et al., 1987, Rendl and Wolkowicz, 1992]. However, we will see below that these transformations actually come about from adding redundant constraints and taking the Lagrangian dual. Therefore, the best transformations are automatically chosen when using the semidefinite relaxation and there is no need for choosing any transformations.

2.3   PROJECTED EIGENVALUE BOUND

We saw above that we can solve the orthogonal relaxation of the homogeneous QAP using Lagrangian duality. We can then improve the resulting bound using perturbations. However, this results in a linear term. The next obvious question is how to handle this linear term. In addition, can we improve the bound by including the linear row and column sum constraints?

Since Lagrangian duality was so successful, it appears to make sense to use this now. However, The linear constraints have to be handled in a special way. We cannot just bring them into the Lagrangian with Lagrange multipliers as they will be ignored, since they have a zero contribution to the Hessian of the Lagrangian, see [Poljak et al., 1995]. There are several ways to overcome this problem. One way is to eliminate the linear constraint. However, one would then drastically change the orthogonality constraint. Instead, we can use a special substitution and elimination. Let V be an $n \times (n-1)$ orthogonal matrix with e in the null space of V^T, i.e.

$$V^T e = 0, \quad V^T V = I.$$

Let $u := e/\|e\|$. Therefore

$$P := [u \mid V] \tag{7.19}$$

is a square orthogonal matrix. Then the following holds, see [Hadley et al., 1992].

Lemma 1 *Let P be defined as in (7.19); let X be $n \times n$ and Y be $(n-1) \times (n-1)$. Suppose that X and Y satisfy*

$$X = P \begin{pmatrix} 1 & 0 \\ 0 & Y \end{pmatrix} P^T. \tag{7.20}$$

Then, the following three statements hold.

1. $X \in \mathcal{E}$;
2. $X \in \mathcal{N} \iff VYV^T \geq -uu^T$;
3. $X \in \mathcal{O}_n \iff Y \in \mathcal{O}_{n-1}$.

Conversely, if $X \in \mathcal{E}$, then there is a Y such that (7.20) holds.

Lemma 1 let's us substitute for X and the linear equality constraints without damaging the orthogonality constraints.

$$
\begin{aligned}
q(X) &= \text{Trace}\, [A(vv^T + VYV^T)B^T + C](vv^T + VY^TV^T) \\
&= \text{Trace}\, \{Avv^T B^T vv^T + Avv^T B^T VY^TV^T + AVYV^T B^T vv^T + \\
&\qquad AVYV^T B^T VY^TV^T + Cvv^T + CVY^TV^T\} \\
&= \text{Trace}\, \{(V^T AV)Y(V^T B^T V) + V^T CV + \tfrac{2}{n}V^T r(A)r^T(B)V\}Y^T + \\
&\qquad \tfrac{s(A)s(B)}{n^2} + \tfrac{s(C)}{n}.
\end{aligned}
$$

Let $\hat{A} := V^T AV$, $\hat{B} := V^T BV$, $\hat{C} := V^T CV$ and $\hat{D} := \tfrac{2}{n}V^T r(A)r^T(B)V + \hat{C}$. We now define the *projected* problem PQAP.

$$
\text{PQAP} \quad
\begin{aligned}
\min \quad & \text{Trace}\, \hat{A}Y\hat{B}^T Y^T + \text{Trace}\, D[vv^T + VY^TV^T] \\
& \qquad - \tfrac{1}{n^2}s(A)s(B) \\
\text{s.t.} \quad & Y \in \mathcal{O}, VYV^T \geq -vv^T,
\end{aligned}
\tag{7.21}
$$

i.e. we have a very similar problem to QAP. The variable is still a square matrix, Y (though one dimension smaller). The constraints are still orthogonality and nonnegativity, though the nonnegativity is not just on the matrix variable Y itself.

We have derived the following equivalence and bound from [Hadley et al., 1992].

Theorem 2 *Let X and Y be related by (7.20). Then X solves QAP $\iff$ Y solves PQAP.*

Theorem 3 *Let a symmetric QAP with matrices A,B and C be given. Then, using the notation from above*

$$QAP(A, B, C) \geq \left\langle \lambda(\hat{A}), \lambda(\hat{B}) \right\rangle_{-} + LAP(D) - s(A)s(B)/n^2.$$

Though there are special cases where one can minimize both the quadratic and the resulting term together, this is not true in general. This results in the deterioration of the bound since we are using the sum of the minimum of two functions rather than the minimum of the sum. However, this bound is a definite improvement over the eigenvalue bound without increasing the evaluation cost. We First note that the constant row and column transformations are not needed.

Proposition 2

$$PB(A, B, C) = PB(A(g, 0), B(f, 0), C(g, f, 0, 0)) \quad \forall g, f \in \Re^n.$$

2.4   STRENGTHENED PROJECTED EIGENVALUE BOUND

The above bound and proposition are proved in [Hadley et al., 1992]. In [Anstreicher, 1999] (numerical tests in [Anstreicher and Brixius, 1999]), a strengthened version of the projected eigenvalue bound is presented. This bound is an attempt to handle both the quadratic and linear terms together while maintaining convexity, or equivalently, tractability of the bound. We now look at a Lagrangian dual approach.

As mentioned above, we have to be careful how we handle linear constraints when taking the Lagrangian dual. Above, we substituted for X and eliminated the linear constraints. Another, equivalent, approach is to change the linear constraint to a quadratic constraint, e.g. to $||X^T e - e||^2 + ||Xe - e||^2 = 0$. This was the approach used in [Zhao et al., 1998]. We now look at the Lagrangian dual of the problem with orthogonal and linear constraints, i.e.

$$\text{QAP}_{\mathcal{EE}} \quad \mu^O := \quad \min \ \text{Trace } AXBX^T - 2CX^T$$
$$\text{s.t.} \quad X^T X = I, XX^T = I, \quad (7.22)$$
$$Xe = e, X^T e = e.$$

In the case that $C = 0$, if we take the Lagrangian dual of this, then we get the bound that is equivalent to the projected eigenvalue bound. However, if we do the elimination or use the form $||X^T e - e||^2 + ||Xe - e||^2 = 0$ for the linear constraints, then the relaxation is equivalent to the elimination. Therefore, we can add redundant constraints that involve the linear constraints and not change the bound. For example, we can add the constraint, $XBX^T e = XBe$. After adding a Lagrange multiplier, we get Trace $ev^T (XBX^T - XB) = 0$. This is equivalent to the constant row transformation $A + ev^T$ which results in a linear term Trace $ev^T XB =$ Trace $r(B)v^T X$. Similarly, we get the other constant row and column sum transformations. So we see that the reason for the lack of improvement is due to the considerations of adding redundant constraints in the Lagrangian dual. In fact, we can get many other transformations (perturbations) in this way, though they are no longer needed.

2.5   TRUST REGION TYPE BOUND

The Rayleigh quotient result for the minimum eigenvalue can be expressed using inequality, i.e.

$$\lambda_{\min}(A) = \min_{x^T x \leq 1} x^T Ax.$$

Just as above, this can be expressed using strong duality and semidefinite programming. (In fact, we can add a linear term here as well and get the TRS.) The extension to matrices for QAP would involve the constraint $XX^T \preceq I$. A further relaxation of the above orthogonal relaxation is the trust region relaxation studied in [Karisch et al., 1994, Anstreicher et al., 1999],

$$\mu^T := \min \quad \text{Trace } AXBX^T \\ \text{s.t.} \quad XX^T \preceq I. \tag{7.23}$$

Though using the constraints $XX^T \preceq I$ in place of $XX^T = I$ weakens the bound on QAP; i.e. $\mu^T \leq \mu^O$, the constraints $XX^T \preceq I$ are convex, and so it is hoped that solving this problem would be useful in obtaining bounds for QAP and improve eigenvalue bounds in the case that $C \neq 0$. We first present the solution to the problem.

Theorem 4 *Let* $V^T AV = \Sigma$, $U^T BU = \Lambda$, *where* $U, V \in \mathcal{O}$, $\Sigma = \text{Diag}(\sigma)$, $\Lambda = \text{Diag}(\lambda)$, $\sigma_1 \geq \sigma_2 \geq \cdots \geq \sigma_n$, $\lambda_1 \geq \lambda_2 \geq \cdots \geq \lambda_n$. *Then for any* X *with* $XX^T \preceq I$ *we have*

$$\sum_{i=1}^{n} \min\{0, \lambda_i \sigma_{n-i+1}\} \leq \text{Trace } AXBX^T \leq \sum_{i=1}^{n} \max\{0, \lambda_i \sigma_i\}.$$

The upper bound is attained for $X = V\text{Diag}(\epsilon)U^T$, *where* $\epsilon_i = 1$ *if* $\sigma_i \lambda_i \geq 0$, *and* $\epsilon_i = 0$ *otherwise. The lower bound is attained for* $X = V\text{Diag}(\epsilon)JU^T$, *where* $\epsilon_i = 1$ *if* $\sigma_i \lambda_{n+1-i} \leq 0$, *and* $\epsilon_i = 0$ *otherwise,* $J = (e_n, e_{n-1}, \cdots, e_1)$ *and* e_i *is the ith element unit vector.*

As above we use the following problem with the redundant constraint added.

$$\text{QAPT} \qquad \mu^T = \quad \min \quad \text{Trace } AXBX^T$$
$$\text{s.t.} \quad XX^T \preceq I, \quad X^T X \preceq I.$$

The dual program is

$$\text{DQAPT} \qquad \mu^T \geq \mu^{DT} := \quad \max \quad -\text{Trace } S - \text{Trace } T$$
$$\text{s.t.} \quad (B \otimes A) + (I \otimes S) + (T \otimes I) \succeq 0$$
$$S \succeq 0, \ T \succeq 0.$$

The following strong duality result is presented in [Anstreicher et al., 1999].

Theorem 5 *Strong duality holds for QAPT and DQAPT, i.e.* $\mu^T = \mu^{DT}$ *and both primal and dual values are attained.*

No numerical tests have yet been done with this relaxation. However, it is interesting to observe that strong duality holds in this case even though the objective function is not convex. It is still unknown what happens if the objective function is not homogeneous.

3.   SDP RELAXATIONS

We now present the SDP relaxation of QAP studied in [Zhao et al., 1998]. (The missing details can be found there.) This relaxation arose by adding many redundant constraints and taking the Lagrangian dual of the Lagrangian dual and then removing redundant constraints at the end. Since we add many redundant constraints at the start, it can be shown that this is the strongest of the bounds that we have looked at so far. In addition, the final bound is greatly simplified by the application of a so-called *gangster operator*. This illustrates the strength of the Lagrangian dual approach to finding the SDP relaxation. One can add many redundant constraints at the start which contribute to the final SDP relaxation. This appears to create a very large SDP relaxation. However, many linear constraints can be shown to be redundant in the final relaxation.

Additional strengthening can be obtained by adding linear inequalities. This is discussed below in Section 3.4.

The QAP can be *lifted* into a higher dimensional space of symmetric matrices so as to obtain a tractable (convex) relaxation. (See e.g. [Grötschel et al., 1988, Lovász and Schrijver, 1991].) Suppose we represent the QAP using binary vectors $x := \text{vec}(X)$. Then the embedding in $\mathcal{S}^{n^2+1}$ is obtained by

$$Y_X := \begin{pmatrix} x_0 \\ \text{vec}(X) \end{pmatrix} (x_0, \text{vec}(X)^T), \ x_0^2 = 1,$$

which results in Y_X being a symmetric and positive semidefinite matrix.

We now outline this for the quadratic constraints that arise from the fact that X is a $(0,1)$, orthogonal matrix . Let $X \in \Pi$ be a permutation matrix and, again, let $x = \text{vec}\,(X)$, $x_0^2 = 1$, and $c = \text{vec}\,(C)$. Then the objective function for QAP (by abuse of notation we add x_0) is

$$
\begin{aligned}
q(X, x_0) &= \text{Trace } AXBX^T - 2CX^T x_0 \\
&= x^T (B \otimes A)x - 2c^T x x_0 \\
&= \text{trace} x x^T (B \otimes A) - 2c^T x x_0 \\
&= \text{trace} L_Q Y_X,
\end{aligned}
$$

where we define the $(n^2 + 1) \times (n^2 + 1)$ matrices

$$
L_Q := \begin{bmatrix} 0 & -\text{vec}\,(C)^T \\ -\text{vec}\,(C) & B \otimes A \end{bmatrix}, \tag{7.24}
$$

and

$$
Y_X := \begin{bmatrix} 1 & x^T \\ x & xx^T \end{bmatrix}. \tag{7.25}
$$

This shows how the objective function of QAP is transformed into a linear function in the SDP relaxation. Note that if we denote $Y = Y_X$, then the element $Y_{(i,j),(k,l)}$ corresponds to $x_{ij} x_{kl}$.

We already have three constraints on the matrix Y, i.e. it is positive semidefinite, the top-left component $Y_{00} = 1$, and it is rank-one. The first two constraints are tractable constraints; while the rank-one constraint is too hard to satisfy and is discarded in the SDP relaxation.

In order to guarantee that the matrix Y, in the case that it is rank one, arises from a permutation matrix X, we need to add additional constraints. For example, the $(0,1)$-constraints $X_{ij}^2 - X_{ij} = 0$ are equivalent to the restriction that the diagonal of Y is equal to its first row (or column). This results in the arrow constraint, see (7.38) below. Similarly, the orthogonality constraint, $XX^T = I, X^T X = I$ can be written using the block diagonal constraint, see (7.39). and the block off diagonal constraints, see (7.40). The SDP relaxation with these constraints, as well as the ones arising from the row and column sums equal 1, is given below in (7.37).

3.1   LAGRANGIAN RELAXATION

Though we can derive the SDP relaxations directly as above, it is interesting and useful to know that the relaxation comes from the dual of the (homogenized) Lagrangian dual. Thus SDP relaxation is equivalent to Lagrangian relaxation for an appropriately constrained problem. In the process we see several of the interesting operators that arise in the relaxation and add the gangster operator which results in a great simplification of the relaxation.

Remark 2 *Note that it could be important to know where the relaxation comes from in order to recover good approximate feasible solutions. More precisely, we can use the optimal solution of the dual of the SDP in the Lagrangian relaxation and then find the optimal matrix X where this Lagrangian attains its minimum. This X is then a good approximation for the original QAP, see e.g. [Kruk and Wolkowicz, 1998, Fujie and Kojima, 1997].*

As we saw above (and also in [Poljak et al., 1995, Zhao et al., 1998, Anstreicher and Wolkowicz, 1999, Anstreicher et al., 1999, Anjos and Wolkowicz, 1999]), adding, possibly redundant, quadratic constraints often tightens the SDP relaxation obtained through the Lagrangian dual. Using the fact that Π can be characterized as the intersection of $(0,1)$-matrices with $\mathcal{E}$ and $\mathcal{O}$, i.e.

$$\Pi = \mathcal{E} \cap \mathcal{Z} = \mathcal{O} \cap \mathcal{Z}, \tag{7.26}$$

we can rewrite QAP as

$$(QAP_{\mathcal{E}}) \quad \begin{aligned} \mu^* := \quad &\min \quad \text{Trace}\, AXBX^T - 2CX^T \\ &\text{s.t.} \quad XX^T = X^TX = I \\ &\qquad Xe = X^Te = e \\ &\qquad X_{ij}^2 - X_{ij} = 0, \ \forall i,j. \end{aligned} \tag{7.27}$$

We can see that there are a lot of redundant constraints in $(QAP_{\mathcal{E}})$. However, as we show below, they are not necessarily redundant in the SDP relaxations. Additional redundant (but useful in the relaxation) constraints will be added below, e.g. we can use the fact that the rank-one matrices formed from the columns of X, i.e. $X_{:i}X_{:j}^T$, are diagonal matrices if $i = j$; while their diagonals are 0 if $i \neq j$. This is equivalent to the fact that the Hadamard products

$$(XP) \circ X = 0,$$

for all permutations P that do not leave any of the columns not permuted, i.e. the permutation does not have a 1-cycle. This latter constraint implies that there are a lot of zeros in the lifted matrices. This is essentially the ingredient for the gangster operator.

We now apply the recipe for the relaxations. We have added the, possibly redundant, constraints to the model. We now continue with the homogenization and taking Lagrangian duals. After changing the row and column sum constraints into $\|Xe - e\|^2 + \|X^Te - e\|^2 = 0$, we consider the following equivalent problem to QAP.

$$\mu_O := \begin{array}{ll} \min & \text{Trace } AXBX^T - 2CX^T \\ \text{s.t.} & XX^T = I \end{array}$$

$$\text{(QAP}_O) \qquad\qquad X^TX = I \qquad\qquad (7.28)$$
$$\|Xe - e\|^2 + \|X^Te - e\|^2 = 0$$
$$X_{ij}^2 - X_{ij} = 0, \quad \forall i, j.$$

We used the approach that changes the linear constraint to a quadratic constraint e.g. to $\|Xe - e\|^2 = 0$. This was the approach used in [Zhao et al., 1998]. The result is an SDP relaxation where the Slater constraint qualification (strict feasibility) fails. (To overcome this problem, in order to successfully apply interior point methods, one projects the problem onto the so-called minimal face of the problem. Following this one can also remove redundant constraints. This is outlined below in Section 3.2.)

$$\mu_O \geq \mu_{\mathcal{L}} := \max_{W, u_0} \min_{XX^T = X^T X = I, x_0^2 = 1} \{\text{Trace}\,[AXBX^T + W(X \circ X)^T$$
$$+ u_0(\|Xe\|^2 + \|X^Te\|^2) - x_0(2C + W)X^T]$$
$$- 2x_0u_0e^T(X + X^T)e + 2nu_0x_0^2\}. \qquad (7.29)$$

Introducing a Lagrange multiplier w_0 for the constraint on x_0 and Lagrange multipliers S_b for $XX^T = I$ and S_o for $X^TX = I$, we get the lower bound μ_R

$$\mu_O \geq \mu_{\mathcal{L}} \geq \mu_R :=$$
$$\max_{W, S_b, S_o, u_0, w_0} \min_{X, x_0} \{\text{Trace}\,[AXBX^T + u_0(\|Xe\|^2 + \|X^Te\|^2)$$
$$+ W(X \circ X)^T + w_0x_0^2 + S_bXX^T + S_oX^TX] \qquad (7.30)$$
$$- \text{Trace } x_0(2C + W)X^T - 2x_0u_0e^T(X + X^T)e$$
$$- w_0 - \text{Trace } S_b - \text{Trace } S_o + 2nu_0x_0^2\}.$$

Both inequalities can be strict, i.e. there can be duality gaps in each of the Lagrangian relaxations. Following is an example of a duality gap that arises from the Lagrangian relaxation of the orthogonality constraint (see [Zhao et al., 1998]).

Example 1 *Consider the the pure quadratic, orthogonally constrained problem*

$$\mu^* := \begin{array}{ll} \min & \text{Trace } AXBX^T \\ \text{s.t.} & XX^T = I, \end{array} \qquad (7.31)$$

with 2×2 matrices

$$A = \begin{pmatrix} 1 & 0 \\ 0 & 2 \end{pmatrix}, \qquad B = \begin{pmatrix} 3 & 0 \\ 0 & 4 \end{pmatrix}.$$

The dual problem is

$$\mu^D := \max \quad -\text{Trace } Ss$$
$$\text{s.t.} \quad (B \otimes A + I \otimes Ss) \succeq 0 \qquad (7.32)$$
$$S = S^T.$$

Then $\mu^* = 10$. But is the dual optimal value μ^D also 10? We have

$$B \otimes A = \begin{pmatrix} 3 & 0 & 0 & 0 \\ 0 & 6 & 0 & 0 \\ 0 & 0 & 4 & 0 \\ 0 & 0 & 0 & 8 \end{pmatrix}.$$

Then in order to satisfy dual feasibility, we must have $S_{11} \geq -3$ and $S_{22} \geq -6$. In order to maximize the dual, equality must hold. Therefore $-\text{Trace } Ss = 9$ in the optimum. Thus we have a duality gap for this simple example.

One can also easily construct a counterexample in the pure linear case, i.e. the case where $A = B = 0$. The 2×2 example with $C = \begin{pmatrix} 1 & 1 \\ 1 & 1 \end{pmatrix}$ provides such an example, see [Wolkowicz, 2000a].

In (7.30), we grouped the quadratic, linear, and constant terms together. We now define $x := \text{vec}(X)$, $y^T := (x_0, x^T)$ and $w^T := (w_0, \text{vec}(W)^T)$ and get

$$\mu_R =$$
$$\max_{w, S_b, S_o, u_0} \min_y \{ y^T [L_Q + \text{Arrow}(w) + \text{B}^0\text{Diag}(S_b) + \text{O}^0\text{Diag}(S_o) + u_0 D] y$$
$$- w_0 - \text{Trace } S_b - \text{Trace } S_o \},$$
$$(7.33)$$

where L_Q is as above and the linear operators

$$\text{Arrow}(w) := \begin{bmatrix} w_0 & -\frac{1}{2}w_{1:n^2}^T \\ -\frac{1}{2}w_{1:n^2} & \text{Diag}(w_{1:n^2}) \end{bmatrix}, \qquad (7.34)$$

$$\text{B}^0\text{Diag}(S) := \begin{bmatrix} 0 & 0 \\ 0 & I \otimes S_b \end{bmatrix}, \qquad (7.35)$$

$$\text{O}^0\text{Diag}(S_o) := \begin{bmatrix} 0 & 0 \\ 0 & S_o \otimes I \end{bmatrix}, \qquad (7.36)$$

and

$$D := \begin{bmatrix} n & -e^T \otimes e^T \\ -e \otimes e & I \otimes E \end{bmatrix} + \begin{bmatrix} n & -e^T \otimes e^T \\ -e \otimes e & E \otimes I \end{bmatrix}.$$

There is a hidden semidefinite constraint in (7.33), i.e. the inner minimization problem is bounded below only if the Hessian of the quadratic form is positive semidefinite. In this case the quadratic form has minimum value 0. This yields the following SDP.

$$(D_{\mathcal{O}}) \quad \begin{array}{ll} \max & -w_0 - \text{Trace } S_b - \text{Trace } S_o \\ \text{s.t.} & L_Q + \text{Arrow }(w) + \text{B}^0\text{Diag }(S_b) + \text{O}^0\text{Diag }(S_o) + u_0 D \succeq 0. \end{array}$$

We now obtain our desired SDP relaxation of $(QAP_{\mathcal{O}})$ as the Lagrangian dual of $(D_{\mathcal{O}})$. This dual is derived just as in linear programming. The similarities are very noticeable. We introduce the $(n^2 + 1) \times (n^2 + 1)$ dual matrix variable $Y \succeq 0$ and derive the dual program to the SDP $(D_{\mathcal{O}})$.

$$(SDP_{\mathcal{O}}) \quad \begin{array}{ll} \min & \text{Trace } L_Q Y \\ \text{s.t.} & \text{b}^0\text{diag }(Y) = I, \quad \text{o}^0\text{diag }(Y) = I \\ & \text{arrow }(Y) = e_0, \quad \text{Trace } DY = 0 \\ & Y \succeq 0, \end{array} \qquad (7.37)$$

where the *arrow operator*, acting on the $(n^2 + 1) \times (n^2 + 1)$ matrix Y, is the adjoint operator to Arrow $(\cdot)$ and is defined by

$$\text{arrow }(Y) := \text{diag }(Y) - \left(0, (Y_{0,1:n^2})^T\right), \qquad (7.38)$$

i.e. the arrow constraint guarantees that the diagonal and 0-th row (or column) are identical.

The *block-0-diagonal operator* and *off-0-diagonal operator* acting on Y are defined by

$$\text{b}^0\text{diag }(Y) := \sum_{k=1}^{n} Y_{(k,\cdot),(k,\cdot)} \qquad (7.39)$$

and

$$\text{o}^0\text{diag }(Y) := \sum_{k=1}^{n} Y_{(\cdot,k),(\cdot,k)}. \qquad (7.40)$$

These are the adjoint operators of $\text{B}^0\text{Diag }(\cdot)$ and $\text{O}^0\text{Diag }(\cdot)$, respectively. The block-0-diagonal operator guarantees that the sum of the diagonal blocks equals the identity. The off-0-diagonal operator guarantees that the trace of each diagonal block is 1, while the trace of the off-diagonal blocks is 0. These constraints come from the orthogonality constraints, $XX^T = I$ and $X^T X = I$, respectively.

We have expressed the orthogonality constraints with both $XX^T = I$ and $X^T X = I$. It is interesting to note that this redundancy adds extra constraints into the relaxation which are not redundant. These constraints reduce the size of the feasible set and so tighten the bounds.

3.2   GEOMETRY OF THE RELAXATION

Proposition 1 *Suppose that Y is feasible for the SDP relaxation (7.37). Then Y is singular.*

Proof. Note that $D \neq 0$ and both D, Y are positive semidefinite. Therefore Y has to be singular in order to satisfy the constraint Trace $DY = 0$.   ∎

Thus the feasible set of the primal problem (SDP_O) has no strictly feasible points. However, it is easy to see that the dual problem (D_O), does satisfy the Slater's constraint qualification (strict feasibility). This means that there is no duality gap between this SDP dual pair but, interior-point algorithms will have difficulty because, as formulated in this way, we have an ill-posed problem, e.g. the dual may not be attained. (See Example 2 below.) Fortuitously, one can use this to advantage, i.e. when Slater's condition fails one can project onto the so-called minimal face of the problem, see [Borwein and Wolkowicz, 1981] and also [Ramana et al., 1997]. Moreover, in our case here we can do this analytically and gain advantages without losing anything to numerical instability.

Example 2 *Consider the SDP pair*

$$
\begin{array}{lll}
\text{min} & 2X_{12} & \\
(P) \quad \text{s.t.} & \text{diag}(X) = \begin{pmatrix} 0 \\ 1 \end{pmatrix} \\
& X \succeq 0
\end{array}
\qquad
\begin{array}{ll}
\text{max} & y_2 \\
(D) \quad \text{s.t.} & \begin{bmatrix} y_1 & 0 \\ 0 & y_2 \end{bmatrix} \preceq \begin{bmatrix} 0 & 1 \\ 1 & 0 \end{bmatrix}
\end{array}
$$

Slater's condition holds for the dual but not for the primal. The optimal value for both is 0. The primal is attained, but the dual is not.

3.2.1   The Minimal Face.

In order to overcome the above difficulties, we need to explore the geometrical structure of F_O. We project the feasible set into a smaller dimensional space so that strict feasibility is satisfied. To do this we need to characterize the so-called minimal face of the problem, see [Borwein and Wolkowicz, 1981].

The points

$$
Y_X := \begin{pmatrix} 1 \\ \text{vec}(X) \end{pmatrix} (1 \ \text{vec}(X)^T), \quad X \in \Pi
$$

are feasible. Moreover, these points are rank-one matrices and are, therefore, contained in the set of extreme points of F_O, see e.g. [Pataki, 2000]. We need only consider faces of F_O which contain all of these extreme points. To do this, we take a closer look at the assignment (row and column sums) constraints defined by $\mathcal{E}$. Surprisingly, it is only these constraints that are needed to define the minimal face. (This is not true in general, see Example 3 below.)

Define the following $(n^2 + 1) \times ((n-1)^2 + 1)$ matrix.

$$\hat{V} := \left[\begin{array}{c|c} 1 & 0 \\ \hline \frac{1}{n}(e \otimes e) & V \otimes V \end{array} \right], \tag{7.41}$$

where V is an $n \times (n-1)$ matrix containing a basis of the orthogonal complement of e, i.e. $V^T e = 0$. Our choice for V is

$$V := \left[\begin{array}{c} I_{n-1} \\ \hline -e_{n-1}^T \end{array} \right].$$

The following theorem characterizes the minimal face by finding the barycenter of the convex hull of the permutation matrices and using the fact that a face of $\mathcal{P}$ can be characterized using the null space (or range space) of any point in its relative interior. We now see that the barycenter has a very simple and elegant structure.

Theorem 6 *Define the barycenter*

$$\hat{Y} := \frac{1}{n!} \sum_{X \in \Pi} Y_X. \tag{7.42}$$

Then:

1. $\hat{Y}$ *has a 1 in the (0,0) position and n diagonal $n \times n$ blocks with diagonal elements $1/n$. The first row and column equal the diagonal. The rest of the matrix is made up of $n \times n$ blocks with all elements equal to $1/(n(n-1))$ except for the diagonal elements which are 0.*

$$\hat{Y} = \left[\begin{array}{c|c} 1 & \frac{1}{n}e^T \\ \hline \frac{1}{n}e & \left[\frac{1}{n^2}E \otimes E\right] + \left[\frac{1}{n^2(n-1)}(nI - E) \otimes (nI - E)\right] \end{array} \right].$$

2. *The rank of $\hat{Y}$ is given by*

$$\mathrm{rank}\,(\hat{Y}) = (n-1)^2 + 1.$$

3. *The $n^2 + 1$ eigenvalues of $\hat{Y}$ are given in the vector*

$$(2, \frac{1}{n-1}e_{(n-1)^2}^T, 0e_{2n-1}^T)^T.$$

4. *The null space and range space are*

$$\mathcal{N}(\hat{Y}) = \mathcal{R}(\hat{T}^T) \text{ and } \mathcal{R}(\hat{Y}) = \mathcal{R}(\hat{V}) \text{ (so that } \mathcal{N}(\hat{T}) = \mathcal{R}(\hat{V})\text{)}.$$

With the above characterization of the barycenter, we can find the minimal face of $\mathcal{P}$ that contains the feasible set of the relaxation SDP. We let $t(n) := \frac{n(n+1)}{2}$.

Corollary 1 *The dimension of the minimal face is $t((n-1)^2 + 1)$. Moreover, the minimal face can be expressed as $\hat{V} \mathcal{S}_{(n-1)^2+1} \hat{V}^T$.*

The above characterization of the barycenter yields a characterization of the minimal face. At first glance it appears that there would be a simpler proof for this characterization, the proof would use only the row and column sums constraints. Finding the barycenter is the key in exploiting the geometrical structure of a given problem with an assignment structure. However, it is not always true that the other constraints in the relaxation are redundant, as the following shows.

Example 3 *Consider the constraints*

$$
\begin{array}{rcll}
x_1 & & & = 1 \\
x_1 & +x_2 & +x_3 & +x_4 & = 1 \\
x_1, & x_2, & x_3, & x_4 & \geq 0
\end{array}
$$

The only solution is $(1, 0, 0, 0)$. Hence the barycenter of the relaxation is the set with only a rank one matrix in it. However, the null space of the above system has dimension 3. Thus the projection using the null space yields a minimal face with matrices of dimension greater than 1.

3.2.2   The Projected SDP Relaxation.   In Theorem 6, we presented explicit expressions for the range and null space of the barycenter, denoted $\hat{Y}$. It is well known, see e.g. [Barker and Carlson, 1975], that the faces of the positive semidefinite cone are characterized by the nullspace of points in their relative interior, i.e. $\mathcal{K}$ is a face if

$$
\mathcal{K} = \{X \succeq 0 : \mathcal{N}(X) \supset S\} = \{X \succeq 0 : \mathcal{R}(X) \subset S^\perp\},
$$

and

$$
\operatorname{relint} \mathcal{K} = \{X \succeq 0 : \mathcal{N}(X) = S\} = \{X \succeq 0 : \mathcal{R}(X) = S^\perp\},
$$

where S is a given subspace. In particular, if $\hat{X} \in \operatorname{relint} \mathcal{K}$, the matrix V is $n \times k$, and $\mathcal{R}(V) = \mathcal{R}(\hat{X})$, then

$$
\mathcal{K} = V \mathcal{P}_k V^T.
$$

Therefore, using $\hat{V}$ in Theorem 6, we can project the SDP relaxation (SDP_O) onto the minimal face. The projected problem is

$$
(QAP_{R1}) \quad
\begin{aligned}
\mu_{R1} := \quad & \min \quad \text{Trace}\,(\hat{V}^T L_Q \hat{V})R \\
& \text{s.t.} \quad b^0 \text{diag}\,(\hat{V} R \hat{V}^T) = I, \qquad o^0 \text{diag}\,(\hat{V} R \hat{V}^T) = I \\
& \qquad \text{arrow}\,(\hat{V} R \hat{V}^T) = e_0, \qquad R \succeq 0.
\end{aligned}
$$

$$(7.43)$$

Note that the constraint $\text{Trace}\,(\hat{V}^T D \hat{V})R = 0$ can be dropped since it is always satisfied, i.e. $D\hat{V} = 0$.

By construction, this program satisfies the generalized Slater constraint qualification for both primal and dual. Therefore there will be no duality gap, the optimal solutions are attained for both primal and dual, and both the primal and dual optimal solution sets are bounded.

3.3   THE GANGSTER OPERATOR

The feasible set of the SDP relaxation is convex but not polyhedral. It contains the set of matrices of the form Y_X corresponding to the permutation matrices $X \in \Pi$. But the SDP relaxations, discussed above, can contain many points that are not in the affine hull of these Y_X. In particular, it can contain matrices with nonzeros in positions that are zero in the affine hull of the Y_X. We can therefore strengthen the relaxation by adding constraints corresponding to these zeros.

Note that the barycenter $\hat{Y}$ is in the relative interior of the feasible set. Therefore the null space of $\hat{Y}$ determines the dimension of the minimal face which contains the feasible set. However, the dimension of the feasible set can be (and is) smaller. We now take a closer look at the structure of $\hat{Y}$ to determine the 0 entries. (These entries are easy to handle with a linear operator constraint.) The relaxation is obtained from

$$
\begin{aligned}
Y_X \;=\; & \begin{pmatrix} 1 \\ \text{vec}\,(X) \end{pmatrix} \begin{pmatrix} 1 & \text{vec}\,(X)^T \end{pmatrix} \\[2mm]
=\; & \begin{pmatrix} 1 \\ X_{:1} \\ X_{:2} \\ \vdots \\ X_{:n} \end{pmatrix} \begin{pmatrix} 1 & X_{:1}^T & X_{:2}^T & \vdots & X_{:n}^T \end{pmatrix}
\end{aligned}
$$

which contains the n^2 blocks

$$(X_{:i} X_{:j}^T).$$

We then have

$$\text{diag}\,(X_{:i}X_{:j}^T) = X_{:i} \circ X_{:j} = 0, \text{ if } i \neq j,$$

and

$$X_{i:} \circ X_{j:} = 0, \text{ if } i \neq j,$$

i.e. the diagonal of the off-diagonal blocks are identically zero and the off-diagonal of the diagonal blocks are identically zero. These are exactly the zeros of the barycenter $\hat{Y}$.

The above description defines the so-called gangster operator, i.e. for $J \subset \{(i,j) : 1 \leq i,j \leq n^2 + 1\}$. $\mathcal{G}_J : \mathcal{S}^{n^2+1} \to \mathcal{S}^{n^2+1}$ is called the *Gangster* operator if

$$(\mathcal{G}_J(Y))_{ij} := \begin{cases} Y_{ij} & \text{if } (i,j) \in J \\ 0 & \text{otherwise.} \end{cases} \tag{7.44}$$

Denote the subspace of matrices

$$\mathcal{S}^J := \{X \in \mathcal{S}^{n^2+1} : X_{ij} = 0 \text{ if } (i,j) \notin J\}.$$

Then the range and null space of $\mathcal{G}_J$ satisfy

$$\mathcal{R}(\mathcal{G}_J) = \mathcal{S}^J$$

and

$$\mathcal{N}(\mathcal{G}_{-J}) = \mathcal{S}^{-J},$$

where $-J$ denotes the complement of the set J. Let $J := \{(i,j) : \hat{Y}_{ij} = 0\}$, be the zeros found above using the Hadamard product; we have

$$\mathcal{G}_J(\hat{Y}) = 0. \tag{7.45}$$

Thus the gangster operator, acting on a matrix Y, shoots holes (zeros) through the matrix Y in the positions where $\hat{Y}$ is not zero. For any permutation matrix $X \in \Pi$, the matrix Y_X has all its entries either 0 or 1; and $\hat{Y}$ is just a convex combination of all these matrices Y_X for $X \in \Pi$. Hence, from (7.45), we have

$$\mathcal{G}_J(Y_X) = 0, \text{ for all } X \in \Pi.$$

Therefore, we can further tighten our relaxation by adding the constraint

$$\mathcal{G}_J(Y) = 0. \tag{7.46}$$

Note that the adjoint equation

$$\text{trace}(\mathcal{G}_J^*(Z)Y) = \text{trace}(Z\mathcal{G}_J(Y)),$$

implies that the gangster operator is self-adjoint, i.e.

$$\mathcal{G}_J^* = \mathcal{G}_J.$$

3.3.1   The Gangster Operator and Redundant Constraints.   The addition of the gangster operator allows makes many constraints redundant. Define the subset $\hat{J}$ of J, of indices of Y, (a union of two sets)

$$
\begin{aligned}
\hat{J} \; := \; & \{(i,j) : i = (p-1)n + q, j = (p-1)n + r, q \neq r\} \; \cup \\
& \{(i,j) : i = (p-1)n + q, j = (r-1)n + q, p \neq r, (p,r \neq n), \\
& ((r,p),(p,r) \neq (n-2, n-1), (n-1, n-2))\}.
\end{aligned}
$$

These are the indices for the 0 elements of the barycenter. (We do not include (up to symmetry) the off-diagonal block $(n-2, n-1)$ or the last column of off-diagonal blocks.) After removing redundant constraints, this results in the following simple projected relaxation.

$$
(QAP_{R2}) \qquad
\begin{aligned}
\mu_{R2} := \quad & \min \quad \mathrm{Trace}\,(\hat{V}^T L_Q \hat{V}) R \\
& \text{s.t.} \quad \mathcal{G}_{\hat{j}}(\hat{V} R \hat{V}^T) = E_{00} \\
& \qquad R \succeq 0.
\end{aligned}
\qquad (7.47)
$$

The dimension of the range space is determined by the cardinality of the set $\bar{J}$, i.e. there are $n^3 - 2n^2 + 1$ constraints.

The dual problem is

$$
\begin{aligned}
\mu_{R2} = \quad & \max \quad -Y_{00} \\
& \text{s.t.} \quad \hat{V}^T(L_Q + \mathcal{G}_{\hat{j}}^*(Y))\hat{V} \succeq 0.
\end{aligned}
$$

Note $\mathcal{R}(\mathcal{G}_{\hat{j}}^*) = \mathcal{R}(\mathcal{G}_{\bar{j}}) = \mathcal{S}^{\bar{J}}$. The dual problem can be expressed as follows

$$
\begin{aligned}
\mu_{R2} = \quad & \max \quad -Y_{00} \\
& \text{s.t.} \quad \hat{V}^T(L_Q + Y)\hat{V} \succeq 0 \\
& \qquad Y \in \mathcal{S}^{\bar{J}}.
\end{aligned}
$$

3.4   INEQUALITY CONSTRAINTS

An important technique that is used to further tighten the derived relaxations is to add generic linear inequality constraints. These constraints come from the relaxation of the (0,1)-constraints of the original problem. For $Y = Y_X$, with $X \in \Pi$, the simplest inequalities are of the type

$$
Y_{(i,j),(k,l)} \geq 0, \quad \text{since } x_{ij} x_{kl} \geq 0,
$$

see e.g. [Jünger and Kaibel, 1995, Rijal, 1995]. In addition, in [Helmberg et al., 1995] the authors show that the so called triangle inequalities of the general integer quadratic programming problem in the (-1,+1)-model are also generic inequalities for the (0,1)-formulation. The basic relaxation (QAP_{R1})

can use both nonnegativity and nonpositivity constraints to approximate the gangster operator, e.g.

$$\mathcal{G}_J(\hat{V}R\hat{V}^T) \leq 0. \tag{7.48}$$

The advantage of this formulation is that the number of inequalities can be adapted so that the model is not too large. The larger the model is the better it approximates the original gangster operator.

Further strengthening can be done using a second lifting, see [Anjos and Wolkowicz, 1999].

3.4.1   A Comparison with Linear Relaxations.

We now look at how our relaxations of QAP compare to relaxations based on linear programming. Adams and Johnson [Adams and Johnson, 1994] derive a linear relaxation providing bounds which are at least as good as other lower bounds based on linear relaxations or reformulations of QAP. Using our notation, their continuous linear program can be written as

$$(QAP_{CLP}) \quad \mu_{CLP} := \min\{\text{Trace } LZ : Z \in \mathcal{F}_{CLP}\} \tag{7.49}$$

where the feasible set is

$$\mathcal{F}_{CLP} := \{ \ Z \in \mathcal{N}; \ Z_{(i,j),(k,l)} = Z_{(k,l),(i,j)}, \ 1 \leq i,j,k,l \leq n, i < k, j \neq l;$$
$$\sum_k Z_{(k,l),(k,l)} = \sum_l Z_{(k,l),(k,l)} = 1;$$
$$\sum_{i \neq k} Z_{(i,j),(k,l)} = Z_{(k,l),(k,l)}, \ 1 \leq j,k,l \leq n, \ j \neq l;$$
$$\sum_{j \neq l} Z_{(i,j),(k,l)} = Z_{(k,l),(k,l)}, \ 1 \leq i,k,l \leq n, \ i \neq k \ \}.$$

We now compare the feasible sets of relaxations (QAP_{R3}) and (QAP_{CLP}). It is easy to see that the elements of Z which are not considered in $\mathcal{F}_{CLP}$ are just the elements covered by the gangster operator, i.e. for which $\mathcal{G}_J(Y) = 0$. In (QAP_{R3}) the gangster operator is replaced by nonnegative and nonpositive constraints. The linear constraints in $\mathcal{F}_{CLP}$ are just the lifted assignment constraints, but they are taken care of by the projection and the arrow operator in (QAP_{R3}). The nonnegativity of the elements is enforced in both feasible sets. Hence the only difference is that we impose the additional constraint $Y \in \mathcal{P}$.

4.   CONCLUSION

In this chapter we have derived and ranked many of the known bounds for QAP. The comparisons between the bounds was done using Lagrangian relaxation. Bounds were derived by using relaxed quadratic models of QAP and taking the Lagrangian relaxation. Thus, stronger quadratic models resulted in stronger bounds. The strongest of these bounds was the SDP relaxation studied in Section 3.2.2. This relaxation had a surprisingly simple form after the addition of the gangster operator. A primal-dual interior-point algorithm that solves this SDP relaxation, along with numerical tests, can be found in [Zhao et al., 1998]. Further testing for these bounds as well as new bounds based on trust region methods is being done in [Rendl and Wolkowicz, 1999].

References

[Adams and Johnson, 1994] Adams, W. and Johnson, T. (1994). Improved linear programming-based lower bounds for the quadratic assignment problem. In *Proceedings of the DIMACS Workshop on Quadratic Assignment Problems*, volume 16 of *DIMACS Series in Discrete Mathematics and Theoretical Computer Science*, pages 43–75. American Mathematical Society.

[Anjos and Wolkowicz, 1999] Anjos, M. and Wolkowicz, H. (1999). A strengthened SDP relaxation via a second lifting for the Max-Cut problem. Technical Report Research Report, CORR 99-55, University of Waterloo, Waterloo, Ontario. 28 pages.

[Anstreicher, 1999] Anstreicher, K. (1999). Eigenvalue bounds versus semidefinite relaxations for the quadratic assignment problem. Technical report, University of Iowa, Iowa City, IA.

[Anstreicher and Brixius, 1999] Anstreicher, K. and Brixius, N. (1999). A new bound for the quadratic assignment problem based on convex quadratic programming. Technical report, University of Iowa, Iowa City, IA.

[Anstreicher et al., 1999] Anstreicher, K., Chen, X., Wolkowicz, H., and Yuan, Y. (1999). Strong duality for a trust-region type relaxation of QAP. *Linear Algebra Appl.*, To appear.

[Anstreicher and Wolkowicz, 1999] Anstreicher, K. and Wolkowicz, H. (1999). On Lagrangian relaxation of quadratic matrix constraints. *SIAM J. Matrix Anal. Appl.*, To appear.

[Assad and Xu, 1985] Assad, A. and Xu, W. (1985). On lower bounds for a class of quadratic $\{0, 1\}$ programs. *Operations Research Letters*, 4:175–180.

[Barker and Carlson, 1975] Barker, G. and Carlson, D. (1975). Cones of diagonally dominant matrices. *Pacific J. of Math.*, 57:15–32.

[Bhatia, 1987] Bhatia, R. (1987). *Perturbation Bounds for Matrix Eigenvalues : Pitman Research Notes in Mathematics Series 162*. Longman, New York.

[Borwein and Wolkowicz, 1981] Borwein, J. and Wolkowicz, H. (1981). Regularizing the abstract convex program. *J. Math. Anal. Appl.*, 83(2):495–530.

[Carraresi and Malucelli, 1992] Carraresi, P. and Malucelli, F. (1992). A new lower bound for the quadratic assignment problem. *Oper. Res.*, 40(suppl. 1):S22–S27.

[Carraresi and Malucelli, 1994] Carraresi, P. and Malucelli, F. (1994). A reformulation scheme and new lower bounds for the QAP. In Pardalos, P. and Wolkowicz, H., editors, *Quadratic assignment and related problems (New Brunswick, NJ, 1993)*, pages 147–160. Amer. Math. Soc., Providence, RI.

[Celis et al., 1984] Celis, M., Dennis Jr., J., and Tapia, R. (1984). A trust region strategy for nonlinear equality constrained optimization. In *Proceedings of the SIAM Conference on Numerical Optimization, Boulder, CO*. Also available as Technical Report TR84-1, Department of Mathematical Sciences, Rice University, Houston, TX.

[Edelman et al., 1999] Edelman, A., Arias, T., and Smith, S. (1999). The geometry of algorithms with orthogonality constraints. *SIAM J. Matrix Anal. Appl.*, 20(2):303–353 (electronic).

[Finke et al., 1987] Finke, G., Burkard, R., and Rendl, F. (1987). Quadratic assignment problems. *Ann. Discrete Math.*, 31:61–82.

[Frieze and Yadegar, 1983] Frieze, A. M. and Yadegar, J. (1983). On the quadratic assignment problem. *Discrete Applied Mathematics*, 5:89–98.

[Fujie and Kojima, 1997] Fujie, T. and Kojima, M. (1997). Semidefinite programming relaxation for nonconvex quadratic programs. *J. Global Optim.*, 10(4):367–380.

[Gilmore, 1962] Gilmore, P. (1962). Optimal and suboptimal algorithms for the quadratic assignment problem. *SIAM Journal on Applied Mathematics*, 10:305–313.

[Grötschel et al., 1988] Grötschel, M., Lovász, L., and Schrijver, A. (1988). *Geometric Algorithms and Combinatorial Optimization*. Springer Verlag, Berlin-Heidelberg.

[Hadley et al., 1990] Hadley, S., Rendl, F., and Wolkowicz, H. (1990). Bounds for the quadratic assignment problems using continuous optimization. In *Integer Programming and Combinatorial Optimization*. University of Waterloo Press.

[Hadley et al., 1992] Hadley, S., Rendl, F., and Wolkowicz, H. (1992). A new lower bound via projection for the quadratic assignment problem. *Math. Oper. Res.*, 17(3):727–739.

[Helmberg et al., 1995] Helmberg, C., Poljak, S., Rendl, F., and Wolkowicz, H. (1995). Combining semidefinite and polyhedral relaxations for integer programs. In *Integer Programming and Combinatorial Optimization (Copenhagen, 1995)*, pages 124–134. Springer, Berlin.

[Jünger and Kaibel, 1995] Jünger, M. and Kaibel, V. (1995). A basic study of the qap-polytope. Technical Report Technical Report No. 96.215, Institut für Informatik, Universität zu Köln, Germany.

[Karisch et al., 1994] Karisch, S., Rendl, F., and Wolkowicz, H. (1994). Trust regions and relaxations for the quadratic assignment problem. In *Quadratic assignment and related problems (New Brunswick, NJ, 1993)*, pages 199–219. Amer. Math. Soc., Providence, RI.

[Koopmans and Beckmann, 1957] Koopmans, T. C. and Beckmann, M. J. (1957). Assignment problems and the location of economic activities. *E-conometrica*, 25:53–76.

[Kruk and Wolkowicz, 1998] Kruk, S. and Wolkowicz, H. (1998). SQ^{2}P, sequential quadratic constrained quadratic programming. In xiang Yuan, Y., editor, *Advances in Nonlinear Programming*, volume 14 of *Applied Optimization*, pages 177–204. Kluwer, Dordrecht. Proceedings of Nonlinear Programming Conference in Beijing in honour of Professor M.J.D. Powell.

[Lawler, 1963] Lawler, E. (1963). The quadratic assignment problem. *Management Sci.*, 9:586–599.

[Li et al., 1994] Li, Y., Pardalos, P., Ramakrishnan, K., and Resende, M. (1994). Lower bounds for the quadratic assignment problem. *Ann. Oper. Res.*, 50:387–410. Applications of combinatorial optimization.

[Lovász and Schrijver, 1991] Lovász, L. and Schrijver, A. (1991). Cones of matrices and set-functions and 0-1 optimization. *SIAM J. Optim.*, 1(2):166–190.

[Pardalos et al., 1994] Pardalos, P., Rendl, F., and Wolkowicz, H. (1994). The quadratic assignment problem: a survey and recent developments. In Pardalos, P. and Wolkowicz, H., editors, *Quadratic assignment and related problems (New Brunswick, NJ, 1993)*, pages 1–42. Amer. Math. Soc., Providence, RI.

[Pardalos and Wolkowicz, 1994] Pardalos, P. and Wolkowicz, H., editors (1994). *Quadratic Assignment and Related Problems*. American Mathematical Society, Providence, RI. Papers from the workshop held at Rutgers University, New Brunswick, New Jersey, May 20–21, 1993.

[Pataki, 2000] Pataki, G. (2000). Geometry of Semidefinite Programming. In Wolkowicz, H., Saigal, R., and Vandenberghe, L., editors, *HANDBOOK OF SEMIDEFINITE PROGRAMMING: Theory, Algorithms, and Applications*. Kluwer Academic Publishers, Boston, MA. To appear.

[Poljak et al., 1995] Poljak, S., Rendl, F., and Wolkowicz, H. (1995). A recipe for semidefinite relaxation for $(0, 1)$-quadratic programming. *J. Global Optim.*, 7(1):51–73.

[Ramana et al., 1997] Ramana, M., Tuncel, L., and Wolkowicz, H. (1997). Strong duality for semidefinite programming. *SIAM J. Optim.*, 7(3):641–662.

[Rendl and Wolkowicz, 1992] Rendl, F. and Wolkowicz, H. (1992). Applications of parametric programming and eigenvalue maximization to the quadratic assignment problem. *Math. Programming*, 53(1, Ser. A):63–78.

[Rendl and Wolkowicz, 1999] Rendl, R. and Wolkowicz, H. (1999). Testing bounds for the quadratic assignment problem. Technical Report in progress, University of Waterloo, Waterloo, Canada.

[Resende et al., 1995] Resende, M., Ramakrishnan, K., and Drezner, Z. (1995). Computing lower bounds for the quadratic assignment problem with an interior point algorithm for linear programming. *Oper. Res.*, 43(5):781–791.

[Rijal, 1995] Rijal, M. (1995). Scheduling, design and assignment problems with quadratic costs. PhD thesis, New York University, New York, USA.

[Shor, 1987] Shor, N. (1987). Quadratic optimization problems. *Izv. Akad. Nauk SSSR Tekhn. Kibernet.*, 222(1):128–139, 222.

[Stern and Wolkowicz, 1995] Stern, R. and Wolkowicz, H. (1995). Indefinite trust region subproblems and nonsymmetric eigenvalue perturbations. *SIAM J. Optim.*, 5(2):286–313.

[Stiefel, 5 36] Stiefel, E. (1935-36). Richtungsfelder und fernparallelismus in n-dimensionalem mannigfaltigkeiten. *Commentarii Math. Helvetici*, 8:305–353.

[Wolkowicz, 2000a] Wolkowicz, H. (2000a). A note on lack of strong duality for quadratic problems with orthogonal constraints. Technical report, University of Waterloo.

[Wolkowicz, 2000b] Wolkowicz, H. (2000b). Semidefinite and Lagrangian relaxations for hard combinatorial problems. In Powell, M., editor, *Proceedings of 19th IFIP TC7 CONFERENCE ON System Modeling and Optimization, July, 1999, Cambridge*, page 35. Kluwer Academic Publishers, Boston, MA. To appear.

[Zhao et al., 1998] Zhao, Q., Karisch, S., Rendl, F., and Wolkowicz, H. (1998). Semidefinite programming relaxations for the quadratic assignment problem. *J. Comb. Optim.*, 2(1):71–109.

Chapter 8

HEURISTICS FOR NONLINEAR ASSIGNMENT PROBLEMS

Stefan Voss
Technische Universität Braunschweig
Allgemeine Betriebswirtschaftslehre
Wirtschaftsinformatik und Informationsmanagement
Abt-Jerusalem-Str. 7, D-38106 Braunschweig, Germany
voss@tu-bs.de

Abstract   Given two sets of elements, assignment problems require a mapping of the elements of one set to those of the other. Distinguishing between bijective and injective mappings we may classify two main areas, i.e., assignment and semi-assignment problems. Nonlinear assignment problems (NAPs) are those (semi-) assignment problems where the objective function is nonlinear.

In combinatorial optimization many NAPs are natural extensions of the linear (semi-) assignment problem and include, among others, the quadratic assignment problem (QAP) and its variants. Due to the intrinsic complexity not only of the QAP and related NAPs heuristics are a primary choice when it comes to the successful solution of these problems in time boundaries deemed practical. In this paper we survey existing heuristic approaches for various NAPs.

1.   INTRODUCTION

Assignment problems, e.g., assigning objects to locations or tasks to workers or jobs to machines are amongst the most important problems in combinatorial optimization. Usually, researchers distinguish linear from nonlinear assignment problems based on the objective function to be considered. More formally, given two sets A and B assignment problems require a bijective mapping between the elements of A and B (where both sets have the same cardinality). Semi-assignment problems refer to an injective mapping. Using binary variables x_{ip} which take the value 1 if $p \in B$ is assigned to $i \in A$ and 0 otherwise, then the problem

P.M. Pardalos and L.S. Pitsoulis (eds.), Nonlinear Assignment Problems, 175–215.
© 2000 *Kluwer Academic Publishers.*

$$\text{Minimize } Z(x) \tag{8.1}$$

subject to

$$\sum_{p \in B} x_{ip} = 1 \qquad \forall i \in A \tag{8.2}$$

$$x_{ip} \in \{0, 1\} \qquad \forall i \in A, \forall p \in B \tag{8.3}$$

may be classified as nonlinear semi-assignment problem once Z is a nonlinear function.

For $|A| = |B|$ model (8.1) - (8.3) together with

$$\sum_{i \in A} x_{ip} = 1 \qquad \forall p \in B \tag{8.4}$$

becomes an assignment problem.

To distinguish assignment problems from semi-assignment problems let us consider an example. The *quadratic assignment problem* (QAP), which probably is the most well-known nonlinear assignment problem (NAP), is to find an assignment of a number of n objects to a number of n locations such that the cumulative product of flow between any two objects and the distance between any two locations is minimized. More formally, we are given a distance matrix $(d_{ij})_{n \times n}$ describing the mutual distances between pairs of locations and a cost matrix $(c_{ij})_{n \times n}$ describing the pairwise connection or flow costs between pairs of objects. Then the QAP is to find a permutation π of the set $\{1, \ldots, n\}$ such that

$$\sum_{i=1}^{n} \sum_{j=1}^{n} c_{ij} \cdot d_{\pi(i)\pi(j)} \tag{8.5}$$

is minimized.

Equivalently, the QAP may be modeled by using binary variables x_{ip} which take the value 1 if object i is assigned to location p and take the value 0 otherwise:

$$\text{Minimize } Z(x) = \sum_{i=1}^{n} \sum_{j=1}^{n} \sum_{p=1}^{n} \sum_{q=1}^{n} c_{ij} \cdot d_{pq} \cdot x_{ip} \cdot x_{jq} \tag{8.6}$$

subject to

$$\sum_{i=1}^{n} x_{ip} = 1 \qquad p = 1, \dots, n \qquad (8.7)$$

$$\sum_{p=1}^{n} x_{ip} = 1 \qquad i = 1, \dots, n \qquad (8.8)$$

$$x_{ip} \in \{0, 1\} \qquad i, p = 1, \dots, n \qquad (8.9)$$

The QAP is $\mathcal{NP}$-hard, encompassing the classical *traveling salesman problem* (TSP), which asks for the shortest closed path (*tour*) through a graph that visits every node exactly once, as a special case. There is a vast amount of literature on the QAP as can be seen from the excellent surveys [Çela, 1998, Pardalos and Wolkowicz, 1994] which also provide a large number of introductory references. (There the reader may also find some carefully developed distinction between the QAP with symmetric and asymmetric matrices which will not be considered in detail here.)

It should be noted, exemplified by comparing model (8.1) - (8.3) with the QAP, that if the set of constraints (8.7) is relaxed we obtain the *quadratic semi-assignment problem* (QSAP), another $\mathcal{NP}$-hard NAP. In the QSAP, we are given a set of n objects that have to be assigned to any of m locations. The difference between the QAP and the QSAP is that in the latter each location may take none, one, or even more than one object, whereas in the QAP each location has to obtain exactly one object, and vice versa.

Besides the QAP and the QSAP as well-known NAPs there is a great variety of more general problems based on the consideration of modified objective functions as well as additional restrictions. Taking into account the complexity of most of these problems heuristics have become a natural choice for their solution. As NAPs mainly differ in the objective function many of them may share a common representation based on the above idea of finding a permutation π of a given set $\{1, \dots, n\}$. Following this fact, in this paper we survey existing heuristic approaches for various NAPs.

The paper is organized as follows. First we present general features for finding initial feasible solutions and for improving them once they are given. This especially concerns local search based concepts as well as modern metaheuristics. In Section 4. the QAP and other NAPs are reviewed with respect to the concepts described. Section 5. considers nonlinear semi-assignment problems. For reason of clarity it should be noted that we do not intend to provide a very detailed description of all possible heuristics but to provide the reader with a rough review with a very comprehensive list of references. The collection of these references in its entirety is the main contribution of this paper. Finally, some ideas for future research are given.

2.   FINDING INITIAL FEASIBLE SOLUTIONS

In this section we survey heuristics to determine feasible solutions for NAPs. It should be noted that usually semi-assignment problems have a higher degree of freedom what might be the reason that a greater variety of ideas may be developed for those problems.

Once a bijective mapping is sought a solution of an NAP requires determining a permutation. To explain the concepts we may distinguish between assignment, partial assignment and complete assignment. That is, while an assignment may refer to combining a specific facility with a specific location and vice versa, a partial assignment refers to the combination of a subset of n facilities with an equal-sized subset of locations and vice versa. Correspondingly, in a complete assignment all facilities are uniquely (bijectively) combined with all locations.

Constructing initial feasible solutions for assignment problems can be accomplished, e.g., by starting with the identity permutation, a random permutation, or some heuristically determined starting solution. One example of the latter approach is to determine a permutation by a *priority rule based* or by a *cheapest insertion heuristic* motivated from TSP-like problems. Starting with $\pi = < i >$ (which constitutes a partial assignment) we successively build a complete assignment or permutation by choosing in each iteration $k = 2, \ldots, n$ the best combination under consideration of the remaining $n - k + 1$ facilities and all k insertion positions. That is, we have to evaluate the objective function increase for all these combinations at each iteration.

Building on this cheapest insertion heuristic, one may also use a corresponding *pilot method* (cf. [Duin and Voß, 1999]). The pilot method is a meta-heuristic that builds primarily on the idea to *look ahead* for each possible local choice (by computing a so-called "pilot"), memorizing the best result, and performing the according move. One may apply this strategy by successively performing a cheapest insertion heuristic for all possible local steps (i.e., starting with all incomplete solutions resulting from adding some not yet included facility at some position to the current incomplete solution). Usually, it is reasonable to restrict the pilot process to a given *evaluation depth*. That is, the pilot method is performed up to an incomplete solution (partial assignment) based on this evaluation depth and then completed by continuing with a conventional cheapest insertion heuristic.

In general we may define some sort of *heuristic measure* which is iteratively followed until a complete solution is build. Usually this heuristic measure is performed in a greedy fashion. To obtain different initial feasible solutions researchers also experiment with random elements, e.g., randomly choosing not necessarily the best but among some best elements according to the heuristic measure.

Additional strategies for determining feasible solutions resort to almost any heuristic procedure known in combinatorial optimization. Examples are regret based approaches (see, e.g., Section 5.1) or truncated branch and bound methods. The question which of these strategies should be recommended usually has to be decided by experimental computations.

3. IMPROVEMENT PROCEDURES

3.1 LOCAL SEARCH

For an excellent general survey on local search see the collection of Aarts and Lenstra [Aarts and Lenstra, 1997].

TSP-related, a permutation π of n objects can be represented as a chain of objects, each connected by an edge, with two dummy objects that represent a virtual fixed first (0) and last ($n+1$) object. With this, moves from a solution to a neighbor solution are composed of attributes defined by giving the edges deleted and inserted.

A 2-exchange move is defined for a pair (p_1, p_2) with $0 \leq p_1$ and $p_1 + 2 \leq p_2 \leq n$ as the deletion of the edges (π_{p_1}, π_{p_1+1}) and (π_{p_2}, π_{p_2+1}) and the inclusion of the edges (π_{p_1}, π_{p_2}) and $(\pi_{p_1+1}, \pi_{p_2+1})$. Such a move represents the inversion of the partial sequence between position $p_1 + 1$ and position p_2. Correspondingly, a 3-exchange neighborhood is defined by the replacement of three edges by three other edges such that a feasible permutation is obtained.

The idea for a shift move is to allow the shift of some object to another position. A shift move for the object at current position p_1 before the object currently at position p_2 with $1 \leq p_1 \leq n$, $1 \leq p_2 \leq n + 1$, $p_1 \neq p_2$, $p_1 + 1 \neq p_2$, may thus be defined as the deletion of the edges (π_{p_1-1}, π_{p_1}), (π_{p_1}, π_{p_1+1}), and (π_{p_2}, π_{p_2+1}) and the inclusion of the edges $(\pi_{p_1-1}, \pi_{p_1+1})$, (π_{p_2}, π_{p_1}) and (π_{p_1}, π_{p_2+1}).

Correspondingly, we may define a swap move for a pair (p_1, p_2) with $1 \leq p_1 < p_2 \leq n$ as the swap of the objects π_{p_1} and π_{p_2} while the remaining permutation remains identical. Thus, a swap move corresponds to the deletion of the edges (π_{p_1-1}, π_{p_1}), (π_{p_1}, π_{p_1+1}), (π_{p_2-1}, π_{p_2}) and (π_{p_2}, π_{p_2+1}) and the inclusion of the edges (π_{p_1-1}, π_{p_2}), (π_{p_2}, π_{p_1+1}), (π_{p_2-1}, π_{p_1}) and (π_{p_1}, π_{p_2+1}). Of course, a swap move may also be performed by the execution of two respective shift moves.

In the sense of *improvement* we may compare solutions before and after a transformation and evaluate the better one based on an objective function. Then, a neighborhood search can look for the best possible improvement, i.e., that move leading to the neighbor with the best objective function value among all neighbors. We call this search *best-fit* or steepest descent. If one is content with the first neighbor providing some improvement, we refer to *first-fit*. The latter approach may be performed in a cyclic manner. That is, whenever an improvement is found it is accepted and the next iteration starts where this improved neighboring solution is found. With respect to the QAP this approach is attributed to Heider (see, e.g., [Çela, 1998]).

3.2   GRASP

Replicating a search procedure to determine a local optimum multiple times with different starting points has been acronymed as *greedy randomized adaptive search* (GRASP) and investigated with respect to different applications, e.g., by [Feo et al., 1991]. In GRASP the different initial solutions or starting points are found by a greedy procedure that incorporates a probabilistic component. Given a list of candidates to choose from, GRASP randomly chooses one of the best candidates from this list in a greedy manner, but not necessarily the best choice possible. The underlying principle is to investigate many good starting points through the greedy procedure and thereby to increase the possibility of finding a global optimum or at least a good local optimum on at least one replication.

3.3   SIMULATED ANNEALING

Simulated annealing extends basic local search by allowing moves to inferior solutions [Kirkpatrick et al., 1983, Dowsland, 1993]. The basic algorithm of simulated annealing may be described as follows: Successively, a candidate move is randomly selected; this move is accepted if it leads to a solution with a better objective function value than the current solution, otherwise the move is accepted with a probability that depends on the deterioration Δ of the objective function value. The probability of acceptance is computed according to the Boltzmann function as $e^{-\Delta/T}$, using a temperature T as control parameter.

Various authors describe a robust concretization of this general simulated annealing procedure. Following [Johnson et al., 1989] the value of T is initially high, which allows many inferior moves to be accepted, and is gradually reduced through multiplication by a parameter *coolingFactor* according to a geometric cooling schedule. At each temperature *sizeFactor* $\times |N|$ move candidates are tested ($|N|$ denotes the current neighbourhood size). The starting temperature is determined as follows: Given a parameter *initialAcceptanceFraction* and based on an abbreviated trial run, the starting temperature is set so that the fraction of accepted moves is approximately *initialAcceptanceFraction*. A further parameter, *frozenAcceptanceFraction* is used to decide whether the annealing process is *frozen* and should be terminated. Every time a temperature is completed with less than *frozenAcceptanceFraction* of the candidate moves accepted, a counter is increased by one. This counter is reset every time a new best solution is found. The procedure is terminated when the counter reaches 5.

Threshold accepting is a modification (or simplification) of simulated annealing with the essential difference between the two methods being the acceptance rules. Threshold accepting accepts every move that leads to a new solution which is 'not much worse' than the older one. The principle idea of threshold accepting was first proposed by [Dueck and Scheuer, 1990].

3.4   TABU SEARCH

The basic paradigm of *tabu search* is to use information about the search history to guide local search approaches to overcome local optimality [Glover and Laguna, 1997]. In general, this is done by a dynamic transformation of the local neighborhood. As for simulated annealing, this may lead to performing deteriorating moves when all improving moves of the current neighborhood are set tabu. For example, at each iteration a best admissible neighbor may be selected. A neighbor, respectively a corresponding move, is called admissible, if it is not tabu or if an aspiration criterion is fulfilled. Below, we briefly describe various tabu search methods that differ especially in the way in which the tabu criteria are defined, taking into consideration the information about the search history (performed moves, traversed solutions). An aspiration criterion may overwrite a possibly unreasonable tabu status of a move. For example, a move that leads to a neighbor with a better objective function value than encountered so far should be considered as admissible.

The most commonly used tabu search method is based on a *recency-based* memory that stores moves, more exactly move attributes, of the recent past (static tabu search, tabu navigation method, TNM). The basic idea of such static tabu search approaches is to prohibit an appropriately defined inversion of performed moves for a given period. For example, one may store the solution attributes that have been created by a performed move in a tabu list. To obtain the current tabu status of a move to a neighbor, one may check whether (or how many of) the solution attributes that would be destroyed by this move are contained in the tabu list.

Strict tabu search embodies the idea of preventing cycling to formerly traversed solutions. That is, the goal of strict tabu search is to provide necessity and sufficiency with respect to the idea of not revisiting any solution. Accordingly, a move is classified as tabu if and only if it leads to a neighbor that has already been visited during the previous part of the search. There are two primary mechanisms to accomplish the tabu criterion: First, we may exploit logical interdependencies between the sequence of moves performed throughout the search process, as realized by the reverse elimination method (REM) and the cancellation sequence method (CSM) (cf., e.g., [Glover and Laguna, 1997, Voß, 1993, Voß, 1996]). Second, we may store information about all solutions visited so far. This may be carried out either exactly or, for reasons of efficiency, approximately (e.g., by using hash codes).

Reactive tabu search aims at the automatic adaptation of the tabu list length of static tabu search [Battiti, 1996]. The basic idea is to increase the tabu list length when the tabu memory indicates that the search is revisiting formerly traversed solutions. A possible concrete algorithm can be described as follows: One starts with a tabu list length l of 1 and increases it to $\min\{\max\{l+1, l \times 1.1\}, u\}$ every time a solution has been repeated, taking into account an appropriate upper bound u (to guarantee at least one admissible move). If there has been no repetition for some iterations, we decrease it appropriately to $\max\{\min\{l - 1, l \times 0.9\}, 1\}$. To accomplish the detection of a repetition of a solution, one may apply a trajectory based memory using hash codes as described for strict tabu search.

As noted above, and especially for reactive tabu search, noticed by [Battiti, 1996], it may be appropriate to include means for diversifying moves whenever the tabu memory indicates that we may be trapped in a certain region of the search space. As a corresponding trigger mechanism, we may use the combination of at least two solutions each having been traversed three times. A very simple escape strategy is to perform randomly a number of moves (in dependence on the moving average of the number of iterations between solution repetitions); more advanced strategies may take into account some long-term memory information (like the frequencies of specific solution attributes in the search history).

Of course there is a great variety of additional ingredients that may make tabu search work successful as, e.g., restricting the number of neighbor solutions to be evaluated (so-called candidate list strategies).

3.5   GENETIC ALGORITHMS

Genetic algorithms are a class of adaptive search procedures based on the principles derived from the dynamics of natural population genetics (see, e.g., [Holland, 1975]).

One of the most crucial ideas for a successful implementation of a genetic algorithm is the representation of an underlying problem by a suitable scheme.

A genetic algorithm starts (for example) with a randomly created initial population of artificial creatures (strings), found, for example, by flipping a 'fair' coin. These strings in whole and in part are the base set for all subsequent populations. They are copied and information is exchanged between the strings in order to find new solutions of the underlying problem. The mechanisms of a simple GA essentially consist of copying strings and exchanging partial strings. A simple GA requires three operators which are named according to the corresponding biological mechanisms: reproduction, crossover, and mutation.

Performing an operator may depend on a so-called *fitness function* or its value (fitness), respectively. This function defines a means of measurement for the profit or the quality of the coded solution for the underlying problem and may depend on the objective function of the given problem.

The QAP as a permutation type problem needs a careful restructuring mechanism when crossover is applied (see, e.g., [Goldberg, 1989]). Here we present the partially matched crossover (PMX) operator and the CX operator.

Contrary to a simple crossover, the PMX operator works with two crossover positions which define a crossover area. Through PMX information is not only exchanged but the strings are also restructured. The best way to understand the mechanism of PMX is to regard an example of two integer strings (permutations):

$$A=4 \quad 3 \quad 1 \mid 9 \quad 6 \quad 10 \mid 5 \quad 2 \quad 8 \quad 7$$
$$B=5 \quad 9 \quad 10 \mid 2 \quad 4 \quad 1 \mid 7 \quad 3 \quad 6 \quad 8$$

The crossover area is marked through the vertical lines. If a simple crossover would be performed in this area, we would obtain two infeasible strings. String A would contain double identifiers 2, 4, and 1 whereas the identifiers 9, 6, and 10 would be missing. To avoid this problem the two strings are restructured. Looking at the first identifier in the crossover area of string A, we find 9. At the same position on string B we find identifier 2. Identifier 2 is now searched on string A and both are exchanged. This process is performed for each identifier in the crossover area of string A. String B is treated analogously. The result is shown as follows:

$$A'=6 \quad 3 \quad 10 \mid 2 \quad 4 \quad 1 \mid 5 \quad 9 \quad 8 \quad 7$$
$$B'=5 \quad 2 \quad 1 \mid 9 \quad 6 \quad 10 \mid 7 \quad 3 \quad 4 \quad 8$$

Next we refer to the presentation of the *CX operator*. The easiest way to show how it works is to use the same example of two integer strings A and B as above:

$$A=4 \quad 3 \quad 1 \quad 9 \quad 6 \quad 10 \quad 5 \quad 2 \quad 8 \quad 7$$
$$B=5 \quad 9 \quad 10 \quad 2 \quad 4 \quad 1 \quad 7 \quad 3 \quad 6 \quad 8$$

Beginning with string A, an object is randomly chosen, say object 9. The object on string B with the same position as the regarded object on string A (the 2) is fixed on string A. Proceeding with 2 on string A, object 3 is the next one which has to be fixed. Now the procedure stops because the object on string B corresponding to the 3 is object 9 and so a cycle

$$(\quad - \quad 3 \quad - \quad 9 \quad - \quad - \quad - \quad 2 \quad - \quad - \quad)$$

has been built. In the next step all empty positions on string A are filled with the corresponding objects of string B and vice versa. In our example the following strings are obtained after applying CX:

$$A = 5 \quad 3 \quad 10 \quad 9 \quad 4 \quad 1 \quad 7 \quad 2 \quad 6 \quad 8$$
$$B = 4 \quad 9 \quad 1 \quad 2 \quad 6 \quad 10 \quad 5 \quad 3 \quad 8 \quad 7$$

Genetic algorithms are closely related to *evolutionary strategies*. Whereas the mutation operator in a genetic algorithm serves to protect the search from premature loss of information, evolution strategies may incorporate some sort of local search procedure (a hill climbing strategy like steepest descent or steepest ascent) with self adapting parameters involved in the procedure.

Recently it appeared that so-called *scatter search* ideas may establish a link between early ideas from various sides – evolutionary strategies, tabu search and genetic algorithms (see, e.g., [Glover, 1995]). Scatter search is designed to operate on a set of points, called reference points, that constitute good solutions obtained from previous solution efforts. The approach systematically generates linear combinations of the reference points to create new points, each of which is mapped into an associated point that yields integer values for discrete variables.

3.6   ANT SYSTEMS

One of the recently explored concepts within intelligent search is the *ant system*, a dynamic optimization process reflecting the natural interaction between ants searching for food (see, e.g., [Taillard, 2000] for a short introduction with connections to the QAP). The ants' ways are influenced by two different kinds of search criteria. The first one is the local visibility of food, i.e., the attractiveness of food in each ant's neighborhood. Additionally, each ant's way through its food space is affected by the other ants' trails as indicators for possibly good directions. The intensity of trails itself is time-dependent: With time going by, parts of the trails ´are gone with the wind´, meanwhile the intensity may increase by new and fresh trails. With the quantities of these trails changing dynamically an autocatalytic optimization process is started forcing the ants' search into most promising regions. This process of interactive learning can easily be modeled for most kinds of optimization problems by using simultaneously and interactively processed search trajectories.

4.   NONLINEAR ASSIGNMENT PROBLEMS

4.1   THE QUADRATIC ASSIGNMENT PROBLEM

The QAP was already introduced in Section 1.. It has a large variety of applications mainly in location and layout planning and in machine scheduling (see, e.g., [Domschke and Krispin, 1997, Heragu, 1997] for recent surveys on layout planning and [Çela, 1998, Burkard et al., 1998, Pardalos and Wolkowicz, 1994] specifically dealing with the QAP).

For the QAP a large number of construction as well as improvement procedures have been proposed in the literature. Numerical results for these methods are usually described for a number of benchmark problems compiled by [Burkard et al., 1991, Burkard et al., 1997]. They describe a collection of known problem instances with sizes $n \geq 8$. Most of these instances are taken from the literature while others were generated by some researchers for their own testing purposes. When referring to numerical results below, we assume this library although a considerable number of studies has developed an independent stream of testing and results. One of the ideas in this respect deserving to be mentioned are problem instances with known optimal solutions (see, e.g., [Li and Pardalos, 1992]).

4.1.1   Finding Initial Feasible Solutions. The development of ideas for construction heuristics for the QAP dates back to Gilmore [Gilmore, 1962]. Construction heuristics usually perform n iterations and in each iteration exactly one object is assigned to one location (see the cheapest insertion idea above; of course the corresponding method may be modified to consider two objects together, cf. [Block, 1978, Edwards et al., 1970, Land, 1963, Müller-Merbach, 1970]). If the classification needs to be more specific then we may distinguish methods that consider the objects and their interaction while building a sequence (e.g. [Gilmore, 1962]) and those that explicitly alternate between choosing a specific object and choosing a specific location this object is assigned to (e.g. CORELAP: COmputerized RElationship Layout Planning from [Lee and Moore, 1967] and modifications [Seehof and Evans, 1967]). Finally these steps can be performed simultaneously; a classification of rules can be found, e.g., in [Domschke and Drexl, 1996].

For additional construction heuristics based on priority rules or heuristic measure see, e.g., [Burkard and Derigs, 1980, Müller-Merbach, 1970]. Devising simple look ahead features within construction heuristics for the QAP may be done following the regret mechanism (see, e.g., the ideas in [Hillier and Conners, 1966, Gilmore, 1962]).

Furthermore, exact algorithms may be used in a truncated version (sometimes also called hybridized or *limited enumeration methods*) to determine good initial feasible solutions that are considered for further improvement (see, e.g., [Bazaraa and Kirca, 1983, Bazaraa and Sherali, 1980, Burkard and Bönniger, 1983, Burkard and Stratmann, 1978]). For instance, based on the Benders' decomposition principle one may solve a sequence of transportation problems to optimality together with some cuts to derive good results for the QAP, however, with considerable memory requirements [Bazaraa and Sherali, 1980]. Additional ideas for limited enumeration methods can be found, e.g., in [Graves and Whinston, 1970]. Applying general non-convex programming ideas is performed by [Torki et al., 1996].

4.1.2   Local Search. Among the improvement procedures for the QAP the most widely known is CRAFT (Computerized Relative Allocation of Facilities Technique) with its development dating back to the early 1960´s [Buffa et al., 1964]. Its principle is to start with a feasible solution and to try to improve it by successive interchanges of single assignments as long as improvements in the objective function are obtained (cf. the swap neighbourhood described above). Early references on variants of the CRAFT algorithm can be found in [Hillier and Conners, 1966, Nugent et al., 1968, Vollmann et al., 1968]. CRAFT is still relevant as being incorporated in many practical facility layout software tools. Furthermore, its results may be quite competitive (see, e.g., [Kusiak and Heragu, 1987]).

Necessarily, local search approaches need initial feasible solutions. When dealing with simple heuristics, one important question is whether it is worth the effort of finding good initial feasible solutions based on the fact that most local search approaches do not seem to be very sensitive with respect to the initial feasible solution [Bruijs, 1984]. Nevertheless, a great variety of construction methods is combined and tested in combination with local search approaches (see, e.g., [Elshafei, 1977, Lashari and Jaisingh, 1980, Murtagh et al., 1982]).

Combining simple look ahead features from limited enumeration methods with local search may yield good results, too [West, 1983].

Modern heuristic search concepts have been applied to the QAP in an almost uncountable number of studies. Most important in most of these approaches is the idea of local search. Considering formulation (8.6) - (8.9), a transition from one feasible solution to another one needs two exchanges within the binary matrix $(x_{ip})_{n \times n}$ such that the assignment conditions remain valid. Accordingly, moves may be visualized as paired-attribute moves, in which attributes denote both the assignments and the type of exchange, i.e., selection for or exclusion from the (actual) solution. More specifically, a move may occur by swapping or exchanging two objects such that each is placed on the location formerly occupied by the other. Given a permutation π of $(1, \ldots, n)$, swapping objects i and j results in a permutation π', i.e., $\pi'(i) := \pi(j)$, $\pi'(j) := \pi(i)$, $\pi'(k) := \pi(k)$ for all $k \neq i, j$. With respect to the binary variables, the move corresponds to setting $x_{ip} := 0$, $x_{jq} := 0$ and $x_{jp} := 1$, $x_{iq} := 1$ indicating that object i is moved from position p to position q and object j from position q to position p, correspondingly. In short, a move may be expressed by (iq, jp).

Most local search methods customarily select best admissible moves (best-fit), and a value $val_{\pi(i,j)}$ for swapping i and j of a given permutation π has to be examined for each move (resulting in permutation π'). For the QAP with symmetric matrices and zero diagonals this change in cost may easily be calculated as:

$$val_{\pi(i,j)} \;:=\; \sum_{h=1}^{n}\sum_{k=1}^{n}\left(c_{hk}d_{\pi(h)\pi(k)} - c_{hk}d_{\pi'(h)\pi'(k)}\right)$$

$$= \; 2\cdot\sum_{k\neq i,j}\left(c_{jk} - c_{ik}\right)\cdot\left(d_{\pi(j)\pi(k)} - d_{\pi(i)\pi(k)}\right)$$

Further improvements in the effort of calculating move values may be obtained by storing all $val_{\pi(i,j)}$ values and updating them in constant time. Therefore, the overall effort for evaluating the neighborhood of a given solution is $O(n^2)$. For similar ideas see also [Frieze et al., 1989].

Further ideas for developing neighborhoods may use, e.g., cyclic 3-optimal exchanges. Once the number of exchanges is limited appropriately, the $O(n^2)$ effort may still hold. For another neighborhood definition see [Murthy et al., 1992]. If the swap neighborhood is extended from two exchanges within the binary matrix $(x_{ip})_{n\times n}$ to three or four but the overall number of steps is kept the same it might be questionable whether the swap neighborhood can be outperformed [Cheh et al., 1991].

4.1.3   Meta-Heuristics.

The main drawback of algorithms like CRAFT is their inability to continue the search upon becoming trapped in a local optimum. This suggests consideration of modern meta-heuristics. As the QAP belongs to the most widely known combinatorial optimization problems it has served as a vehicle for the application of almost any meta-heuristic. Table 8.1 provides a list of references for the application of various meta-heuristics for the QAP[1].

A careful comparison of different meta-heuristics often depends on the understanding and implementation of the method. Therefore, it might be helpful to compare different approaches within a common framework (see, e.g., the heuristic optimization framework HOTFRAME of [Fink and Voß, 1999, Fink and Voß, 1998]). For the QAP a corresponding (yet not fully comprehensive) comparison is undertaken under the Algodesk system [Maniezzo et al., 1995] which has been implemented to provide given QAP data for different algorithms and let them run under identical surroundings (time based termination criterion of one hour, selected data from QAPLIB with $n \leq 30$, the same PC). The authors tested the following methods: a multi-greedy approach (random restart), simulated annealing and a slight variation based on the Boltzmann machine model [Chakrapani and Skorin-Kapov, 1992], tabu search (following a random tabu list length variation from [Taillard, 1991]), a genetic algorithm (incorporating the PMX operator described above), different evolution strategies, sampling and clustering [Boender et al., 1982], and an approach called immune network algorithm [Bersini and Varela, 1991]. To summarize the results of this study it is notable that on the limited testbed considered no method was able to really outperform a multi-greedy or random restart method. The population

Approach	References
Ant System	[Colorni and Maniezzo, 1999, Dorigo et al., 1999, Gambardella et al., 1999] [Maniezzo, 1999, Stützle and Hoos, 1999, Taillard, 1998, Taillard, 2000]
Genetic Algorithms	[Ahuja et al., 1995, Brown et al., 1989, Derigs et al., 1999] [Huntley and Brown, 1991, Mühlenbein, 1989, Nissen, 1993] [Nissen, 1994, Tate and Smith, 1995]
GRASP	[Li et al., 1994, Pardalos et al., 1995, Pardalos et al., 1997] [Resende et al., 1996]
Scatter Search	[Cung et al., 1996]
Simulated Annealing	[Bölte, 1994, Burkard and Rendl, 1984, Cheh et al., 1991] [Connolly, 1990, de Abreu et al., 1999, Golden and Skiscim, 1986] [Heragu and Alfa, 1992, Jajodia et al., 1992, Laursen, 1993] [Lutton and Bonomi, 1986, Paulli, 1993, Peng et al., 1996] [Wilhelm and Ward, 1987]
Tabu Search	[Battiti and Tecchiolli, 1992, Battiti and Tecchiolli, 1994a] [Chakrapani and Skorin-Kapov, 1992, Chakrapani and Skorin-Kapov, 1993] [Chiang and Chiang, 1998, Fiechter et al., 1992] [Paulli, 1993, Skorin-Kapov, 1990, Skorin-Kapov, 1994] [Sondergeld and Voß, 1993, Taillard, 1991, Talbi et al., 1999, Voß, 1995]
Threshold Accepting	[Nissen and Paul, 1993]

Table 8.1   Meta-heuristics for the QAP

based approaches did not behave as good as other strategies like random restart, tabu search, simulated annealing, and sampling and clustering. Of course we have to admit that different researchers even come to different conclusions when trying to compare several methods (see, e.g., the discussion of [Paulli, 1993] and [Battiti and Tecchiolli, 1994b]). A comparison of different improvement procedures can also be found in [Sinclair, 1993, Taillard, 1995].

The meta-heuristic approaches that have been applied in a very large number of studies with great success to the QAP are simulated annealing and tabu search. Here we focus more on the latter.

Tabu search as a metastrategy for guiding a heuristic, such as CRAFT, heavily depends on the underlying tabu list management. Most approaches in the literature are using the TNM and simple variations. One of the first papers in this respect is [Skorin-Kapov, 1990]. The same approach is also followed in [Heragu, 1997]. Taillard [Taillard, 1991] performs a random modification of the tabu list length with considerably more success. With respect to the solution quality over time as a measure, however, the most efficient tabu search method seems to be the reactive tabu search approach [Battiti, 1996, Battiti and Tecchiolli, 1992, Battiti and Tecchiolli, 1994a].

In the mid-nineties it became some sort of sport to "win the horserace", i.e., to provide new best solutions for some of the benchmark instances now available in QAPLIB. Two successful tabu search approaches of that time not yet mentioned are [Fleurent and Ferland, 1994, Voß, 1995] who independently presented new best results for some of those data with $n = 100$. The first deserves special mention as it very successfully combines tabu search and genetic algorithms into a hybrid approach. It should be noted that various other hybrid methods may be developed successfully. For instance, short and long term memory considerations in tabu search may be combined with a simulated annealing approach being applied as a means for an intensification phase [Talbi et al., 1999].

In contrast to the previously mentioned references, [Sondergeld and Voß, 1993, Voß, 1995] investigate methods that dynamically exploit logical inter-dependencies between attributes used to determine tabu status, i.e., CSM and REM. Furthermore, a clustering approach is proposed in [Voß, 1995] that helps to reduce the computational effort within the procedure. As the latter clustering approach may be of interest in other applications, too, and also refers to the interplay of intensification and diversification within tabu search as well as some of the ideas in scatter search (e.g. elite solutions, cf. [Glover, 1995, Glover and Laguna, 1997]) we describe it in more detail. Some of the ideas may also be related to recent restart concepts of [Fleurent and Glover, 1999]. In a certain sense GRASP may be related to the latter extreme version of a one-sided strategic oscillation approach (temporarily overcoming the limitations of feasibility in heuristic search).

Whenever search intensification is performed based on a promising solution the idea is to perform moves that keep the permutations within the near vicinity of the corresponding solution. If search intensification is performed several times throughout the search it may be advantageous for the starting solutions of the intensification phase to be different from each other. In order to diversify the different phases of search intensification, some solutions that have been visited within the neighborhood search may be stored as possible candidates for a restart in future iterations. To obtain reasonable differences in new starting solutions, we proposed a clustering approach.

For any two solutions π and π' we may define some measure of similarity $\Delta(\pi, \pi')$. Whenever $\Delta(\pi, \pi')$ is less than or equal to a given threshold value, these solutions will be considered to belong to the same class of solutions, and otherwise will be considered to belong to different classes. Any time the search is restarted, a starting solution is selected from a class that has not previously been used for intensification purposes, eliminating the chosen solution from the respective class.

Consider the following measure of similarity between two permutations π and π', based on the hamming metric:

$$\Delta(\pi, \pi') := \sum_{i=1}^{n} \delta_i \qquad (8.10)$$

where $\delta_i := 1$ if $\pi(i) \neq \pi'(i)$ and $\delta_i := 0$ otherwise. To initialize these classes each class C consists of at most c_{max} elements and the number of classes is bounded by a number tc. Each nonempty class is represented by a permutation π_C with the best objective function value among all elements of that class. Such a solution may be referred to as an elite solution. Any solution π of the search is a candidate for being included into one of the solution classes.

A simple diversification strategy different from ours has been proposed in [Kelly et al., 1994] as follows. Among all moves that reduce the coincidence between the current solution and the most recent local optimum the most improving or least deteriorating one is chosen (which corresponds to the idea of maximizing the hamming distance between two solutions while following some second criterion).

To end our discussion of the tabu search approaches for the QAP it should be noted that a considerable number of those studies have been performed in parallel environments [Chakrapani and Skorin-Kapov, 1992, Chakrapani and Skorin-Kapov, 1993, Taillard, 1991, Talbi et al., 1999].

With respect to the *ant system* idea a large number of authors have provided numerical results for the QAP (see Table 8.1 and [Taillard, 2000]). Usually these approaches memorize solutions which are generated throughout the search in an $n \times n$ matrix indicating how often a facility was placed on a specific location, i.e., giving a frequency based memory. Some interesting features of the various ant system approaches are as follows. To some extent the different ideas are defining various starting solutions in a constructive way combined with some local search mechanism [Stützle and Hoos, 1999, Taillard, 1998] eventually incorporating some means of randomness, e.g., by a probabilistic local search [Gambardella et al., 1999]. [Maniezzo, 1999] even uses a limited enumeration approach for this purpose. Similar to, e.g., the elite strategy in [Voß, 1995] the approaches also incorporate some intelligent way to change between intensification and diversification phases (in the spirit of an adaptive memory programming concept). Most importantly, however, is the way how different processes are distinguished. An ant process may be defined which receives problem data, a memory state and perhaps other parameters. Incorporating random elements a new solution is determined and resubmitted to the second and main process called queen process. The queen process initializes the memory and manages all ant processes by providing parameter and receiving and distributing solutions from and to the ant processes in a cooperative way. The results reported in the latest ant system papers are stimulating in two ways. First, with respect to the QAP they are competitive to other approaches discussed earlier and second they show a very successful way meta-heuristics might take in general (independent from nonlinear assignment problems) in the sense that ingredients from different meta-heuristics will be merged together leading to more successful solution approaches in the future.

We know that our description is by no means complete and while we have concentrated more on the description of tabu search approaches and less on others we admit that a great variety of additional concepts have been developed over time and applied successfully to the QAP. Examples are scatter search [Cung et al., 1996] or so-called manifold search techniques [Gavish, 1991] which elaborate different candidate directions of a search procedure such as the TNM within tabu search. Also genetic algorithms have not been studied to a full extent besides summarizing some of the relevant references. In this respect it may be of interest to relate the early biased sampling idea of Nugent et al. [Nugent et al., 1968] to some evolutionary programming ideas.

Of interest are also hybrid methods. Besides the examples mentioned above this also circumvents, e.g., the use of genetic algorithms for improving the parameter settings within simulated annealing [Bölte, 1994]. On the other side simulated annealing may be used to improve solutions within a genetic algorithm [Huntley and Brown, 1991]. Finally, we should mention the possibility to apply ideas from *neural networks* to solve the QAP and related facility layout problems (see, e.g., [Tsuchiya et al., 1996]), although we have not yet seen any neural network results that are able to outperform some of the previously presented meta-heuristics. For a survey on neural networks and a somewhat different view see, e.g., [Smith, 1999, Smith et al., 1998]. Whether general problem solvers in this respect (see e.g. the neural network genetic algorithm hybrid approach of [Gen et al., 1999]) may be helpful needs further consideration.

4.1.4   Some Theoretical Results.   For the QAP even the problem of finding a feasible solution which is guaranteed to approximate the optimal objective function value by some $\epsilon > 0$ is $\mathcal{NP}$-hard [Sahni and Gonzalez, 1976]. This result may be modified even for special classes of QAP problem instances (e.g., when the distance matrix satisfies the triangle inequality [Queyranne, 1986]).

For the QAP we also find theoretical investigations on some heuristics described above as well as on the structure of problem instances. For instance, it may be shown that under certain assumptions, as a QAP instance grows large, the relative error of a heuristic algorithm grows small [Burkard and Fincke, 1983]. The results as well as the proof techniques regarding this type of convergence have been improved over time [Dyer et al., 1986, Frenk et al., 1985, Rhee, 1988a] and also applied to different objective functions [Rhee, 1988b]. Experimentally this was confirmed in some studies as, e.g., [Bonomi and Lutton, 1986]. One of the consequences of these results refers to the way randomly generated problem instances are considered in the literature as they tend to be easily solvable by almost any heuristic. (Some of the QAPLIB instances are of this type.)

With respect to complexity theory a great amount of research has been undertaken to define problem classes that are easy to solve or which seem intractable. Some authors also pose the question whether for a given problem determining locally optimal solutions is possible in polynomial time (introducing the notion of PLS-completeness [Johnson et al., 1988]). In this respect it may be of interest to investigate how good locally optimal solutions are at an average to determine whether it is worth to search for a local optimum once this cannot be assured in polynomial time.

For the QAP with the 2-exchange neighborhood it can be seen from its special case, the graph bipartitioning problem [Schäffer and Yannakakis, 1991], that it is PLS-complete (for additional discussion see also [Pardalos et al., 1994]). Based on this observation it may be shown for the QAP that the cost of any locally optimal solution with respect to the 2-exchange neighborhood is at least as small as the average cost over all possible solutions/permutations [Angel and Zissimopoulos, 1998].

4.1.5   Some Applications.

The problem of *balancing mechanical parts* refers to n mechanical parts of weight w_i $(i = 1, \ldots, n)$ that have to be fixed on a support (see, e.g., [Taillard and Voß, 1999]). Each part may bijectively occupy any of n positions on the support, given by their co-ordinates $(x_i, y_i, \ldots)$ $(i = 1, \ldots, n)$, where typically the problem is considered in up to three dimensions $(x, y, \ldots)$. The goal is to position each mechanical part on the support in such a way that the center of gravity of all the parts is as close as possible to an ideal center of gravity, say $(0, \ldots, 0)$. Mathematically, the objective is to find a permutation π of n elements, that minimizes the function $f(\pi)$:

$$min_\pi f(\pi) = \frac{1}{\sum_{i=1}^n w_i} \cdot \left\| \left(\sum_{i=1}^n w_i \cdot x_{\pi_i}, \sum_{i=1}^n w_i \cdot y_{\pi_i}, \ldots \right) \right\|$$

The balancing problem is of major practical importance. Typical applications are the balancing of rotating parts such as turbine rotors. A turbine rotor is composed of an axis around which a number of (almost) identical blades are fixed. Due to mechanical, thermic and chemical constraints, the blades are made in a special alloy that is difficult to manufacture. The result is that the blades are not identical. The weight difference of the blades induces vibrations in the turbine that strongly depend on the position of the center of mass of the blades. The more the center of mass is away from the center of the axis of the turbine, the higher the vibrations are. Therefore, it is very important to find an arrangement of the blades around the axis in such a way that the center of gravity is as close to the rotation center of the rotor as possible. Another practical application of the problem is the balancing of the load of boats, airplanes or lorries.

In the literature [Laporte and Mercure, 1988, Sinclair, 1993, Sondergeld and Voß, 1996, White, 1996, Mason and Rönnqvist, 1997, Pitsoulis et al., 2000, Choi et al., 1999], the balancing problem is commonly explored as a special case of the QAP. The problem has been heuristically solved by several authors [Fathi and Ginjupalli, 1993, Laporte and Mercure, 1988, Sinclair, 1993]. For the description of a well-known greedy heuristic assume, without loss of generality, that the weights $w_i (i = 1, ..., n)$ are ordered according to their respective weights and that the heaviest blade is placed into location 1. In each of the remaining $n - 1$ iterations, say iteration k, the k-th blade is placed into that not yet occupied location l^* such that

$$\sum_{i=1}^{k-1} w_i \cos(2\pi(l^* - loc(i))/n) \qquad (8.11)$$

is minimum among all still available locations.

Contrary to this basic greedy heuristic where at each iteration k the k-th blade is permanently assigned to its final location, two extensions involve a look ahead mechanism by temporarily placing the k-th blade and further exploring the placement of the remaining $n - k$ blades (compare the idea of the pilot method). In iteration k the k-th blade is successively placed in each of the not yet occupied locations, whereas the simple greedy procedure determines the placement of the remaining blades. The assignment of the k-th blade to a location yielding a configuration with a minimum deviation value for all temporary placements is taken as permanent. An additional and more complex look ahead extension of the basic greedy heuristic determines the placement of the remaining blades in iteration k by using the first look ahead procedure.

We have applied simple heuristic solution schemes including different tabu search approaches to the turbine runner balancing problem [Sondergeld and Voß, 1996]. For some benchmark problem instances taken from the literature we had provided better solutions than previously reported. The main conclusion with respect to the tabu search approach is that for the REM two simple and easily implemented methods for scattering the search trajectories over the solution space - elongation of tabu tenure and using distance measures - prove to be more effective than a more complex and even more time-consuming approach. It should be noted that improved results have been obtained with the Popmusic (partial optimization meta-heuristic under special intensification conditions) meta-heuristic of Taillard and Voß [Taillard and Voß, 1999] based on the above mentioned original modeling approach.

Optimizing gate assignments at airport terminals for arriving aircrafts is another interesting application related to the QAP where the objective is to minimize walking distances of arriving passengers (who eventually change planes). Of course the problem is dynamic (and therefore a generalization of the QAP) but even the non-dynamic case has received considerable attention (see, e.g., [Haghani and Chen, 1998] and the references therein). For the present study a versatile priority rule based heuristic is proposed [Haghani and Chen, 1998].

For the QAP there is a great variety of additional applications some of which have even received considerable attention with respect to the development of specialized heuristics (see, e.g., the references in [Burkard et al., 1998, Domschke and Drexl, 1996] as well as [Bartolomei-Suarez and Egbelu, 2000] as just one example among a dime a dozen).

Genetic algorithms have also been applied to a version of the *QAP with associated area requirements* by [Tam, 1992].

4.2   THE MULTI-INDEX ASSIGNMENT PROBLEM

In his seminal work on the QAP, Lawler [Lawler, 1963] showed a linearization technique which is based on using an additional set of variables y_{ijp}. Based on these ideas it comes as a natural extension to consider three-index and multi-index assignment problems in the context of NAPs. (For an annotated bibliography on the QAP and three-index assignment problems see [Burkard and Çela, 1998].)

From a motivational point of view the problem refers to simultaneously assigning, e.g., objects, workers and locations. Formally, the *three-dimensional* or *three-index assignment problem* may be modeled by using binary variables y_{ijp} which take the value 1 if object i is assigned to worker j and location p and take the value 0 otherwise:

$$\text{Minimize } Z(x) = \sum_{i=1}^{n}\sum_{j=1}^{n}\sum_{p=1}^{n} c_{ijp} y_{ijp} \qquad (8.12)$$

subject to

$$\sum_{i=1}^{n}\sum_{j=1}^{n} y_{ijp} = 1 \qquad p = 1,\ldots,n \qquad (8.13)$$

$$\sum_{i=1}^{n}\sum_{p=1}^{n} y_{ijp} = 1 \qquad j = 1,\ldots,n \qquad (8.14)$$

$$\sum_{j=1}^{n}\sum_{p=1}^{n} y_{ijp} = 1 \qquad i = 1,\ldots,n \qquad (8.15)$$

$$y_{ijp} \in \{0,1\} \qquad i,j,p = 1,\ldots,n \qquad (8.16)$$

This problem requires each plane of the cube (y_{ijp}) to have exactly one non-vanishing element. It is $\mathcal{NP}$-hard, and obviously becomes not easier once the dimension is enlarged to yield the multi-index case.

Heuristics for three-index assignment problems can be found, e.g., as part of branch and bound procedures [Magos and Miliotis, 1994] including also improvement procedures. A simple static tabu search approach incorporating frequency based memory ideas has been tested on problem instances of size up to 14. The most interesting feature of this approach is the neighborhood definition based on the set of latin squares [Magos, 1996]. For an early simulated annealing implementation see [Burkard and Rendl, 1984]. Additional branch and bound approaches that may serve as good limited enumeration methods (see Section 4.1.1) are given in [Frieze and Yadegar, 1981, Balas and Saltzman, 1991].

Recently an interesting heuristic has been proposed to solve the three-dimensional assignment problem as a dynamical system by means of coupled selection equations [Starke et al., 1999]. The system is derived as equations of motion penalizing deviations from feasibility similar to the idea of defining energy functions in some neural network approaches. Limited results are presented but the approach needs more detailed testing in comparison to other heuristics.

While for other NAPs polynomial approximation schemes cannot be given there are special cost structures for which this is possible for multi-dimensional assignment problems [Bandelt et al., 1994, Crama and Spieksma, 1992].

Some ideas for heuristics for the multidimensional case can be found in [Poore and Rijavec, 1994]. Meta-heuristics are applied to these problems, too (e.g., GRASP [Murphey et al., 1998]).

One of the special cases of the three-index assignment problem which has been defined in the literature is the *cumulative assignment problem* [Dell´Amico et al., 1999]. Formally the problem is a special case of the three-dimensional case by replacing the objective function coefficients of $C = (c_{ijp})$ in (8.12) by $b_p \times c_{ij}$ for given values b_p. The problem asks for an assignment of each object to a location and for an ordering of the elements of C involved such that the scalar product of $b = (b_1, ..., b_n)$ times the ordered vector of chosen elements from C is minimized (while assuming that without loss of generality $b_1 \geq b_2 \geq \cdots \geq b_n$). This problem is a relaxed version of the time-dependent TSP (or continuous flowshop problem [Fink and Voß, 1998]) which itself is known to be a special case of the QAP by defining appropriate cost values. The cumulative assignment problem asks for two permutations π and π' of the set $\{1, \ldots, n\}$ such that

$$\sum_{i=1}^{n} b_i c_{\pi(i), \pi'(\pi(i))} \tag{8.17}$$

is minimized. The authors define a swap based neighborhood and devise an $O(n_2 \log n)$ implementation. Based on this neighborhood a random restart method combined with a local search mechanism, simulated annealing and tabu search are compared. The tabu search approach, which combines several features of intensification and diversification on three different levels, seems to be most appropriate when compared to some Lagrangean lower bounds.

4.3   THE BIQUADRATIC ASSIGNMENT PROBLEM

Generalizing the QAP in minimizing the weighted sum of products of four factors x_{ij} subject to assignment constraints yields the *biquadratic assignment problem* (BiQAP). Starting from model (8.5) this may be accomplished by minimizing

$$\sum_{i=1}^{n}\sum_{j=1}^{n}\sum_{p=1}^{n}\sum_{q=1}^{n} c_{ijpq} \cdot d_{\pi(i)\pi(j)\pi(p)\pi(q)} \qquad (8.18)$$

for two given arrays $(d_{ijpq})_{n \times n \times n \times n}$ and $(c_{ijpq})_{n \times n \times n \times n}$.

As the calculation of the objective function of the BiQAP takes $O(n^4)$ time one of the crucial considerations for any local search approach is a reduction in calculating the cost effect of a move. Let us assume a 2-exchange neighborhood. While a straightforward implementation requires an effort of $O(n^6)$ as $O(n^2)$ neighbors have to be considered similar ideas as presented above for the QAP (when calculating values $val_{\pi(i,j)}$) may be applied to the BiQAP to reduce the computational effort of each move to $O(n^4)$ [Burkard and Çela, 1995]. Nevertheless, the effort for meta-heuristics which explore a complete 2-exchange neighborhood (i.e., without using any candidate list strategy) may be too burdensome to perform a reasonable number of iterations.

Improvement procedures for the BiQAP are presented in [Burkard and Çela, 1995]. In particular they explore local search and variants of simulated annealing and tabu search. Numerical results are performed for problem instances with $n \leq 32$ for different versions of simulated annealing and tabu search favoring one specific simulated annealing implementation with a cooling schedule of $t_i = 0.8 \cdot t_0$.

A GRASP implementation for the BiQAP is described in [Mavridou et al., 1998] and numerical results are presented for almost the same problem instances as in [Burkard and Çela, 1995] with $n \leq 36$. The results are quite convincing as a comparison with the optimal solution values shows, which are known from the data generator [Burkard et al., 1994].

4.4   MISCELLANEOUS

Very general nonlinear assignment problems arise, e.g., when we consider some sort of time dependency. One such example is the *dynamic facility layout problem*. Different from the QAP in this problem we consider the situation where the layout may change over time and costs have to be considered not only for assignments made but also for relocating objects over time within changing environments or modified requirements. For such problems different ideas have been investigated such as genetic algorithms [Conway and Venkataramanan, 1994] or GRASP as well as random restart [Urban, 1998].

In the literature one may also find a modification of the QAP where m instead of n locations and values b_p indicating the demand of objects for location p with $\sum_{p=1}^{m} b_p = n$ are considered. Then constraints (8.7) are replaced by

$$\sum_{i=1}^{n} x_{ip} = b_p \qquad p = 1, \dots, m \qquad (8.19)$$

A neural network based approach for this problem can be found in [Smith, 1995].

One of the topics not mentioned here and also mainly neglected in the literature are multicriteria assignment problems (see, e.g., [Rosenblatt, 1979, Dutta and Sahu, 1982] for generating the efficient frontier of multicriteria layout problems). Furthermore, in the light of applications with some constraints not explicitly formulated decision support may also consider interactive planning incorporating especially priority rules and other concepts described above. For the QAP and related layout problems a great variety of interactive approaches has been developed over the years (see, e.g., [Ernst, 1978, Sharpe et al., 1985, Wäscher and Chamoni, 1987]).

Generalizing the QAP has been done in various ways in the literature starting in the early 1960s by defining, e.g., the cubic assignment problem [Lawler, 1963] or the bottleneck QAP where the sum objective is replaced by a maximum [Steinberg, 1961]. It should be noted that many of the above approaches may be modified and adapted to these problems (e.g., [Burkard and Fincke, 1982]).

5.   NONLINEAR SEMI-ASSIGNMENT PROBLEMS

5.1   THE QUADRATIC SEMI-ASSIGNMENT PROBLEM

In this section we report successful applications of heuristics to the QSAP. In addition, this includes a more application oriented problem description as the one presented in the introduction as well as information on some real-world applications in the area of schedule synchronization.

5.1.1   Problem Description.   Assigning items to sets such that a quadratic function is minimized may be referred to as the QSAP. This problem indeed is a relaxed version of the well-known QAP where one set of constraints is missing, i.e., feasible solutions need not be permutations but are injective functions. The QSAP may be represented in a mathematical model as follows (which is slightly different from the one in Section 1.). Given sets $A = \{1, \ldots, m\}$ and $B = \{1, \ldots, n\}$ and a (not necessarily) symmetric cost matrix (c_{ihjk}). With binary variables $x_{ih} = 1$ if $h \in B$ is assigned to $i \in A$ and $x_{ih} = 0$ otherwise we derive the model:

$$\text{Minimize } Z(x) = \sum_{i=1}^{m} \sum_{h=1}^{n} \sum_{j=1}^{m} \sum_{k=1}^{n} c_{ihjk} \cdot x_{ih} \cdot x_{jk} \tag{8.20}$$

subject to

$$\sum_{h=1}^{n} x_{ih} = 1 \qquad \forall i = 1, \ldots, m \tag{8.21}$$

$$x_{ih} \in \{0, 1\} \qquad \forall i = 1, \ldots, m, \ h = 1, \ldots, n \tag{8.22}$$

Although the QSAP is $\mathcal{NP}$-hard some special cases may be solved by polynomial time algorithms [Chhajed and Lowe, 1992, Voß, 1992, Simeone, 1986].

As we have seen in the mathematical model, the QSAP is a relaxed version of the QAP. Nevertheless, it turns out that the QAP is a special case of the QSAP. This relationship can be accomplished by defining appropriate cost coefficients c_{ihjk} (cf. [Malucelli, 1993]).

The QSAP has been independently introduced by several authors with respect to specific problem situations (e.g., [Dutta et al., 1982, Chhajed and Lowe, 1992]). We choose the name QSAP although it might be slightly too general. Other authors [Moretti Tomasin et al., 1988], e.g., consider the following scheduling problem under the name QSAP which indeed is a special case of our problem: Given a set of activities and a limited number of facilities, exactly one facility has to be assigned to each activity and every combination of two activities i and j scheduled on the same facility implies some interaction cost c_{ij}. The objective is to minimize the sum of all interaction costs, i.e.

$$\text{Minimize } Z(x) = \sum_{i=1}^{m} \sum_{j=1}^{m} \sum_{h=1}^{n} c_{ij} \cdot x_{ih} \cdot x_{jh} \qquad \text{subject to (8.21) and (8.22).}$$

Applications may be found in floor layout planning, in certain median problems with mutual communication, and in certain scheduling problems with cost-oriented objective functions and problems where the deviation from due dates is penalized by a quadratic function. Furthermore, multiprocessor environments may incorporate corresponding problems [Magirou and Milis, 1989].

Another application of the QSAP arises in schedule synchronization in public mass transit networks where the objective is to minimize the total transfer waiting times of passengers in a mass transit system which is expressed as the sum of individual waiting times within given operation hours [Klemt and Stemme, 1988]. Let there be a number of m lines or routes (Fig. 8.1 shows an example for a transit network with three routes; note that here a line, in contrast to the common sense, is defined for one direction only). With each route i we join a set $N(i)$ of possible departure times. Given a cycle time of t_i minutes for route i then $N(i)$ corresponds with these times within the given cycle time, i.e. $N(i) = \{1, \dots, t_i\}$ is a set with each element giving a specific departure time. This time may be viewed as starting time of i at its first station so that all arrival and departure times further on may be easily calculated resulting in a complete timetable.

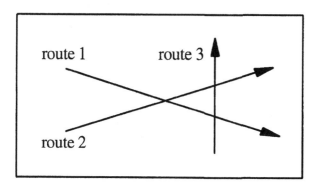

Figure 8.1   Schedule synchronization in public mass transit

The objective is to schedule all routes, i.e., to fix a starting time for each route. Sets A and B denote traffic lines and possible departure times, respectively. The number of allowed departure times for each line equals its cycle time (in time units, e.g., minutes) and the problem size may be referred to as "number of lines × cycle time" with identical cycle times for all lines. Therefore, the QSAP is a suitable model for schedule synchronization.

5.1.2   Existing Heuristics.   Initial solutions for the QSAP may be calculated by a *regret heuristic* which may be sketched as follows [Domschke, 1989]: For every assignment of an element of set B to an element of set A a minimal interaction cost between this one and the other assignments can be calculated. For any element of set A a regret value may be defined as the cost increase from the best assignment to the second best. An element with a maximum regret value is marked and its best assignment is fixed (i.e., included into the solution). For all unmarked elements of set A their minimal interaction costs and furthermore their regret values will be updated according to the fixed assignment and the algorithm continues until all elements of set A are marked. Variants of this algorithm need not choose elements with maximum regret or the best assignment but, e.g., one of the three best at random (which would lead to a GRASP once a local search approach is integrated).

Ideas for local search approaches may be based on two unidirectional improvement procedures. The first searches for any exchange in assignments that decrease the objective function value. That is, the starting time of any route is changed while all other routes stay the same. This will be repeated as long as this strategy gives further improvements. The second one searches for a simultaneous interchange of assignments for any two distinct elements of set A as long as it results in a decrease of the objective function value. That is, the starting times of two routes are changed simultaneously.

Whenever one of these strategies stops we get stuck in a local optimum which need not be global. Correspondingly, meta-heuristics have been considered. For instance, simulated annealing procedures for the QSAP have been described by [Domschke, 1989, Voß, 1992]. Additional approaches may be found in the literature comparing different tabu search methods (TNM, CSM, REM, REM2; REM2 refers to a modification of the REM or strict tabu search approach where revisiting of previously evaluated neighbors of already visited solutions are defined as tabu, too) with each other as well as with the simulated annealing strategy [Domschke et al., 1992, Voß, 1992, Voß, 1993]. Initial feasible solutions were calculated randomly or by successive application of the regret heuristic with two exchange procedures. With simulated annealing as well as the different tabu search methods these initial solutions were tried to be improved within a given time limit.

Results are presented for randomly generated data based on real-world data of a part of (West-)Berlin's subway system [Domschke, 1989, Voß, 1992]. The presented results are taken for the following problem sizes p (among others): 10, 11, and 12 lines with 10 minutes cycle time each. These rather small problem sizes were chosen in order to calculate the optimal solution values in reasonable time, too (see the branch and bound approach of [Domschke, 1989]).

To sum up the main results with respect to the QSAP it can be stated that tabu search generally operates better than simulated annealing. Also simulated annealing seems to be more influenced by the initial solution. There is not much difference between the three tested tabu search methods in solution quality if parameters are chosen well. The TNM asks for most exact determination of the tabu list size whereas the CSM seems to be more robust according to nonoptimal parameters. CSM proved to be most independent from the starting solution quality, too. The REM performed slightly worse but steadily improved a given solution with increasing computation time. It is the method that prevents from revisiting solutions best. Therefore, a combination of a simple starting heuristic, the CSM for fast improvement and probably the REM for further thorough improvement may be a good choice to solve even larger QSAP's.

Based on the procedures developed by [Domschke et al., 1992] as presented above additional computational testing on large scale real-world problems has been performed by [Voß, 1993] including REM2 (as well as sequential testing of a look ahead method). For some real-world problem specific considerations compare also [Voß, 1992]. The data represent schedule synchronization problems from three German cities with 14, 24, and 27 routes operating in two directions each (i.e. $m \leq 54$). A modification with respect to the above mentioned problem description concerns different cycle times for different routes as well as variable cycle times with $10 \leq t_i \leq 80$ minutes over the day.

The results are quite astonishing in the sense that an analysis of the case studies reveals very specialized data, i.e., the underlying traffic networks have a special shape. In one case it might be characterized as a star network with a central station in the downtown part of the respective city. In the other two cases different lines partly use the same tracks implicating that security distances have to be observed. Motivated by the modeling process above we may still use the QSAP model with slight modifications [Voß, 1992].

The special structure of the problems leads to the following results. Nearly in all cases the different tabu search methods are able to find only slight improvements over the initialized solutions obtained with the regret and the simple 2-optimal exchange procedures. Careful analysis of the smallest of the three examples gives us some reasoning (based on the branch and bound approach of [Domschke, 1989]) that the tabu search methods do not fail but that the initial feasible solutions are close to the optimum or even equal to it. In addition, even local optima close to the optimum have a great number of neighborhood solutions each with the same objective function value.

The algorithms have been applied within an overall solution approach for introducing three new bus lines into the public transit network of a major German city (cf. [Daduna and Voß, 1995]). In this case study the objective was not to schedule all lines to minimize the waiting time for the passengers but a hybrid with cost considerations. Therefore, we aimed in providing a large number of 'reasonably good' solutions that could be analyzed due to the cost criteria involved and to choose one solution that satisfies the tradeoff between the different goals. Finally, one of the solutions obtained by all different tabu search methods (lying at most 6% from the best solution found) was taken as a reference solution being modified by a planner in an interactive manner. Compared to the best possible solution with our approach the 'worse' solution provided the ability to save three buses, a cost reduction that led to the acceptance of the slightly worse result with respect to the schedule synchronization objective.

5.2   MISCELLANEOUS

In machine scheduling semi-assignment problems play a major role when it comes to assigning tasks to machines or jobs to processors. Once nonlinear penalty functions are observed they become relevant for our purposes. For instance, when assigning program modules to processors in a distributed computer system with communication needs and capacitated computer resources we derive an interesting generalization of the QSAP. The objective may be changed to a quadratic minimax objective similarly as in the schedule synchronization environment described above. Recently for such problems a simulated annealing approach implemented in Matlab is proposed [Hamam and Hindi, 2000].

6.   FUTURE DIRECTIONS

We presented the application of heuristics for nonlinear assignment problems. Some of the methods described still need more careful consideration as, e.g., the pilot meta-heuristic which has to be examined by further theoretical and experimental research, including a corresponding look-ahead strategy for improvement methods. While meta-heuristics are able to obtain high quality results it still needs to be tested under which circumstances which approach seems to be better suited. Regarding NAPs, one of the crucial aspects in this respect is the (re-)calculation of the objective function values in a local search environment as this greatly influences the number of iterations that may be performed in time limits deemed practical.

While theoretical investigations with respect to the QAP seem to be quite advanced, other problems still deserve consideration. For instance, the complexity of local search for semi-assignment problems as the QSAP may be treated. From different points of view we should strive to extend this survey to general nonlinear combinatorial optimization problems, especially those which may neither be represented as a permutation or by means of binary vectors or matrices. Finally, in connection with the goal to build systems applicable under real-world settings, research should also strive to transfer the findings to, e.g., decision support systems. Respective work may confirm the generality of research on nonlinear assignment problems as various problem types may be subsumed under an appropriate framework given by mainly changing the objective function and not the constraints.

Notes

1. Note that we do not explicitly distinguish between genetic algorithms and evolutionary strategies in this table

References

[Aarts and Lenstra, 1997] Aarts, E. and Lenstra, J., editors (1997). *Local Search in Combinatorial Optimization*. Wiley, Chichester.

[Ahuja et al., 1995] Ahuja, R., Orlin, J., and Tivari, A. (1995). A greedy genetic algorithm for the quadratic assignment problem. Working paper 3826-95, Sloan School of Management, MIT.

[Angel and Zissimopoulos, 1998] Angel, E. and Zissimopoulos, V. (1998). On the quality of local search for the quadratic assignment problem. *Discrete Applied Mathematics*, 82:15–25.

[Balas and Saltzman, 1991] Balas, E. and Saltzman, M. (1991). An algorithm for the three-index assignment problem. *Operations Research*, 39:150–161.

[Bandelt et al., 1994] Bandelt, H.-J., Crama, Y., and Spieksma, F. (1994). Approximation algorithms for multidimensional assignment problems with decomposable costs. *Discrete Applied Mathematics*, 49:25–50.

[Bartolomei-Suarez and Egbelu, 2000] Bartolomei-Suarez, S. and Egbelu, P. (2000). Quadratic assignment problem with adaptable material handling devices. *International Journal of Production Research*, 38:855–874.

[Battiti, 1996] Battiti, R. (1996). Reactive search: Toward self-tuning heuristics. In Rayward-Smith, V., Osman, I., Reeves, C., and Smith, G., editors, *Modern Heuristic Search Methods*, pages 61–83, Chichester. Wiley.

[Battiti and Tecchiolli, 1992] Battiti, R. and Tecchiolli, G. (1992). Parallel biased search for combinatorial optimization: genetic algorithms and tabu. *Microprocessors and Microsystems*, 16:351–367.

[Battiti and Tecchiolli, 1994a] Battiti, R. and Tecchiolli, G. (1994a). The reactive tabu search. *ORSA Journal on Computing*, 6:126–140.

[Battiti and Tecchiolli, 1994b] Battiti, R. and Tecchiolli, G. (1994b). Simulated annealing and tabu search in the long run: a comparison on QAP tasks. *Computer and Mathematical Applications*, 28:1–8.

[Bazaraa and Kirca, 1983] Bazaraa, M. and Kirca, O. (1983). A branch-and-bound heuristic for solving the quadratic assignment problem. *Naval Research Logistics Quarterly*, 30:287–304.

[Bazaraa and Sherali, 1980] Bazaraa, M. and Sherali, H. (1980). Bender's partitioning scheme applied to a new formulation of the quadratic assignment problem. *Naval Research Logistics Quarterly*, 27:29–41.

[Bersini and Varela, 1991] Bersini, H. and Varela, F. (1991). The immune recruitment mechanism: A selective evolutionary strategy. In *Proceedings of the 4th International Conference on Genetic Algorithms*, San Mateo. Morgan Kaufmann.

[Block, 1978] Block, T. (1978). A new construction algorithm for facilities layout. *Journal of Engineering Production*, 2:111–126.

[Boender et al., 1982] Boender, C., Kan, A. R., Timmer, A., and Stougie, L. (1982). A stochastic method for global optimization. *Mathematical Programming*, 22:125–140.

[Bölte, 1994] Bölte, A. (1994). *Modelle und Verfahren zur innerbetrieblichen Standortplanung*. Physica, Heidelberg.

[Bonomi and Lutton, 1986] Bonomi, E. and Lutton, J.-L. (1986). The asymptotic behavior of quadratic sum assignment problems: A statistical mechanics approach. *European Journal of Operational Research*, 26:295–300.

[Brown et al., 1989] Brown, E., Huntley, L., and Spillance, R. (1989). A parallel genetic heuristic for the quadratic assignment problem. In Schaffer, J., editor, *Proceedings of the 3rd International Conference on Genetic Algorithms, Arlington*, pages 406–415, San Mateo. Morgan Kaufmann.

[Bruijs, 1984] Bruijs, P. (1984). On the quality of heuristic solutions to a 19x19 quadratic assignment problem. *European Journal of Operational Research*, 17:21–30.

[Buffa et al., 1964] Buffa, E., Armour, G., and Vollmann, T. (1964). Allocating facilities with CRAFT. *Harvard Business Review*, 42 (2):136–158.

[Burkard and Bönniger, 1983] Burkard, R. and Bönniger, T. (1983). A heuristic for quadratic boolean programs with applications to quadratic assignment problems. *European Journal of Operational Research*, 13:374–386.

[Burkard and Çela, 1995] Burkard, R. and Çela, E. (1995). Heuristics for biquadratic assignment problems and their computational comparison. *European Journal of Operational Research*, 83:283–300.

[Burkard and Çela, 1998] Burkard, R. and Çela, E. (1998). Quadratic and three-dimensional assignments: An annotated bibliography. In Dell'Amico, M., Maffioli, F., and Martello, S., editors, *Annotated Bibliographies in Combinatorial Optimization*, pages 373–392, Chichester. Wiley.

[Burkard et al., 1994] Burkard, R., Çela, E., and Klinz, B. (1994). On the biquadratic assignment problem. In Pardalos, P. and Wolkowicz, H., editors, *Quadratic Assignment and Related Problems*, pages 117–146, Providence. AMS.

[Burkard et al., 1998] Burkard, R., Çela, E., Pardalos, P., and Pitsoulis, L. (1998). The quadratic assignment problem. In Pardalos, P. and Du, D.-Z., editors, *Handbook of Combinatorial Optimization*, pages 241–338, Boston. Kluwer.

[Burkard and Derigs, 1980] Burkard, R. and Derigs, U. (1980). *Assignment and matching problems: Solution methods with FORTRAN programs*, volume 184 of *Lecture Notes in Economics and Mathematical Systems*. Springer, Berlin.

[Burkard and Fincke, 1982] Burkard, R. and Fincke, U. (1982). On random quadratic bottleneck assignment problems. *Mathematical Programming*, 23:227–232.

[Burkard and Fincke, 1983] Burkard, R. and Fincke, U. (1983). The asymptotic probabilistic behavior of quadratic sum assignment problems. *Zeitschrift für Operations Research*, 27:73–81.

[Burkard et al., 1991] Burkard, R., Karisch, S., and Rendl, F. (1991). QAPLIB - a quadratic assignment problem library. *European Journal of Operational Research*, 55:115 – 119.

[Burkard et al., 1997] Burkard, R., Karisch, S., and Rendl, F. (1997). QAPLIB - a quadratic assignment problem library. *Journal of Global Optimization*, 10:391 – 403.

[Burkard and Rendl, 1984] Burkard, R. and Rendl, F. (1984). A thermodynamically motivated simulation procedure for combinatorial optimization problems. *European Journal of Operational Research*, 17:169–174.

[Burkard and Stratmann, 1978] Burkard, R. and Stratmann, K.-H. (1978). Numerical investigations on quadratic assignment problems. *Naval Research Logistics Quarterly*, 25:129–148.

[Çela, 1998] Çela, E. (1998). *The Quadratic Assignment Problem: Theory and Algorithms*. Kluwer, Dordrecht.

[Chakrapani and Skorin-Kapov, 1992] Chakrapani, J. and Skorin-Kapov, J. (1992). A connectionist approach to the quadratic assignment problem. *Computers & Operations Research*, 19:287 – 295.

[Chakrapani and Skorin-Kapov, 1993] Chakrapani, J. and Skorin-Kapov, J. (1993). Massively parallel tabu search for the quadratic assignment problem. *Annals of Operations Research*, 41:327–342.

[Cheh et al., 1991] Cheh, K., Goldberg, J., and Askin, R. (1991). A note on the neighborhood structure in simulated annealing. *Computers & Operations Research*, 18:537–547.

[Chhajed and Lowe, 1992] Chhajed, D. and Lowe, T. (1992). M-median and m-center problems with mutual communication: solvable special cases. *Operations Research*, 40:56 – 66.

[Chiang and Chiang, 1998] Chiang, W.-C. and Chiang, C. (1998). Intelligent local search strategies for solving facility layout problems with the quadratic assignment problem formulation. *European Journal of Operational Research*, 106:457–488.

[Choi et al., 1999] Choi, W., Kang, H., and Baek, T. (1999). A turbine-blade balancing problem. *International Journal of Production Economics*, 60-61:405–410.

[Colorni and Maniezzo, 1999] Colorni, A. and Maniezzo, V. (1999). The ant system applied to the quadratic assignment problem. *IEEE Transactions on Knowledge and Data Engineering*, 11:769–778.

[Connolly, 1990] Connolly, D. (1990). An improved annealing scheme for the QAP. *European Journal of Operational Research*, 46:93–100.

[Conway and Venkataramanan, 1994] Conway, D. and Venkataramanan, M. (1994). Genetic search and the dynamic facility layout problem. *Computers & Operations Research*, 21:955–960.

[Crama and Spieksma, 1992] Crama, Y. and Spieksma, F. (1992). Approximation algorithms for three-dimensional assignment problems with triangle inequalities. *European Journal of Operational Research*, 60:273–279.

[Cung et al., 1996] Cung, V.-D., Mautor, T., Michelon, P., and Tavares, A. (1996). A scatter search based approach for the quadratic assignment problem. Working paper 96/037, PRiSM Laboratory, University of Versailles.

[Daduna and Voß, 1995] Daduna, J. and Voß, S. (1995). Practical experiences in schedule synchronization. In Daduna, J., Branco, I., and Paixao, J., editors, *Computer-Aided Transit Scheduling*, volume 430 of *Lecture Notes in Economics and Mathematical Systems*, pages 39–55. Springer, Berlin.

[de Abreu et al., 1999] de Abreu, N., Querido, T., and Boaventura-Netto, P. (1999). REDINV-SA: A simulated annealing for the quadratic assignment problem. *RAIRO Recherche Operationelle*, 33:249–274.

[Dell'Amico et al., 1999] Dell'Amico, M., Lodi, A., and Maffioli, F. (1999). Solution of the cumulative assignment problem with a well-structured tabu search method. *Journal of Heuristics*, 5:123–143.

[Derigs et al., 1999] Derigs, U., Kabath, M., and Zils, M. (1999). Adaptive genetic algorithms for dynamic autoconfiguration of genetic search algorithms. In Voss, S., Martello, S., Osman, I., and Roucairol, C., editors, *Meta-Heuristics: Advances and Trends in Local Search Paradigms for Optimization*, pages 231–248. Kluwer, Boston.

[Domschke, 1989] Domschke, W. (1989). Schedule synchronization for public transit networks. *OR Spektrum*, 11:17 – 24.

[Domschke and Drexl, 1996] Domschke, W. and Drexl, A. (1996). *Logistik: Standorte, fourth ed.* Oldenbourg, München.

[Domschke et al., 1992] Domschke, W., Forst, P., and Voß, S. (1992). Tabu search techniques for the quadratic semi-assignment problem. In Fandel, G., Gulledge, T., and Jones, A., editors, *New Directions for Operations Research in Manufacturing*, pages 389 – 405. Springer, Berlin.

[Domschke and Krispin, 1997] Domschke, W. and Krispin, G. (1997). Location and layout planning - a survey. *OR Spektrum*, 19:181 – 202.

[Dorigo et al., 1999] Dorigo, M., Maniezzo, V., and Colorni, A. (1999). Ant system: Optimization by a colony of cooperating agents. *IEEE Transactions on Systems, Man and Cybernetics*, 26:29–41.

[Dowsland, 1993] Dowsland, K. (1993). Simulated annealing. In Reeves, C., editor, *Modern Heuristic Techniques for Combinatorial Problems*, pages 20–69. Blackwell: Halsted Pr.

[Dueck and Scheuer, 1990] Dueck, G. and Scheuer, T. (1990). Threshold accepting: a general purpose optimization algorithm appearing superior to simulated annealing. *Journal of Computational Physics*, 90:161 – 175.

[Duin and Voß, 1999] Duin, C. and Voß, S. (1999). The pilot method, a strategy for heuristic repetition with application to the Steiner problem in graphs. *Networks*, 34:181 – 191.

[Dutta et al., 1982] Dutta, A., Koehler, G., and Whinston, A. (1982). On optimal allocation in a distributed processing environment. *Management Science*, 28:839 – 853.

[Dutta and Sahu, 1982] Dutta, K. and Sahu, S. (1982). A multi-goal heuristic for facilities design problem: MUGHAL. *International Journal of Production Research*, 20:147–154.

[Dyer et al., 1986] Dyer, M., Frieze, A., and McDiarmid, C. (1986). On linear programs with random costs. *Mathematical Programming*, 35:3–16.

[Edwards et al., 1970] Edwards, H., Gillett, B., and Hale, M. (1970). Modular allocation technique (MAT). *Management Science*, 17:161–169.

[Elshafei, 1977] Elshafei, A. (1977). Hospital layout as a quadratic assignment problem. *Operational Research Quarterly*, 28:167–179.

[Ernst, 1978] Ernst, W. (1978). *Verfahren zur Fabrikplanung im Mensch-Rechner-Dialog am Bildschirm*. Krausskopf, Mainz.

[Fathi and Ginjupalli, 1993] Fathi, Y. and Ginjupalli, K. (1993). A mathematical model and a heuristic procedure for the turbine balancing problem. *European Journal of Operational Research*, 63:336–342.

[Feo et al., 1991] Feo, T., Venkatraman, K., and Bard, J. (1991). A grasp for a difficult single machine scheduling problem. *Computers & Operations Research*, 18:635 – 643.

[Fiechter et al., 1992] Fiechter, C., Rogger, A., and de Werra, D. (1992). Basic ideas of tabu search with an application to traveling salesman and quadratic assignment. *Ricerca Operativa*, 62:5 – 28.

[Fink and Voß, 1998] Fink, A. and Voß, S. (1998). Applications of modern heuristic search methods to continuous flow-shop scheduling problems. Working paper, Technische Universität Braunschweig, Germany.

[Fink and Voß, 1999] Fink, A. and Voß, S. (1999). Generic metaheuristics application to industrial engineering problems. *Computers & Industrial Engineering*, 37:281–284.

[Fleurent and Ferland, 1994] Fleurent, C. and Ferland, J. (1994). Genetic hybrids for the quadratic assignment problem. In Pardalos, P. and Wolkowicz, H., editors, *Quadratic Assignment and Related Problems*, volume 16 of *DIMACS Series in Discrete Mathematics and Theoretical Computer Science*, pages 173–187. American Mathematical Society, Providence.

[Fleurent and Glover, 1999] Fleurent, C. and Glover, F. (1999). Improved constructive multistart strategies for the quadratic assignment problem using adaptive memory. *INFORMS Journal on Computing*, 11:198–204.

[Frenk et al., 1985] Frenk, J., van Houweninge, M., and Kan, A. R. (1985). Asymptotic properties of the quadratic assignment problem. *Mathematics of Operations Research*, 16:223–239.

[Frieze and Yadegar, 1981] Frieze, A. and Yadegar, J. (1981). An algorithm for solving 3-dimensional assignment problems with applications to scheduling a teaching practice. *Operations Research*, 32:989–995.

[Frieze et al., 1989] Frieze, A., Yadegar, J., El-Horbaty, S., and Parkinson, D. (1989). Algorithms for assignment problems on an array processor. *Parallel Computing*, 11:151–162.

[Gambardella et al., 1999] Gambardella, L., Taillard, E., and Dorigo, M. (1999). Ant colonies for the QAP. *Journal of the Operational Research Society*, 50:167–176.

[Gavish, 1991] Gavish, B. (1991). Manifold search techniques applied to quadratic assignment problems (QAP). Working paper, Owen Graduate School of Management, Vanderbilt University, Nashville.

[Gen et al., 1999] Gen, M., Ida, K., and Lee, C.-Y. (1999). Hybridized neural network and genetic algorithms for solving nonlinear integer programming problem. In *SEAL '98*, volume 1585 of *Lecture Notes in Computer Science*, pages 421–429. Springer, Berlin.

[Gilmore, 1962] Gilmore, P. (1962). Optimal and suboptimal algorithms for the quadratic assignment problem. *J. SIAM*, 10:305–313.

[Glover, 1995] Glover, F. (1995). Scatter search and star-paths: beyond the genetic metaphor. *OR Spektrum*, 17:125–137.

[Glover and Laguna, 1997] Glover, F. and Laguna, M. (1997). *Tabu Search*. Kluwer, Dordrecht.

[Goldberg, 1989] Goldberg, D. (1989). *Genetic Algorithms in Search, Optimization, and Machine Learning*. Addison-Wesley, Reading.

[Golden and Skiscim, 1986] Golden, B. and Skiscim, C. (1986). Using simulated annealing to solve routing and location problems. *Naval Research Logistics Quarterly*, 33:261 – 279.

[Graves and Whinston, 1970] Graves, G. and Whinston, A. (1970). An algorithm for the quadratic assignment problem. *Management Science*, 17:453–471.

[Haghani and Chen, 1998] Haghani, A. and Chen, M.-C. (1998). Optimizing gate assignments at airport terminals. *Transportation Research A*, 32:437–454.

[Hamam and Hindi, 2000] Hamam, Y. and Hindi, K. (2000). Assignment of program modules to processors: A simulated annealing approach. *European Journal of Operational Research*, 122:509–513.

[Heragu, 1997] Heragu, S. (1997). *Facilities Design*. PWS, Boston.

[Heragu and Alfa, 1992] Heragu, S. and Alfa, A. (1992). Experimental analysis of simulated annealing based algorithms for the facility layout problem. *European Journal of Operational Research*, 57:190–202.

[Hillier and Conners, 1966] Hillier, F. and Conners, M. (1966). Quadratic assignment problem algorithms and the location of indivisable facilities. *Management Science*, 13:42–57.

[Holland, 1975] Holland, J. (1975). *Adaptation in Natural and Artificial Systems*. The University of Michigan Press, Ann Arbor.

[Huntley and Brown, 1991] Huntley, C. and Brown, D. (1991). A parallel heuristic for quadratic assignment problems. *Computers & Operations Research*, 18:275–289.

[Jajodia et al., 1992] Jajodia, S., Minis, I., Harhalakis, G., and Proth, J. (1992). CLASS: Computerized layout solutions using simulated annealing. *International Journal of Production Research*, 30:95–108.

[Johnson et al., 1989] Johnson, D., Aragon, C., McGeoch, L., and Schevon, C. (1989). Optimization by simulated annealing: an experimental evaluation; part 1, graph partitioning. *Operations Research*, 37:865–892.

[Johnson et al., 1988] Johnson, D., Papadimitriou, C., and Yannakakis, M. (1988). How easy is local search? *J. Comput. System Sci.*, 37:79–100.

[Kelly et al., 1994] Kelly, J., Laguna, M., and Glover, F. (1994). A study of diversification strategies for the quadratic assignment problem. *Computers & Operations Research*, 21:885–893.

[Kirkpatrick et al., 1983] Kirkpatrick, S., Jr., C. G., and Vecchi, M. (1983). Optimization by simulated annealing. *Science*, 220:671 – 680.

[Klemt and Stemme, 1988] Klemt, W.-D. and Stemme, W. (1988). Schedule synchronization for public transit networks. In Daduna, J. and Wren, A., editors, *Computer-Aided Transit Scheduling*, volume 308 of *Lecture Notes in Economics and Mathematical Systems*, pages 327 – 335. Springer, Berlin.

[Kusiak and Heragu, 1987] Kusiak, A. and Heragu, S. (1987). The facility layout problem. *European Journal of Operational Research*, 29:229–251.

[Land, 1963] Land, A. (1963). A problem of assignment with interrelated costs. *Operational Research Quarterly*, 14:185–198.

[Laporte and Mercure, 1988] Laporte, G. and Mercure, H. (1988). Balancing hydraulic turbine runners: A quadratic assignment problem. *European Journal of Operational Research*, 35:378–381.

[Lashari and Jaisingh, 1980] Lashari, R. and Jaisingh, S. (1980). A heuristic approach to quadratic assignment problems. *Journal of the Operational Research Society*, 31:845–850.

[Laursen, 1993] Laursen, P. (1993). Simulated annealing for the QAP - optimal tradeoff between simulation time and solution quality. *European Journal of Operational Research*, 69:238 – 243.

[Lawler, 1963] Lawler, E. (1963). The quadratic assignment problem. *Management Science*, 9:586 – 599.

[Lee and Moore, 1967] Lee, R. and Moore, J. (1967). CORELAP - computerized relationship layout planning. *Journal of Industrial Engineering*, 18:195–200.

[Li and Pardalos, 1992] Li, Y. and Pardalos, P. (1992). Generating quadratic assignment test problems with known optimal permutations. *Computational Optimization and Applications*, 1:163–184.

[Li et al., 1994] Li, Y., Pardalos, P., and Resende, M. (1994). A greedy randomized adaptive search procedure for the quadratic assignment problem. In Pardalos, P. and Wolkowicz, H., editors, *Quadratic Assignment and Related Problems*, volume 16 of *DIMACS Series in Discrete Mathematics and Theoretical Computer Science*, pages 237–261. American Mathematical Society, Providence.

[Lutton and Bonomi, 1986] Lutton, J. and Bonomi, E. (1986). The asymptotic behavior of a quadratic sum assignment problems: a sta-tistical mechanics approach. *European Journal of Operational Research*, 26:295 –300.

[Magirou and Milis, 1989] Magirou, V. and Milis, J. (1989). An algorithm for the multiprocessor assignment problem. *Operations Research Letters*, 8:351–356.

[Magos, 1996] Magos, D. (1996). Tabu search for the planar three-index assignment problem. *Journal of Global Optimization*, 8:35–48.

[Magos and Miliotis, 1994] Magos, D. and Miliotis, P. (1994). An algorithm for the planar three-index assignment problem. *European Journal of Operational Research*, 77:141–153.

[Malucelli, 1993] Malucelli, F. (1993). *Quadratic assignment problems: solution methods and applications*. PhD thesis, Universita di Pisa.

[Maniezzo, 1999] Maniezzo, V. (1999). Exact and approximate nondeterministic tree-search procedures for the quadratic assignment problem. *INFORMS Journal on Computing*, 11:358–369.

[Maniezzo et al., 1995] Maniezzo, V., Dorigo, M., and Colorni, A. (1995). Algodesk: An experimental comparison of eight evolutionary heuristics applied to the quadratic assignment problem. *European Journal of Operational Research*, 81:188–204.

[Mason and Rönnqvist, 1997] Mason, A. and Rönnqvist, M. (1997). Solution methods for the balancing of jet turbines. *Computers & Operations Research*, 24:153–167.

[Mavridou et al., 1998] Mavridou, T., Pardalos, P., Pitsoulis, L., and Resende, M. (1998). A GRASP for the biquadratic assignment problem. *European Journal of Operational Research*, 105:613–621.

[Moretti Tomasin et al., 1988] Moretti Tomasin, E., Pianca, P., and Sorato, A. (1988). Heuristic algorithms for the quadratic semi-assignment problem. *Ricerca Operativa*, 18:65 – 89.

[Mühlenbein, 1989] Mühlenbein, H. (1989). Varying the probability of mutation in the genetic algorithm. In Schaffer, J., editor, *Proceedings of the Third International Conference on Genetic Algorithms*, pages 416 – 421, San Mateo. Morgan Kaufmann.

[Müller-Merbach, 1970] Müller-Merbach, H. (1970). *Optimale Reihenfolgen*. Springer, Berlin.

[Murphey et al., 1998] Murphey, R., Pardalos, P., and Pitsoulis, L. (1998). A parallel GRASP for the data association multidimensional assignment problem. In Pardalos, P., editor, *Parallel Processing of Discrete Problems*, volume 106 of *IMA Volumes in Mathematics and its Applications*, pages 159–180. Springer, Berlin.

[Murtagh et al., 1982] Murtagh, B., Jefferson, T., and Sornprasit, V. (1982). A heuristic procedure for solving the quadratic assignment problem. *European Journal of Operational Research*, 9:71–76.

[Murthy et al., 1992] Murthy, K., Pardalos, P., and Li, Y. (1992). A local search algorithm for the quadratic assignment problem. *Informatica*, 3:524–538.

[Nissen, 1993] Nissen, V. (1993). A new efficient evolutionary algorithm for the quadratic assignment problem. In Hansmann, K.-W., Bachem, A., Jarke, M., Katzenberger, W., and Marusev, A., editors, *Operations Research Proceedings 1992*, pages 259 – 267, Berlin. Springer.

[Nissen, 1994] Nissen, V. (1994). *Evolutionäre Algorithmen*. DUV, Wiesbaden.

[Nissen and Paul, 1993] Nissen, V. and Paul, H. (1993). A modification of threshold accepting and its application to the quadratic assignment problem. *OR Spektrum*, 17:205–210.

[Nugent et al., 1968] Nugent, C., Vollmann, T., and Ruml, J. (1968). An experimental comparison of techniques for the assignment of facilities to locations. *Operations Research*, 16:150–173.

[Pardalos et al., 1994] Pardalos, P., Burkard, R., and Wolkowicz, H. (1994). The quadratic assignment problem: A survey and recent developments. In Pardalos, P. and Wolkowicz, H., editors, *Quadratic Assignment and Related Problems*, pages 1–42, Providence. AMS.

[Pardalos et al., 1995] Pardalos, P., Pitsoulis, L., and Resende, M. (1995). A parallel grasp implementation for solving the quadratic assignment problem. In Ferreira, A. and Rolim, J., editors, *Parallel Algorithms for Irregular Problems: State of the Art*, pages 115–133. Kluwer, Boston.

[Pardalos et al., 1997] Pardalos, P., Pitsoulis, L., and Resende, M. (1997). Fortran subroutines for approximate solution of sparse quadratic assignment problems using grasp. *ACM Transactions on Mathematical Software*, 23:196–208.

[Pardalos and Wolkowicz, 1994] Pardalos, P. and Wolkowicz, H., editors (1994). *Quadratic Assignment and Related Problems*, volume 16 of *DIMACS Series in Discrete Mathematics and Theoretical Computer Science*. American Mathematical Society, Providence.

[Paulli, 1993] Paulli, J. (1993). Information utilization in simulated annealing and tabu search. *Committee on Algorithms Bulletin*, 22:28 – 34.

[Peng et al., 1996] Peng, T., Huanchen, W., and Dongme, Z. (1996). Simulated annealing for the quadratic assignment problem: A further study. *Computers & Industrial Engineering*, 31:925–928.

[Pitsoulis et al., 2000] Pitsoulis, L., Pardalos, P., and Hearn, D. (2000). Approximate solutions to the turbine balancing problem. *European Journal of Operational Research*, (to appear)

[Poore and Rijavec, 1994] Poore, A. and Rijavec, N. (1994). A parallel heuristic for quadratic assignment problems. *Journal of Computing and Information Technology*, 2:25–37.

[Queyranne, 1986] Queyranne, M. (1986). Performance ratio of polynomial algorithms for triangle-inequality quadratic assignment problems. *Operations Research Letters*, 4:231–234.

[Resende et al., 1996] Resende, M., Pardalos, P., and Li, Y. (1996). Fortran subroutines for approximate solution of dense quadratic assignment problems using grasp. *ACM Transactions on Mathematical Software*, 22:104–118.

[Rhee, 1988a] Rhee, W. (1988a). A note on asymptotic properties of the quadratic assignment problem. *Operations Research Letters*, 7:197–200.

[Rhee, 1988b] Rhee, W. (1988b). Stochastic analysis of the quadratic assignment problem. *Mathematics of Operations Research*, 16:223–239.

[Rosenblatt, 1979] Rosenblatt, M. (1979). The facilities layout problem: A multi-goal approach. *International Journal of Production Research*, 17:323–332.

[Sahni and Gonzalez, 1976] Sahni, S. and Gonzalez, T. (1976). P-complete approximation problems. *Journal of the ACM (Association for Computing Machinery)*, 23:555–565.

[Schäffer and Yannakakis, 1991] Schäffer, A. and Yannakakis, M. (1991). Simple local search problems that are hard to solve. *SIAM Journal on Computing*, 20:56–87.

[Seehof and Evans, 1967] Seehof, J. and Evans, W. (1967). Automated layout design program. *Journal of Industrial Engineering*, 18:690–695.

[Sharpe et al., 1985] Sharpe, R., Marksjo, B., Mitchell, J., and Crawford, J. (1985). An interactive model for the layout of buildings. *Applied Mathematical Modeling*, 9:207–214.

[Simeone, 1986] Simeone, B. (1986). An asymptotically exact algorithm for equipartition problems. *Discrete Applied Mathematics*, 14:283 – 293.

[Sinclair, 1993] Sinclair, M. (1993). Comparison of the performance of modern heuristics for combinatorial optimization on real data. *Computers & Operations Research*, 20:687 – 695.

[Skorin-Kapov, 1990] Skorin-Kapov, J. (1990). Tabu search applied to the quadratic assignment problem. *ORSA Journal on Computing*, 2:33 – 45.

[Skorin-Kapov, 1994] Skorin-Kapov, J. (1994). Extensions of a tabu search adaptation to the quadratic assignment problem. *Computers & Operations Research*, 21:855–866.

[Smith, 1995] Smith, K. (1995). Solving the generalized quadratic assignment problem using a self-organizing process. In *Proceedings of the IEEE International Conference on Neural Networks (ICNN)*, pages 1876–1879. Vol. 4, Perth.

[Smith, 1999] Smith, K. (1999). Neural networks for combinatorial optimization: A review of more than a decade of research. *INFORMS Journal on Computing*, 11:15–34.

[Smith et al., 1998] Smith, K., Palaniswami, M., and Krishnamoorthy, M. (1998). Neural techniques for combinatorial optimization with applications. *IEEE Transactions on Neural Networks*, 9:1301–1318.

[Sondergeld and Voß, 1993] Sondergeld, L. and Voß, S. (1993).   Solving quadratic assignment problems using the cancellation sequence method. Working paper, TH Darmstadt.

[Sondergeld and Voß, 1996] Sondergeld, L. and Voß, S. (1996). A star-shaped diversification approach in tabu search. In Osman, I. and Kelly, J., editors, *Meta-Heuristics: Theory and Applications*, pages 489–502. Kluwer, Boston.

[Starke et al., 1999] Starke, J., Schanz, M., and Haken, H. (1999). Treatment of combinatorial optimization problems using selection equations with cost terms. part ii. NP-hard three-dimensional assignment problems. *Physica D*, 134:242 – 252.

[Steinberg, 1961] Steinberg, L. (1961).   The backboard wiring problem: A placement algorithm. *SIAM Review*, 3:37–50.

[Stützle and Hoos, 1999] Stützle, T. and Hoos, H. (1999).   The max-min ant system and local search for combinatorial optimization problems. In Voss, S., Martello, S., Osman, I., and Roucairol, C., editors, *Meta-Heuristics: Advances and Trends in Local Search Paradigms for Optimization*, pages 313–329. Kluwer, Boston.

[Taillard, 1991] Taillard, E. (1991).   Robust taboo search for the quadratic assignment problem. *Parallel Computing*, 17:443 – 455.

[Taillard, 1995] Taillard, E. (1995).   Comparison of iterative searches for the quadratic assignment problem. *Location Science*, 3:87–105.

[Taillard, 1998] Taillard, E. (1998).   FANT: Fast ant system. Working paper, IDSIA, Lugano.

[Taillard, 2000] Taillard, E. (2000).   An introduction to ant systems. In Laguna, M. and Gonzalez-Velarde, J., editors, *Computing Tools for Modeling, Optimization and Simulation*, pages 131–144. Kluwer, Boston.

[Taillard and Voß, 1999] Taillard, E. and Voß, S. (1999). Popmusic. Working paper, University of Applied Sciences of Western Switzerland.

[Talbi et al., 1999] Talbi, E.-G., Hafidi, Z., and Geib, J.-M. (1999). Parallel tabu search for large optimization problems. In Voss, S., Martello, S., Osman, I., and Roucairol, C., editors, *Meta-Heuristics: Advances and Trends in Local Search Paradigms for Optimization*, pages 345–358. Kluwer, Boston.

[Tam, 1992] Tam, K. (1992). Genetic algorithms, function optimization, and facility layout design. *European Journal of Operational Research*, 63:322 – 346.

[Tate and Smith, 1995] Tate, D. and Smith, A. (1995). A genetic approach to the quadratic assignment problem. *Computers & Operations Research*, 22:73–83.

[Torki et al., 1996] Torki, A., Yajima, Y., and Enkawa, T. (1996). A low-rank bilinear programming approach for sub-optimal solution of the quadratic assignment problem. *European Journal of Operational Research*, 94:384–391.

[Tsuchiya et al., 1996] Tsuchiya, K., Bharitkar, S., and Takefuji, Y. (1996). A neural network approach to facility layout problems. *European Journal of Operational Research*, 89:556–563.

[Urban, 1998] Urban, T. (1998). Solution procedures for the dynamic facility layout problem. *Annals of Operations Research*, 76:323–342.

[Vollmann et al., 1968] Vollmann, T., Nugent, C., and Zartler, R. (1968). A computerized model for office layout. *Journal of Industrial Engineering*, 19:321–327.

[Voß, 1992] Voß, S. (1992). Network design formulations in schedule synchronization. In Desrochers, M. and Rousseau, J.-M., editors, *Computer-Aided Transit Scheduling*, volume 386 of *Lecture Notes in Economics and Mathematical Systems*, pages 137 – 152. Springer, Berlin.

[Voß, 1993] Voß, S. (1993). Tabu search: applications and prospects. In Du, D.-Z. and Pardalos, P., editors, *Network Optimization Problems*, pages 333 – 353. World Scientific, Singapore.

[Voß, 1995] Voß, S. (1995). Solving quadratic assignment problems using the reverse elimination method. In Nash, S. and Sofer, A., editors, *The Impact of Emerging Technologies on Computer Science and Operations Research*, pages 281 – 296. Kluwer, Dordrecht.

[Voß, 1996] Voß, S. (1996). Observing logical interdependencies in tabu search: Methods and results. In Rayward-Smith, V., Osman, I., Reeves, C., and Smith, G., editors, *Modern Heuristic Search Methods*, pages 41–59, Chichester. Wiley.

[Wäscher and Chamoni, 1987] Wäscher, G. and Chamoni, P. (1987). MICRO-LAY: An interactive computer program for factory layout planning on microcomputers. *European Journal of Operational Research*, 31:185–193.

[West, 1983] West, D. (1983). Algorithm 608: Approximate solution of the quadratic assignment problem. *ACM Transactions on Mathematical Software*, 9:461–466.

[White, 1996] White, D. (1996). A lagrangean relaxation approach for a turbine design quadratic assignment problem. *Journal of the Operational Research Society*, 47:766–775.

[Wilhelm and Ward, 1987] Wilhelm, M. and Ward, T. (1987). Solving quadratic assignment problems by 'simulated annealing'. *IIE Transactions*, 19:107 – 119.

Chapter 9

SYMBOLIC SCHEDULING OF PARAMETERIZED TASK GRAPHS ON PARALLEL MACHINES

Michel Cosnard

LORIA INRIA Lorraine
615, rue du Jardin Botanique
54602 Villers les Nancy
France
Michel.Cosnard@loria.fr

Emmanuel Jeannot

LaBRI, University of Bordeaux I
351, cours de la Liberation
33405 Talence Cedex
France
ejeannot@labri.u-bordeaux.fr

Tao Yang

CS Dept. UCSB
Engr Building I
Santa Barbara, CA 93106
USA
tyang@cs.ucsb.edu

Abstract   In this chapter we address the problem of allocating parallel tasks on a distributed memory machine for coarse-grain applications represented by *parameterized task graphs* (PTG). A PTG is a new computation model for representing directed acyclic task graphs (DAG) symbolically. The size of a PTG is independent of the problem size and its parameters can be instantiated at run time. Parameter-independent optimization is important for exploiting non-static parallelism in scientific computing programs with varying problem sizes and the previous DAG scheduling algorithms are not able to handle such cases. We present and study a

P.M. Pardalos and L.S. Pitsoulis (eds.), Nonlinear Assignment Problems, 217–243.

symbolic scheduling algorithm called SLC (Symbolic Linear Clustering) which derives task clusters from a PTG using affine piecewise mapping functions and then evenly assigns clusters to processors. Our experimental results show that the proposed method is effective for a number of compute-intensive problems in scientific applications.

1.   INTRODUCTION

Directed acyclic task graphs (DAGs) have been used in modeling parallel applications and performing performance prediction and optimization [Adve and Vernon, , Chong et al., 1995, Deelman et al., 1998, Fu and Yang, 1996, Gerasoulis et al., 1995, Sarkar, 1989]. There are a number of algorithms which have been proposed to perform static task graph mapping on distributed memory machines [El-Rewini et al., 1994, Kwok and Ahmad, 1996, Liou and Palis, 1998, Palis et al., 1996, Papadimitriou and Yannakakis, 1990, Yang and Gerasoulis, 1994]. Because a task graph is obtained statically and the number of processors must be given in advance before scheduling, these methods present two major drawbacks:

- Static scheduling is not adaptive. Each time the problem parameter values change or the number of processors of the target machine changes, a scheduling solution has to be recomputed.

- Static scheduling is not scalable. For large problem sizes, the corresponding task graph may contain a large number of tasks and dependence edges, and a static scheduler might fail due to a memory constraint.

The previous work in parallelizing compilers [Anderson and Lam, 1993, Feautrier, 1994, Mongenet, 1997] has studied the compact parallelism representation based on fine-grain level dependence analysis and their model normally deals with DOALL parallelism. These works cannot be applied directly to our problem because they mainly deal with unitary communications and computations and a major extension is needed for handling coarse-grain parallelism. The class of allocation/mapping functions derived in these works are at most affine functions. Our goal is to extend these results for exploiting coarse-grain DAG parallelism in a symbolic manner, in order to overcome the two drawbacks mentioned above.

In this chapter we use a parameterized task graph (PTG) [Cosnard and Loi, 1995, Loi, 1996] to model task computation symbolically. A PTG can be considered as a DAG, and is augmented by a set of parameters. Thus such a graph is symbolic and its size does not vary when problem parameters change. In our previous work [Cosnard et al., 1998, Cosnard and Jeannot, 1999] we have presented a dynamic scheme for scheduling parameterized task graphs. Thus, memory requirement is very low but still depends on the parameter values. This chapter focuses on symbolic scheduling of parameterized task graphs on work-station/SMP clusters or distributed memory machines where multi-threading is available on each node and inter-node communication requires latency-aware optimization in task scheduling. Preliminary versions of this work can be found in [Cosnard et al., 1999, Jeannot, 1999].

In the rest of this chapter, we use term "processor" or "node" to denote a computing unit in a parallel machine or a cluster. For an SMP cluster, term "processor" in our algorithm setting can be interpreted as a machine SMP machine containing multiple CPUs.

Our algorithm, called SLC (Symbolic Linear Clustering), first performs symbolic clustering and assigns selected tasks in a dependence path to the same processor in order to reduce inter-task communication while still preserving available parallelism. Then it assigns symbolic clusters uniformly to processors. Since each node may own several symbolic task clusters, our runtime scheme executes clusters on each processor using a multi-threading fashion to overlap computation with communication.

The rest of this book chapter is organized as follows. Section 2. gives the definition of parameterized task graphs. Section 3. provides an overview of our approach. Section 4. presents our symbolic clustering method. Section 5. discusses how clusters can be numbered explicitly so that symbolic clusters can be evenly mapped to physical processors. Section 6. describes the runtime execution of clusters using threads on each processor. Section 7. presents simulation and experimental results. Section 8. presents related work on symbolic computation allocation. Section 9. concludes the chapter. In the appendix we list the kernel code for our experiments as well as the clustering function found by SLC.

2. DEFINITIONS AND NOTATIONS

Parameterized Task Graphs. A parameterized task graph (PTG) is a compact model for parallel computation [Cosnard and Loi, 1995]. It contains a set of symbolic tasks and each symbolic task is represented by a name and an iteration vector. A PTG also contains a set of communication rules that describe data items transfered among tasks. Since each task can contain a set of instructions executed sequentially, this model mainly is targeted at coarse-grain parallelism.

There are two types of communication rules which are either emission or reception rules to model how tasks send or receive data. Reception and emission rules are dual forms of each other. Reception rules describe a set of parents that a task depends on. Emission rules describe a set of children that a task needs to send data. An emission rule R (a reception rule) has the form:

$$R : Ta(\vec{u}) \longrightarrow Tb(\vec{v}) : D(\vec{y})|P$$

where $\vec{u}$ and $\vec{v}$ are the iteration vectors of tasks Ta and Tb. Rule R means that if predicate P is true, task $Ta(\vec{u})$ sends data $D(\vec{y})$ to task $Tb(\vec{v})$. Vector $\vec{y}$ describes which part of the data D is sent to task Tb. P is a polyhedron which describes valid values of vectors $\vec{u}$, $\vec{v}$ and $\vec{y}$.

The number of components in $\vec{y}$ depends on the dimension of data variable D. For example, it is 0 if D is a scalar, and 1 if D is a 1D vector. We assume that the variables $\vec{y}$ do not appear in the predicates describing the variables of $\vec{u}$ and $\vec{v}$. Hence, the data part of a rule can be removed easily without getting *holes* in the polyhedron.

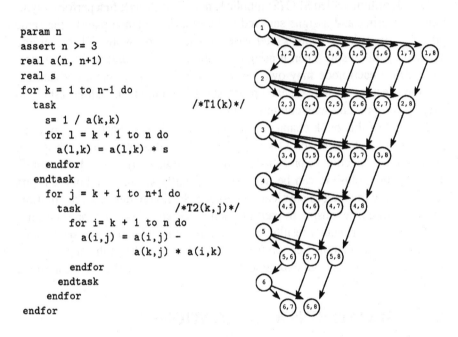

```
param n
assert n >= 3
real a(n, n+1)
real s
for k = 1 to n-1 do
  task                    /*T1(k)*/
    s= 1 / a(k,k)
    for l = k + 1 to n do
      a(l,k) = a(l,k) * s
    endfor
  endtask
    for j = k + 1 to n+1 do
      task                /*T2(k,j)*/
        for i= k + 1 to n do
          a(i,j) = a(i,j) -
                   a(k,j) * a(i,k)
        endfor
      endtask
    endfor
endfor
```

Figure 9.1   Gaussian elimination and expanded task graph(n=7)

Derivation of a Parameterized Task Graph.   We have used the PlusPyr software tool [Loi, 1996] in order to construct a PTG from a sequential program. PlusPyr is able to derive a parameterized task graph for a sequential program if a user also provides an annotation to describe how this program is partitioned into tasks. Figure 9.1 illustrates an annotated program for Gaussian Elimination without pivoting. It contains two generic tasks $T1(k)$ and $T2(k,j)$. Key words "task" and "endtask" specify the beginning and end of a task and "param" specifies symbolic parameters used in this program, which could vary based on the actual problem size. Figure 9.1 also illustrates an instantiated Gaussian Elimination task graph when program parameter n is 7.

The current implementation of PlusPyr supports dependence analysis and communication synthesis among tasks if the input program has static control and only deals with matrix vector or scalar computation. The techniques for dependence analysis and communication summarization are based on the work done by Feautrier and others on integer parametric programming (see [Feautrier, 1991, Schrijver, 1986] for an introduction and [Cosnard and Loi, 1995, Cosnard and Loi, 1996] for more details).

For the input GE program shown in Figure 9.1, the emission rules of the corresponding parameterized task graph, computed by our tool PlusPyr, are shown in Figure 9.2. For example, the third emission rule indicates that for any k and j such that $1 \leq k \leq n - 2$ and $k + 2 \leq j \leq n + 1$, task $T2(k,j)$ sends part of column j to task $T2(k+1,j)$.

1. $T1(k) \longrightarrow T2(k,j) : A(i,k)|1 \leq k \leq n-1, k+1 \leq j \leq n+1, k+1 \leq i \leq n$

2. $T2(k,j) \longrightarrow T1(k+1) : A(i,k+1)|1 \leq k \leq n-2, j = k+1, k+1 \leq i \leq n$

3. $T2(k,j) \longrightarrow T2(k+1,j) : A(i,j)|1 \leq k \leq n-2, k+2 \leq j \leq n+1, k+1 \leq i \leq n$

Figure 9.2   Emission rules for gaussian elimination

Bijective Emission Rules.   Our symbolic clustering algorithm identifies a type of rules called *bijection rules* to cluster and we present its definition as follows. Given an emission rule of format $R : Ta(\vec{u}) \longrightarrow Tb(\vec{v}) : D(\vec{y})|P$, this rule is **bijective** if for each instance of iteration vector $\vec{u}$, there exists exactly one corresponding instance of iteration vector $\vec{v}$. For example, in Figure 9.2, the first emission rule is not bijective because this rule describes a dependence from task $T1(k)$ to many of its children $T2(k, k + 1), T2(k, k + 2) \cdots$. The second and third rules are bijective.

We further discuss how to judge if an emission rule is bijective. We assume that a task contains nested loops and its iteration vector is an affine function of program parameters and loop indices. Let $\vec{i}$ be the vector composed of all the problem parameters and loop indices in the lexicographic order. Thus we can express vectors $\vec{u}$ and $\vec{v}$ as:

$$\vec{u} = D_1\vec{i} + \vec{k_1} \quad \text{and} \quad \vec{v} = D_2\vec{i} + \vec{k_2}.$$

where $\vec{k_1}$ and $\vec{k_2}$ are constant integer vectors. $D1$ and $D2$ are constant integer matrices. Notice that if there are equalities in P, we should substitute variables to remove some variables before constructing D_1 and D_2.

To find if this rule is bijective, we find E_1 and $\vec{w}_1$ such that $\vec{\imath} = E_1(\vec{u} - \vec{k}_1) + \vec{w}_1$. There could exist many E_1 and $\vec{w}_1$ that satisfy this equation. We impose a condition that $\vec{w}_1$ is composed of the indices that do not appear in $\vec{u}$. Formally, $w_{1_j} = \delta_j i_j$, where the w_{1_j} (resp. i_j) is j^{th} element of $\vec{w}_1$ (resp. $\vec{\imath}$); $\delta_j = 0$ if i_j appears in $\vec{u}$, and $\delta_j = 1$ if i_j does not appear in $\vec{u}$.

Then we have: $\vec{v} = D_2 E_1(\vec{u} - \vec{k}_1) + D_2\vec{w}_1 + \vec{k}_2$. Similarly we can build E_2 and $\vec{w}_2$ such that: $\vec{u} = D_1 E_2(\vec{v} - \vec{k}_2) + D_1\vec{w}_2 + \vec{k}_1$.

The above rule is bijective if and only if $D_2\vec{w}_1$ and $D_1\vec{w}_2$ are both constant. This condition ensures that for each instance of $\vec{u}$ there is only one instance of $\vec{v}$ and for each instance of $\vec{v}$ there is only one instance of $\vec{u}$.

In Figure 9.2, the fact that the first rule is not bijective can be verified as follows. Since $\vec{u} = (k)$, $\vec{v} = \begin{pmatrix} k \\ j \end{pmatrix}$, and the vector of the program parameters and all the loops indices is $\vec{\imath} = (n \; k \; l \; j \; i)^T$, $D_1 = (0, 1, 0, 0, 0)$ and

$$D_2 = \begin{pmatrix} 0 & 1 & 0 & 0 & 0 \\ 0 & 0 & 0 & 1 & 0 \end{pmatrix}.$$

We deduce that: $E_1 = (0\,1\,0\,0\,0)^T$ and $\vec{w}_1 = (n\,0\,l\,j\,i)^T$ so that $\vec{\imath} = E_1\vec{u} + \vec{w}_1$. Then, $D_2\vec{w}_1 = \begin{pmatrix} 0 \\ j \end{pmatrix}$, which is not a constant.

On the other hand, the second emission rule in Figure 9.2 is bijective. In this case, $\vec{u} = \begin{pmatrix} k \\ j \end{pmatrix}$ and $\vec{v} = (k + 1)$. According to the predicate $j = k + 1$, we substitute all the occurrences of j by $k + 1$. Hence we have: $\vec{u} = \begin{pmatrix} k \\ k + 1 \end{pmatrix}$.

Thus, $D_1 = \begin{pmatrix} 0 & 1 & 0 & 0 & 0 \\ 0 & 1 & 0 & 0 & 0 \end{pmatrix}$ and $D_2 = (0\,1\,0\,0\,0)$. We deduce that: $\vec{w}_1 = \vec{w}_2 = (n\,0\,l\,j\,i)^T$. Therefore, both $D_2\vec{w}_1$ and $D_1\vec{w}_2$ are constant.

3.   OVERVIEW OF THE SLC METHOD

The optimization goal of our symbolic clustering algorithm is the same as that of static scheduling algorithms [Sarkar, 1989, Yang and Gerasoulis, 1994]: eliminate unnecessary communication and preserve data locality, map symbolic clusters to processors evenly to achieve load balance, and execute clusters within each processor using the multi-threading technique to overlap computation with communication. The mapping and scheduling process is symbolic in the sense that the change in the problem size and the number of processors does not affect the solution derived by our algorithm.

The main steps involved in our method for scheduling and executing a parameterized task graph are summarized as follows:

1. Given a PTG, we first simplify this graph by merging rules, when possible, in order to reduce the number of rules. Then we extract and sort all bijective rules in the PTG. Then, we perform a linear clustering on these bijection rules, The clustering process is discussed in details in Section 4.. The important aspect of this process is that merged clusters are always linear whatever parameter values are, namely, no independent task are placed in the same cluster. In this way, parallelism is preserved while unnecessary communication is eliminated.

   Clustering is conducted by assigning two end tasks of a dependence edge to the same cluster (so communication between these two tasks is considered zero). This process is called *edge zeroing*. Since each emission rule represents a set of dependence edges, we will call this process as "rule zeroing".

2. Given a linear cluster of a PTG, the second step of SLC is to provide the identification of each cluster, which is a mapping function from a task ID to a cluster number. For example, mapping function $\kappa(T1, (3, 7))$ is the cluster ID of task $T1(3, 7)$. This procedure is trivial if clustering is not done symbolically. In our setting, this mapping function allows us to map a cluster symbolically to a physical processor.

3. The third step is to derive the mapping and packaging of data items used during execution. This is done using communication rules. When a rule is not zeroed this means that data will be sent from a processor to another. Rules with cluster mapping describe which kind of data (scalar, vector, matrix, ...) should be sent. We use these informations for generating message packaging code.

4. The last part is the execution of symbolic clusters assigned to each processor. At runtime, symbolic clusters are mapped to the given number of processors (p) evenly using a cyclic (cluster ID mod p) or block mapping formula (cluster ID $/p$).

   Task execution is asynchronous and is driven by message communication. Each processor maintains a ready queue and if all data items needed for a task are available locally, this task becomes ready. For each processor, we maintain t active threads and each of them can pick up a ready task for execution. The advantage of having multiple threads is that the system can take advantages of SMP nodes if applicable. After a task completes its execution, data items are sent following the emission rules related to this task.

   There is an important issue on how to initiate execution. At the beginning of execution, each processor needs to find starting tasks (i.e. tasks without parents) and put them on its ready task queue. This is done by computing all valid instances of each task assigned to this processor, called $valid(Ta)$. Then this processor computes all the instances of Ta that need to receive data following the reception rules (call it $recv(Ta)$). The difference $valid(Ta) - recv(Ta)$ is a polyhedron that describes all starting task instances of Ta assigned to this processor.

In the next two sections, we will discuss in details how tasks can be clustered and mapped symbolically.

4.   SYMBOLIC LINEAR CLUSTERING

In this step, we allocate tasks to an unbounded number of virtual processors and in the literature this is often call clustering. All the tasks assigned to the same cluster will be executed on the same physical processor.

A cluster is said **linear** if all its task form a path in the instantiated task graph of a PTG for given program parameters. The motivation for linear clustering is based on a study by Yang and Gerasoulis [Gerasoulis and Yang, 1993]. They have proven that a linear clustering provides good performance on an unbounded number of processors for coarse grain graphs. If g is the granularity a task graph G, PT_{lc} is the parallel time of any linear clustering, and PT_{opt} is the parallel time of an optimal clustering applied to G, then $PT_{lc} \leq (1 + 1/g)PT_{opt}$. Thus as long as g is not too small, linear clustering produces a schedule competitive to the optimum when there are a sufficient number of processors.

Our clustering algorithm contains two parts: 1) merge rules together if possible to reduce the searching space for clustering. 2) find bijective clusters and cluster those rules linearly.

Rule merging.   Before clustering the given PTG, we merge a few emission rules if possible. Two emission rules are mergable if each rule describes the same set of dependence edges and their data items communicated can be combined. For example, let us consider the two following rules:

$$R1 : T1(k) \longrightarrow T2(k + 1) : A(k)|1 \leq k \leq n$$

and

$$R2 : T1(k) \longrightarrow T2(k + 1) : A(i)|1 \leq k \leq n, k + 1 \leq i \leq n.$$

$R1$ and $R2$ describe the same set of edges. We see that rule $R1$ sends element $A(k)$ and rule $R2$ sends elements of vector A from $k + 1$ to n. Hence, these two rules can be merged in a rule R that sends elements of vector A from k to n:

$$R : T1(k) \longrightarrow T2(k + 1) : A(i)|1 \leq k \leq n, k \leq i \leq n.$$

Input: R a set of communication rules
Output: The set $\mathcal{Z}$ of zeroed rules
$\mathcal{Z} := \emptyset$
merge_rules(R)
bijection_rule_set:=compute_bijection_rules(R)
sort_rules(bijection_rule_set)
foreach r **in** bijection_rule_set **do**
   if not_in_conflict(r,$\mathcal{Z}$) **then**
     $\mathcal{Z}$+=r
   endif
enddo

Figure 9.3   The rule clustering algorithm

Merging of rules is important because it reduces the number of rules. In that way, the clustering algorithm discussed below can spend less time in searching and zeroing bijective rules.

Rule zeroing.   We give a formal definition of rule zeroing as follows. Given a rule with form $Ta(\vec{u}) \longrightarrow Tb(\vec{v}) : D(\vec{y})|P$, it is *zeroed* if and only if $\kappa(Ta, \vec{u}) = \kappa(Tb, \vec{v})$ for all the valid instances of $\vec{u}$ and $\vec{v}$ in P. The zeroing algorithm is summarized in Figure 9.3.

The zeroing algorithm extracts all bijection rules, using the method described in Section 2.. Then it sorts all bijective emission rules. The sorting is done such that rules implying more communication are to be zeroed first. The rule ordering is done by taking into consideration the dimension of data communicated (i.e. a scalar, a row or a column, a matrix block). Once bijective emission rules are sorted by a decreasing dimension of data communication, *transitive* rules (as defined below) are placed at the end of this sorted list.

An emission rule R is called *transitive* if it sends data from task $T1$ to $T3$ and there exist two rules $R1$ and $R2$ such that $R1$ sends data from $T1$ to $T2$ and $R2$ sends data from $T2$ to $T3$ (see Figure 9.4).

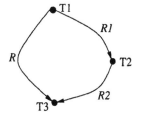

Figure 9.4 R: a transitive rule

As shown in next section, it appears that a clustering decision based on R may be in conflict with that for $R1$ and $R2$. But the clustering decisions for $R2$ and $R1$ do not conflict with each other. Since our goal is to zero as many rules as possible, we rank the zeroing proiority of rules $R1$ and $R2$ higher than that for rule R and thus we place transitive rules at the end of the sorted rule list.

During the zeroing loop shown in Figure 9.3, the algorithm checks if zeroing a selected rule is not in conflict with a rule already zeroed. Two rules are in conflict if, once zeroed, the produced clustering is not linear. If the rule is not in conflict with any other zeroed rules then the algorithm adds this newly-zeroed rule to the set of zeroed rules. The algorithm repeats this process until all rules have been examined once in the sorted order.

Conflict Detection.   Now we describe how to determine if zeroing of two rules is in conflict. Let us consider the two following emission rules called R_1 and R_2:

$$R_1 : T_a(\vec{u}) \longrightarrow T_b(\vec{v})|P_1$$
$$R_2 : T_c(\vec{w}) \longrightarrow T_d(\vec{z})|P_2$$

Suppose that R_1 is already zeroed. We want to check if R_2 is not in conflict with zeroing of R_1. There are two kinds of conflict: the fork conflict and the join conflict.

- We have a *fork conflict* if:

  1. $T_a = T_c$;
  2. There exists an instance of $\vec{u}$ called $\vec{u_1}$ ($\vec{u_1} \in P_1$) and an instance of $\vec{w}$ called $\vec{w_1}$ ($\vec{w_1} \in P_2$) such that $T_a(\vec{u_1}) = T_c(\vec{w_1})$;
  3. For the above v_1 and w_1 values, let $\vec{v_1}$ be the image of $\vec{u_1}$ according to rule R_1 and $\vec{z_1}$ be the image of $\vec{w_1}$ according to rule R_2. $\vec{v_1}$ and $\vec{z_1}$ satisfy condition $T_b(\vec{v_1}) \neq T_d(\vec{z_1})$.

- We have a *join conflict* if:

  1. $T_b = T_d$;
  2. There exists an instance of $\vec{v}$, called $\vec{v_1}$ ($v_1 \in P_1$) and an instance of $\vec{z}$ called $\vec{z_1}$ ($\vec{z_1} \in P_3$) such that $T_b(\vec{v_1}) = T_d(\vec{z_1})$;
  3. For the above v_1 and z_1 values, let $\vec{u_1}$ be the source of $\vec{v_1}$ according to rule R_1 and $\vec{w_1}$ be the source of $\vec{z_1}$ according to rule R_2. $\vec{u_1}$ and $\vec{w_1}$ satisfy condition $T_a(\vec{u_1}) \neq T_c(\vec{w_1})$.

In order to automatically compute if zeroing of two rules is in a *fork conflict* (the *join conflict* case is symmetric) we proceed as follows. For checking the second sufficient condition of a fork conflict, we compute a set I, which is the intersection of all the tasks $T_a(\vec{u})(\vec{u} \in P_1)$ with all the tasks $T_c(\vec{w})(\vec{w} \in P_2)$.

If I is not empty then condition 2 is verified. For checking the third sufficient condition, we need $T_b = T_d$ and we compute if R_1 restricted to I is equal to R_2 restricted to I. If not then zeroing R_1 and R_2 is in a *fork conflict*. We can symbolically compute I and the restriction of R_1 and R_2 to I using the *omega test* software from University of Maryland [Pugh, 1992].

For the example in Figure 9.2, Rule 2 is valid only when $j = k + 1$ and Rule 3 is valid only when $k + 2 \le j \le n + 1$. Hence there is no instance of $T_2(k, j)$ which have two child tasks following Rule 2 and Rule 3, i.e. $T_2(k + 1, j)$ and $T_1(k + 1)$. Thus zeroing these two rules is not in conflict for producing linear clusters.

We can prove the following theorem to show the correctness of our rule zeroing algorithm.

Theorem 1 *If bijective emission rules of a PTG are zeroed using the SLC zeroing algorithm, then the result of clustering is linear.*

  Sketch of Proof: When a bijective rule is zeroed, exactly one edge for each instance of the rule is zeroed in the instantiated task graph.

Since SLC ensures that there are no conflict in zeroing multiple bijective rules. Therefore, zeroing those rules produces a clustering with no independent task instances assigned to the same cluster.

            ■

5.   CLUSTER IDENTIFICATION

The previous algorithm zeros a number of rules and as a result it zeros a number of edges in the instantiated task graph for a given set of program parameter values. However, there is no cluster ID given for each cluster. We need such an ID number explicitly so we can assign clusters evenly to physical processors using a symbolic mapping function. In static scheduling research [Sarkar, 1989, Yang and Gerasoulis, 1994], producing such a number is easy since clusters can be explicitly numbered.

Given a set of zeroed rules, we show how to explicitly build a symbolic function $\kappa(Ta, \vec{u})$ such that for any generic task Ta and any valid instance $\vec{u}$, this function gives a cluster number for this task instance. We call this procedure as *cluster identification* or *cluster numbering*.

Notice that all the tasks in the same cluster will have the same cluster number and tasks from different clusters should not have the same cluster number. More formally, let us consider a zeroed rule:

$$R1 : Ta(\vec{u}) \longrightarrow Tb(\vec{v})|P_1$$

Under constrain P_1, we must have:

$$\kappa(Ta, \vec{u}) = \kappa(Tb, \vec{v}). \tag{9.1}$$

We briefly discuss our method using an example as follows. The basis of our method is the same as the one used by Feautrier in [Feautrier, 1994] for automatic distribution. We assume that our clustering function is affine with respect to program parameters and the iteration vector. Thus we let

$$\kappa(Ta, \vec{u}) = \vec{\alpha_a}\vec{u} + \beta_a + \vec{\gamma_a}\vec{p}$$

where $\vec{\alpha_a}$ and γ_a are vectors of unknowns which have to be found, β_a is a constant and $\vec{p}$ is the vector of the program parameters. We also have

$$\kappa(Tb, \vec{v}) = \vec{\alpha_b}\vec{u} + \beta_b + \vec{\gamma_b}\vec{p}.$$

Then solving Equation 9.1 leads to $\vec{\alpha_a}\vec{u} + \beta_a + \vec{\gamma_a}\vec{p} = \vec{\alpha_b}\vec{v} + \beta_b + \vec{\gamma_b}\vec{p}$.

For example, let us look at the second rule in Figure 9.2: $T2(k, j) \longrightarrow T1(k+1) : A(i, k+1)|1 \leq k \leq n-2, j = k+1, k+1 \leq i \leq n$. By zeroing this rule, we have $\kappa(T2, (k, j)) = \alpha_{2,1}k + \alpha_{2,2}j + \beta_2 + \gamma_{2,1}n$ and $\kappa(T1, (k+1)) = \alpha_{1,1}(k+1) + \beta_1 + \gamma_{1,1}n$. This leads to

$$\begin{cases} \alpha_{2,1}k + \alpha_{2,2}j + \beta_2 + \gamma_{2,1}n = \alpha_{1,1}(k+1) + \beta_1 + \gamma_{1,1}n \\ j = k+1 \end{cases} \quad (9.2)$$

Equation 9.2 is satisfied with $\alpha_{2,1} + \alpha_{2,2} = \alpha_{1,1}$, $\beta_2 + \alpha_{2,2} = \beta_1 + \alpha_{1,1}$ and $\gamma_{2,1} = \gamma_{1,1}$.

For emission Rule 3 in Figure 9.2: $(T2(k, j) \longrightarrow T2(k+1, j) : \{A(i, j)|1 \leq k \leq n-2, k+2 \leq j \leq n+1, k+1 \leq i \leq n\})$, we have:

$$\alpha_{2,1}k + \alpha_{2,2}j + \beta_2 + \gamma_{2,1}n = \alpha_{2,1}(k+1) + \alpha_{2,2}j + \beta_2 + \gamma_{2,1}n. \quad (9.3)$$

Equation 9.3 is satisfied with $\alpha_{2,1} = 0$.

The above analysis for the second and third rules of Figure 9.2 leads to the following linear system:

$$\begin{cases} \alpha_{2,1} + \alpha_{2,2} - \alpha_{1,1} = 0 \\ \beta_2 + \alpha_{2,2} - \beta_1 - \alpha_{1,1} = 0 \\ \gamma_{2,1} = \gamma_{1,1}\alpha_{2,1} = 0 \end{cases} \quad (9.4)$$

System 9.4 has many solutions. In particular,

$$\begin{cases} \alpha_{2,2} = \alpha_{1,1} = 1 \\ \beta_1 = \beta_2 = \alpha_{2,1} = \gamma_{1,1} = \gamma_{2,1} = 0 \end{cases} \quad (9.5)$$

Thus, one explicit clustering function κ for Gaussian Elimination PTG is:

$$\kappa(T2, (k, j)) = j$$

and

$$\kappa(T1, (k)) = k.$$

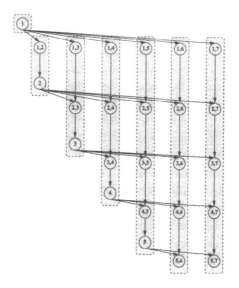

Figure 9.5   A clustering for an instantiated gaussian elimination graph with $n = 6$

This clustering is depicted in Figure 9.5.

The above example looks simple; however there is a complication we need to handle. The mapping derivation from two rules for the same task may lead to a restrictive condition on the $\vec{\alpha}$ vector value in computing the cluster affine function of this task. For instance, given the following two zeroed rules :

$$R1 : T1(k, j) \longrightarrow T1(k+1, j) | 1 \leq k \leq n, 1 \leq j \leq k$$

$$R2 : T1(k, j) \longrightarrow T1(k, j+1) | 1 \leq k \leq n, k+1 \leq j \leq n.$$

Zeroing rule R1 leads to $\alpha_{1,1} = 0$ but zeroing rule R2 leads to $\alpha_{1,2} = 0$. We do not want the $\vec{\alpha}$ solution vector to be zero for all the tasks because this solution maps all task instances into one cluster, which yields no parallelism. In the above case, we know that zeroing these two rules are not in conflict in terms of the linear clustering constraint. To fix this problem, we can use the two mapping functions for task $T1$ with two disjoint polyhedra:

$$\kappa(T1, (k, j)) = \begin{cases} j & \text{if } 1 \leq k \leq n, 1 \leq j \leq k \\ k & \text{otherwise.} \end{cases}$$

In general, we have proposed a technique called graph splitting which uses multiple mapping functions for each selected task if the derived solution is $\vec{\alpha} = \vec{0}$ for all tasks. Given a zeroing of a PTG, we construct a dependence graph G only based on all zeroed bijective rules. We compute the cluster mapping function for each task based on zeroing results. If for all the tasks Ta the mapping solution is $\vec{\alpha} = 0$, we identify the rules that cause setting, we split the G into a few sub-PTGs by dividing the polyhedra domain of Ta into disjoint parts based on these rules. For the above example, the restriction is caused by two rules among the instances of the same task. Then we apply the cluster numbering procedure recursively to each subgraph. The obtained solution for task T_a is no longer an affine function but a *piecewise affine function*. Notice if a task has multiple out-going edges in G, those rules can also potentially impose a restrictive condition on cluster numbering and we can also split based on these rules. Figure 9.6 illustrates splitting of a task $T2$ based on the rules "$T2 \rightarrow T3$" and "$T2 \rightarrow T4$". Notice that all rules in G are bijective, thus subgraphs $G1$ and $G2$ deal with disjoint sets of all task instances. For example, task instances of $T2$ are divided into two parts, one in $G1$ and another in $G2$. Then task instances of $T1$ are also divided in two parts accordingly due to bijection and rule "$T1 \rightarrow T5$" will only need to appear in $G2$.

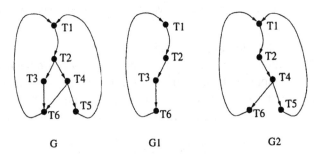

Figure 9.6   Spliting of graph G into G_1 and G_2 at node $T2$ (note that G_2 may be partitioned again at node $T4$)

6.   MULTI-THREADED EXECUTION OF CLUSTERS

So far each processor has been assigned a number of clusters. Assume that there are t clusters at one processor. Each cluster is a linear task chain and is considered as a thread on that processor. Then at most t tasks can be executed concurrently on this processor. Multi-threading is important for hiding communication latency and overlapping computation with communication [Tang and Gao, 1998].

We have shown how to build parameterized polyhedrons that describe starting tasks. Once parameter values are given during runtime, each processor P_x scans all starting tasks. For each starting task, if its processor location is P_x and this task should be executed at processor P_x.

We propose the following method for executing tasks on each processor: we maintain a ready queue for tasks ready to be executed and a waiting queue for tasks that have already received data and are still waiting for additional data.

- When a thread is idle, it picks up a task from the ready queue if available.

- When a task receives a data item:

    1. If this task is not in the waiting queue then we compute the number of its parents and decrement it. This number is called the *number of remaining parents*. If this number becomes zero, we put this task in the ready queue; otherwise we put it in the waiting queue.

    2. If this task is in the waiting queue then we decrement the number of its remaining parent. If this number reaches zero, we put this task in the ready queue.

- When a task completes computation, it sends data to the child tasks described by its emission rules.

Notice that for each generic task, we need to derive a polynomial that computes the number of its parent tasks. This polynomial is derived from the reception rules and can be used for any valid instance of iteration vectors and problem parameters.

7.   SIMULATION AND EXPERIMENTAL RESULTS

Size of the matrix	1000	2000	4000
Number of tasks $(n^2/2 + 3n/2 - 2)$	501 498	2 002 998	8 005 998
Number of edges $(n^2 + n - 4)$	1 000 996	4 001 996	16 003 996
Number of float op $((n^3 + n^2 - 2)/10^6)$	1 001	8 004	64 016
Size in MB for Pyrros	20	89	374

Table 9.1 Characteristics of gaussian elimination DAG for various matrix sizes

We have implemented SLC for compile-time symbolic scheduling of a PTG, which automatically finds the clustering function κ for each task. We have also implemented part of runtime support for executing a PTG using the SLC schedule, which allows us to conduct experiments in assessing performance of an SLC symbolic schedule.

The benchmarks used in this chapter are several compute-intensive kernels found in scientific applications: Gaussian elimination, Cholesky factorization,

Givens algorithm, Jordan diagonalization, and matrix multiplication. The advantage of our algorithm is that it requires small memory space in clustering and produces symbolic solutions independent of problem sizes and the number of processors used. Furthermore, the time taken by our algorithm to compute the allocation function does only depend on the number of rules and hence is independent of the parameter values. In general the time taken by SLC is a few seconds and is never greater than one minute. In this section, we mainly assess if the scheduling performance of our algorithm is still competitive to a static algorithm while retaining symbolic processing advantages.

A Performance Comparison of SLC and DSC.   Table 9.2 compare performance of SLC with the DSC static clustering algorithm [Yang and Gerasoulis, 1994] for some small parameter size : 100, 200 and 400, the matrix dimension. We have not been able to perform tests for larger matrix sizes because task graphs become too large (over 1 million tasks/edges) and DSC cannot schedule them on our machine (see Table 9.1 for Gaussian Elimination characteristics). For larger values, the most useful ones, only SLC can be applied.

The result of matrix multiplication is not shown because SLC and DSC always find the optimal clustering, which yields no difference. The last column of these two tables is the ratio R of the DSC schedule length over the SLC schedule length. If R is greater than 1, it means that SLC outperforms DSC. The results show that despite that DSC computes a clustering suitable to each graph, R is never lower than 0.83. It appears that SLC outperforms DSC for some graph instances. These results indicate that the SLC algorithm delivers a symbolic scheduling solution with performance highly competitive to the DSC static solution in terms of scheduling quality.

We also have simulated the execution of clusters on a bounded number of processors (from 2 to 64). We have used the PYRROS scheduling tool [Yang and Gerasoulis, 1992] to merge clusters and simulate execution of tasks. The results are shown in Figure 9.8. In that case R is between 0.93 and 1.3, thus the scheduling quality of SLC is also good on a bounded number of processors. The results shows that our approach is reasonable in term of solution quality : we find equivalent schedules.

Gaussian Elimination on IBM SP2.   For Gaussian elimination code, we have developed a preliminary version of multi-threaded code to execute symbolically mapped clusters using a multi-threading package called *PM2* [Namyst and Méhaut, 1995]. We have run it on an IBM SP2 with 16 RS6000 processors. Figure 9.7 shows the speedup of the Gaussian elimination code with various matrix sizes. We have a speedup of 12.28 for 16 processors when the matrix size is 4000. The result shows that our Gaussian elimination code scales well.

Graph	n	Sequential time	No. clusters DSC	Sched. length DSC	No. clusters SLC	Sched. length SLC	R
Gauss	100	1009800	101	25093	102	30001	0.84
Gauss	200	8039600	201	100193	202	120001	0.83
Gauss	400	64159200	401	400393	402	480001	0.83
Jordan	100	1509950	101	40002	101	40202	0.99
Jordan	200	12039900	201	160002	201	160402	1.00
Jordan	400	96159800	401	640002	401	640802	1.00
Givens	100	3720750	98	166393	101	175906	0.95
Givens	200	39214956	318	671163	201	755580	0.89
Givens	400	257418656	496	2668550	401	2844880	0.94
Cholesky	100	510150	100	32156	101	30100	1.07
Cholesky	200	4040300	239	121759	201	120200	1.01

Table 9.2   Comparison between SLC and DSC on an unbounded number of processors (relatively slow machine/coarse grain tasks)

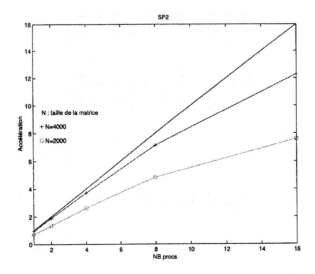

Figure 9.7   Speedups of gaussian elimination on IBM SP2

8.   RELATED WORK

There are a number of papers addressing computation mapping to processors. For loop programs, the SUIF compiler [Amarasinghe et al., 1995] uses a mapping algorithm [Anderson and Lam, 1993] which maximizes parallelism and minimizes communication for `doall` or `forall` parallelism. Both data mapping and computation allocation functions are affine.

In [Darte and Robert, 1993], Darte and Robert have introduced the *"communication graph"* model which describes read/write data dependence in the case of affine loop nests. For the case of one uniform loop nest, they prove that the problem of aligning computation with data in order to minimize communication is NP-Complete. They propose a heuristic for mapping loop nests for aligning computation with data. This work is further extended by Dion and Robert in [Dion and Robert, 1995]. They introduce the *"access graph"* model which is a communication graph whose edges are labeled with the access matrix of data arrays. They prove that aligning computation with data in order to minimize the residual communication is an NP-Complete problem for the case of one uniform loop nest.

Feautrier in [Feautrier, 1994] proposes an algorithm for *"cutting"* edges in a data flow graph. Edges are selected in a greedy manner and a Gaussian elimination is perform to find the allocation function. Data are then mapped according to the owner compute rule. In [Feautrier, 1996], Feautrier extended

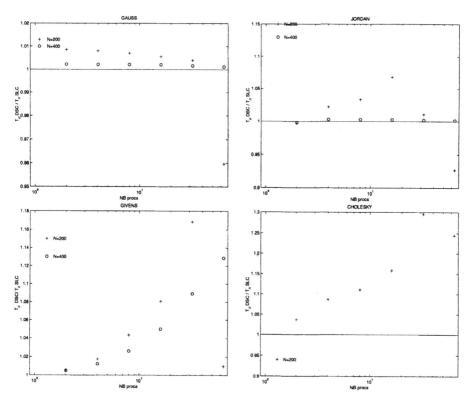

Figure 9.8 Makespan ratio DSC/SLC for a bounded number of processors (using PYRROS cyclic cluster merging). Cluster execution within each processor is done using the PYRROS RCP* algorithm

his previous results in order to map data more efficiently and the owner-compute rule is not always respected.

In [Mongenet, 1997], Mongenet addresses the problem of data allocation and instruction scheduling when programs are modeled as a system of recursive affine functions. Dependences are sorted according to the dimension of transmitted data. She gives a condition on dependences such that they do not imply communication in parallel execution. She shows how point-to-point communication and broadcast can be determined. Once data are mapped, the owner compute rule is applied for allocating computation.

All these works share a common characteristic: allocation/mapping functions are affine. None of these works can address affine piecewise functions. Also these work addresses either fine-grain parallelism or DOALL parallelism. Our setting is more complicated and deals with symbolic coarse-grain DAG parallelism.

9.   CONCLUSIONS

In this chapter we have presented the SLC algorithm which symbolically schedules parameterized task graphs and can perform optimization used in static scheduling, independent of program parameter values and the number of processors.

Our contribution is twofold. First, we have shown how to find a linear clustering for a PTG. Second, we have modified and augmented the Feautrier's algorithm with graph splitting for deriving an explicit clustering function in order to allocate clusters onto the processors. We guarantee that our solution can always been expressed as a piecewise affine function, which is not the case for any other known heuristics. We have demonstrated that symbolically scheduled tasks can be executed efficiently on multiprocessors and can deliver good performance for several compute-intensive kernel benchmarks. Our experimental results indicate that SLC finds symbolic schedules with performance highly competitive to static scheduling algorithms such as DSC.

Currently we are developing a code generator that transforms an annotated sequential program into parallel code using the SLC algorithm and evaluate our method in a wider range of benchmarks.

Acknowledgments

This work is supported in part by the European Community Eurêka EuroTOPS Project, the NSF/CNRS grant 139812 and the NSF grant INT-95113361/CCR-9702640.

Appendix

Various kernels and their task graphs are presented here. The clustering found by SLC is shown in shaded boxes.

References

[Adve and Vernon,] Adve, V. S. and Vernon, M. K. A Deterministic Model for Parallel Program Performance Evaluation. (Submitted for publication).

[Amarasinghe et al., 1995] Amarasinghe, S., Anderson, J. M., Lam, M. S., and Tseng, C. (1995). The SUIF Compiler for Scalable Parallel Machines. In *seventh SIAM Conference on Parallel Processing for Scientific Computing.*

[Anderson and Lam, 1993] Anderson, J. M. and Lam, M. S. (1993). Global Optimizations for Parallelism and Locality on Scalable Parallel Machines. In *ACM SIGPLAN'93 Conference on Programming Language Design and Implementation.*

[Chong et al., 1995] Chong, F. T., Sharma, S. D., Brewer, E. A., and Saltz, J. (1995). Multiprocessor Runtime Support for Fine-Grained Irregular DAGs.

In Kalia, R. K. and Vashishta, P., editors, *Toward Teraflop Computing and New Grand Challenge" Applications.*, New York. Nova Science Publishers.

[Cosnard and Jeannot, 1999] Cosnard, M. and Jeannot, E. (1999). Compact DAG Representation and Its Dynamic Scheduling. *Journal of Parallel and Distributed Computing*, 58(3):487–514.

[Cosnard et al., 1998] Cosnard, M., Jeannot, E., and Rougeot, L. (1998). Low Memory Cost Dynamic Scheduling of Large Coarse Grain Task Graphs. In *IEEE International Parallel Processing Symposium (IPPS'98)*, Orlando, Florida. IEEE.

[Cosnard et al., 1999] Cosnard, M., Jeannot, E., and Yang, T. (1999). SLC: Symbolic Scheduling for Executing Parameterized Task Graphs on Multiprocessors. In *International Conference on Parallel Processing (ICPP'99)*, Aizu Wakamatsu, Japan.

[Cosnard and Loi, 1995] Cosnard, M. and Loi, M. (1995). Automatic Task Graph Generation Techniques. *Parallel Processing Letters*, 5(4):527–538.

[Cosnard and Loi, 1996] Cosnard, M. and Loi, M. (1996). A Simple Algorithm for the Generation of Efficient Loop Structures. *International Journal of Parallel Programming*, 24(3):265–289.

[Darte and Robert, 1993] Darte, A. and Robert, Y. (1993). On the Alignment Problem. *Parallel Processing Letters*, 4(3):259–270.

[Deelman et al., 1998] Deelman, E., Dube, A., Hoisie, A., Luo, Y., Oliver, R., Sunderam-Stukel, D., Wasserman, H., Adve, V., Bagrodia, R., Browne, J., Houstis, E., Lubeck, O., Rice, J., Teller, P., and Vernon, M. (1998). POEMS: End-to-End Performance Design of Large Parallel Adaptive Computational Systems. In *First International Workshop on Software and Performance*, Santa Fe, USA.

[Dion and Robert, 1995] Dion, M. and Robert, Y. (1995). Mapping Affine Loop Nests : New Results. In *Int. Conf. on High Performance Computing and Networking, HPCN'95*, pages 184–189.

[El-Rewini et al., 1994] El-Rewini, H., Lewis, T., and Ali, H. (1994). *Task Scheduling in Parallel and Distributed Systems*. Prentice Hall.

[Feautrier, 1994] Feautrier, P. (1994). Toward Automatic Distribution. *Parallel Processing Letters*, 4(3):233–244.

[Feautrier, 1991] Feautrier, P. (1991). Dataflow analysis of array and scalar references. *International Journal of Parallel Programming*, 20(1):23–53.

[Feautrier, 1996] Feautrier, P. (1996). Distribution automatique des données et des calculs. *T.S.I.*, 15(5):529–557.

[Fu and Yang, 1996] Fu, C. and Yang, T. (1996). Sparse LU Factorization with Partial Pivoting on Distributed Memory Machines. In *ACM/IEEE Supercomputing'96*, Pittsburgh.

[Gerasoulis et al., 1995] Gerasoulis, A., Jiao, J., and Yang, T. (1995). Scheduling of Structured and Unstructured Computation . In Hsu, D., Rosenberg, A., and Sotteau, D., editors, *Interconnections Networks and Mappings and Scheduling Parallel Computation* , pages 139–172. American Math. Society.

[Gerasoulis and Yang, 1993] Gerasoulis, A. and Yang, T. (1993). On the Granularity and Clustering of Direct Acyclic Task Graphs. *IEEE Transactions on Parallel and Distributed Systems*, 4(6):686–701.

[Jeannot, 1999] Jeannot, E. (1999).   *Allocation de graphes de tâches paramétrés et génération de code.* PhD thesis, École Normale Supérieure de Lyon, France. `ftp://ftp.ens-Lyon.fr/pub/LIP/Rapports/PhD/PhD1999/PhD1999-08.ps.Z.`

[Kwok and Ahmad, 1996] Kwok, Y.-K. and Ahmad, I. (1996).   Dynamic Critical-Path Scheduling: An Effective Technique for Allocating Task Graphs to Multiprocessors. *IEEE Transactions on Parallel and Distributed Systems*, 7(5):506–521.

[Liou and Palis, 1998] Liou, J.-C. and Palis, M. A. (1998). A New Heuristic for Scheduling Parallel Programs on Multiprocessor. In *IEEE Intl. Conf. on Parallel Architectures and Compilation Techniques (PACT'98)*, pages 358–365, Paris.

[Loi, 1996] Loi, M. (1996). *Construction et exécution de graphe de tâches acycliques à gros grain.* PhD thesis, Ecole Normale Supérieure de Lyon, France.

[Mongenet, 1997] Mongenet, C. (1997). Affine Dependence Classification for Communications Minimization. *IJPP*, 25(6).

[Namyst and Méhaut, 1995] Namyst, R. and Méhaut, J.-F. (1995). PM2: Parallel Multithreaded Machine. A computing environment for distributed architectures. In *Parallel Computing (ParCo'95)*, pages 279–285. Elsevier Science Publishers.

[Palis et al., 1996] Palis, M., Liou, J.-C., and Wei, D. (1996). Task Clustering and Scheduling for Distributed Memory Parallel Architectures. *IEEE Transactions on Parallel and Distributed Systems*, 7(1):46–55.

[Papadimitriou and Yannakakis, 1990] Papadimitriou, C. and Yannakakis, M. (1990). Toward an Architecture Independent Analysis of Parallel Algorithms. *SIAM Journal on Computing*, 19(2):322–328.

[Pugh, 1992] Pugh, W. (1992). The Omega Test a fast and practical integer programming algorithm for dependence analysis. *Communication of the ACM.* (website: `http://www.cs.umd.edu/projects/omega`).

[Sarkar, 1989] Sarkar, V. (1989). *Partitioning and Scheduling Parallel Program for Execution on Multiprocessors.* MIT Press, Cambridge MA.

[Schrijver, 1986] Schrijver, A. (1986). *Theory of linear and integer programming*. John Wiley & sons.

[Tang and Gao, 1998] Tang, X. and Gao, G. R. (1998). How "Hard" is Thread Partitioning and How "Bad" is a List Scheduling Based Partitioning Algorithm? In *tenth ACM Symposium on Parallel Algorithms and Architectures (SPAA98)*, Puerto Vallarta, Mexico.

[Yang and Gerasoulis, 1992] Yang, T. and Gerasoulis, A. (1992). Pyrros: Static Task Scheduling and Code Generation for Message Passing Multiprocessor. In *Supercomputing'92*, pages 428–437, Washington D.C. ACM.

[Yang and Gerasoulis, 1994] Yang, T. and Gerasoulis, A. (1994). DSC Scheduling Parallel Tasks on an Unbounded Number of Processors. *IEEE Transactions on Parallel and Distributed Systems*, 5(9):951–967.

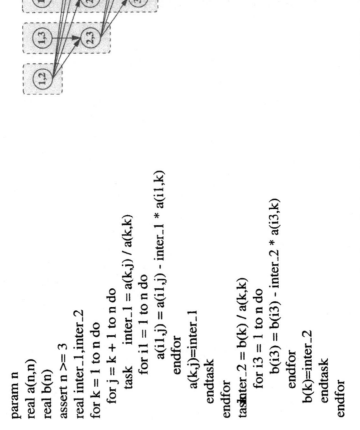

```
param n
real a(n,n)
real b(n)
assert n >= 3
real inter_1,inter_2
for k = 1 to n do
    for j = k + 1 to n do
        task    inter_1 = a(k,j) / a(k,k)
                for i1 = 1 to n do
                    a(i1,j) = a(i1,j) - inter_1 * a(i1,k)
                endfor
                a(k,j)=inter_1
        endtask
    endfor
    task inter_2 = b(k) / a(k,k)
                for i3 = 1 to n do
                    b(i3) = b(i3) - inter_2 * a(i3,k)
                endfor
                b(k)=inter_2
        endtask
endfor
```

Figure 9.A.1   Jordan diagonalization and its task graph (n=6), SLC finds $\kappa(T_1, (k,j)) = j$, $\kappa(T_2, (k)) = k$

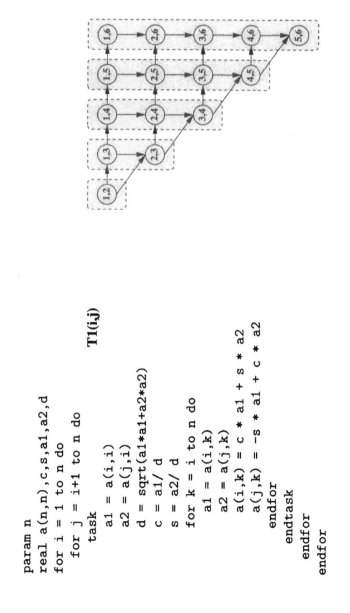

```
param n
real a(n,n),c,s,a1,a2,d
for i = 1 to n do
  for j = i+1 to n do
    task                         T1(i,j)
      a1 = a(i,i)
      a2 = a(j,i)
      d = sqrt(a1*a1+a2*a2)
      c = a1/ d
      s = a2/ d
      for k = i to n do
        a1 = a(i,k)
        a2 = a(j,k)
        a(i,k) = c * a1 + s * a2
        a(j,k) = -s * a1 + c * a2
      endfor
    endtask
  endfor
endfor
```

Figure 9.A.2   Givens algorithm and its task graph (n=6), SLC finds $\kappa(T_1, (i, j)) = j$

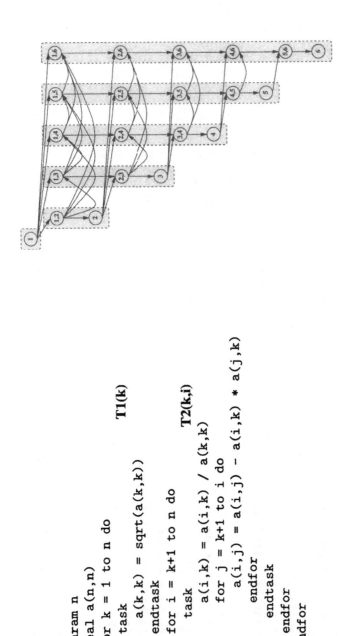

```
param n
real a(n,n)
for k = 1 to n do                    T1(k)
    task
        a(k,k) = sqrt(a(k,k))
    endtask
    for i = k+1 to n do
        task                          T2(k,i)
            a(i,k) = a(i,k) / a(k,k)
            for j = k+1 to i do
                a(i,j) = a(i,j) - a(i,k) * a(j,k)
            endfor
        endtask
    endfor
endfor
```

Figure 9.A.3   Cholesky algorithm and its task graph (n=6), SLC finds $\kappa(T_1, (k)) = k$, $\kappa(T_2, (k, i)) = i$

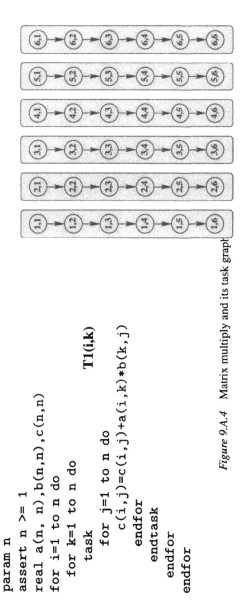

```
param n
assert n >= 1
real a(n, n),b(n,n),c(n,n)
for i=1 to n do
  for k=1 to n do
    task                    T1(i,k)
      for j=1 to n do
        c(i,j)=c(i,j)+a(i,k)*b(k,j)
      endfor
    endtask
  endfor
endfor
```

Figure 9.A.4   Matrix multiply and its task graph

Chapter 10

DECOMPOSITION ALGORITHMS FOR COMMUNICATION MINIMIZATION IN PARALLEL COMPUTING

Ioannis T. Christou
Delta Technology Inc.
Dept. of Operations Research
1001 International Blvd.
Atlanta, GA 30354
ioannis.christou@delta-air.com

Robert R. Meyer
University of Wisconsin - Madison
Computer Sciences Dept.
1201 W. Dayton Str.
Madison, WI 53706
rrm@cs.wisc.edu

Abstract   We present algorithms that minimize the communication overheads in large classes of parallel computing applications of scientific computing and engineering. Communication minimization is essentially equivalent to graph partitioning which in turn can be formulated as quadratic assignment. Decomposition - coordination techniques using genetic algorithms as multi-coordinators have allowed us to solve problems including millions or even billions of variables in a few seconds on a network of workstations. Furthermore, we establish the asymptotically optimal behavior of these algorithms for some regular types of domains.

1.   INTRODUCTION

The recent emergence of parallel computing environments that occurred during the nineties enabled the solution of many scientific and engineering problems whose complexity defeated previous attempts to solve them on serial ma-

P.M. Pardalos and L.S. Pitsoulis (eds.), Nonlinear Assignment Problems, 245–302.
© 2000 *Kluwer Academic Publishers.*

chines. Parallel computing has now advanced both from the hardware as well as from the algorithmic aspect to become a commercially available framework for solving very large scale problems.

With the advent of parallel computing came, however, new challenges; one of them is the problem of decomposition and the related problem of communication overheads. In order to solve a problem in parallel, one must be able to decompose it in an effective manner among the available processors, with two objectives in mind. The first objective is to identify the computations that can be performed in parallel *with as little communication as possible*. The second objective is to obtain an acceptable load balance, which leads to high utilization of all the available processors. Satisfying both these objectives can be a rather difficult task.

1.1   THE MINIMUM PERIMETER PROBLEM

This discussion focuses on the problem of minimizing communication overheads in certain parallel computing applications. This problem eventually leads to the Minimum Perimeter problem (MP), which belongs to the class of NP-hard problems. Minimum perimeter problems arise in various scientific and engineering computations in parallel computing environments. In the solution of Partial Differential Equations using finite difference schemes [Strikwerda, 1989], or the simulation of molecule behavior in Chemical Engineering or in edge detection in image processing [Schalkoff, 1989] and computer vision, or in the solution of max-flow problems over graphs with a grid structure using preflow-push [Ahuja et al., 1993], one must eventually perform a series of computations over a domain consisting of grid cells; the update of each cell requires the previous value of the cell as well as the value of its immediate four neighbors, namely, its northern, southern, eastern and western neighbors. The name *5-point uniform grid* refers to any such computation. Often, some area of the domain is refined to a more detailed level of granularity to obtain better precision in this part of the grid. The grid is then no longer uniform.

In order to efficiently perform such 5-point computations over a discretized domain on a distributed memory parallel computer (like a network of high-performance workstations or the Connection Machine CM-5 [TMC, 1991]) the computational load must be balanced across processors in a way that also minimizes interprocessor communication. This communication will occur at the boundaries of the regions assigned to the processor. It is therefore necessary to partition the grid in such a way so as to incur as small a total perimeter of the partition as possible. As the parallel processing paradigm shifts towards networks of workstations where the communication delays can be very high compared to the processing speed of the machines, it becomes more and more important that high quality partitions of the given domain among the available processors can be found efficiently. So we define the Minimum Perimeter prob-

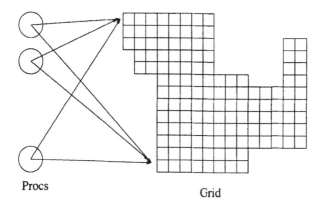

Procs   Grid

Figure 10.1   Network assignment formulation of MP

em MP($\mathcal{G}$, P) as follows: *Given a Grid $\mathcal{G}$ of unit cells, a number of processors P, and an associated load a_i for each processor, find an assignment of the grid cells to the processors so that the perimeter of the partition is minimized while observing the load balancing constraint that processor i is assigned exactly a_i cells.* The perimeter of a partition is the sum of the lengths of the boundaries of the regions that each processor occupies , while the load of each processor is the area of the region it occupies. By considering the graph of the grid, where each grid cell corresponds to a vertex and each border between two neighboring cells corresponds to an edge between the corresponding vertices, the problem becomes a Graph Partitioning problem. Even if we restrict MP to the class of uniform 5-point grids, it remains NP-hard as was shown by Yackel [Yackel, 1993]. MP can easily be formulated as nonlinear assignment, and in particular is a Quadratic Assignment problem (see Pardalos et. al. [Pardalos et al., 1993]), with $|\mathcal{G}|P$ binary variables and $|\mathcal{G}| + P$ constraints. Letting $\mathcal{I}$ denote the set of pairs of adjacent cells, and a_i the area for processor i, the QAP formulation is as follows:

$$\min \sum_{i,j\in\mathcal{G}} \sum_{\substack{p,p'=1\\p\neq p'}}^{P} c_{ij}x_i^p x_j^{p'} \tag{10.1}$$

$$s.t. \begin{cases} \sum_{i\in\mathcal{G}} x_i^p = a_p & p=1\ldots P \\ \sum_{p=1}^{P} x_i^p = 1 & i\in\mathcal{G} \\ x_i^p \in \mathbf{B} = \{0,1\} & i\in\mathcal{G}, p=1\ldots P \end{cases}$$

$$\text{where } c_{ij} = \begin{cases} 1 & \text{if } (i,j)\in\mathcal{I} \\ 0 & \text{else} \end{cases}$$

The objective of this optimization problem is a sum of quadratic terms of binary variables, while the constraints are network constraints as shown in Figure 10.1, where each processor represents a node with supply a_p and each cell in the grid represents a node with demand 1. Many well known methods for graph partitioning attempt to solve this QAP by exploiting certain properties of appropriate relaxations of its discrete nature. Others, exploit properties of relaxations of the assignment constraints of the problem to obtain information about the shape of each subdomain in an optimal solution.

1.2   APPROACHES FOR SOLVING THE MINIMUM PERIMETER PROBLEM

As noted before, MP is a graph partitioning problem (each grid cell representing a vertex and each boundary between two cells representing an edge between the two cells). In the remainder of this section, we review some of the most widely used algorithms that have been proposed before for solving MP. Because of the extreme size of the literature on the subject, we restrict our attention on methodologies that have proved themselves in practice or that are related to the decomposition - coordination framework which we shall analyze in detail. Many of the well established methods, like the spectral methods, the Kernighan-Lin method or the geometric mesh partitioner, work on arbitrary graphs; however, many of them also have the disadvantage of requiring the number of partitions to be a power of two.

1.2.1   Spectral Methods.   The spectral bisection algorithm was originally proposed by Pothen et. al. in [Pothen et al., 1990], and attempts to partition an arbitrary graph into two components of equal size while minimizing the cut edges of the partition. The idea behind the partitioning scheme traces back to the work in algebraic graph theory done by Fiedler [Fiedler, 1973] which relates the second smallest eigenvalue and corresponding eigenvector of the Laplacian matrix of a graph $\mathcal{G}$ to the edge and vertex connectivities of the graph. The

Laplacian matrix of a graph $\mathcal{G}(V, E)$ with n nodes is defined as follows:

$$L_{ij} = \begin{cases} -1 & \text{if } (v_i, v_j) \in E \\ d_i & \text{if } i = j \\ 0 & \text{else} \end{cases}$$

where d_i is the degree of node v_i. It is easy to check that $L = D - A$ where D is a diagonal matrix with diagonal elements the node degrees of the graph and A is the adjacency matrix of the graph. Essentially, spectral bisection attempts to solve the following QAP

$$\min \frac{1}{4} x' L x \qquad (10.2)$$

$$s.t. \begin{cases} x'e = 0 \\ x_i = \pm 1 \end{cases}$$

which is equivalent to solving the general graph partitioning problem exactly. The constraints simply require the partition to be balanced and to indicate to which partition each node belongs. The objective counts cut edges.

Since graph partitioning is NP-complete, there is little hope of solving (10.2) exactly, and thus spectral bisection *relaxes* the problem by solving its continuous counterpart with the integrality constraints replaced by the constraint $x'x = n$. It is exactly this relaxation that is the key approximation in the application of spectral methods. Hendrickson and Leland in [Hendrickson and Leland, 1995b] showed that the relaxed problem has as a solution $x = \sqrt{n}u_2$ where u_2 is the normalized eigenvector corresponding to the second lowest eigenvalue λ_2 of L, and n is the cardinality of V, and this solution is unique if $\lambda_2 \neq \lambda_3$ where λ_3 is the third lowest eigenvalue of L. This is due to a theorem proved in [Hendrickson and Leland, 1995b] stating that the matrix L is symmetric positive semi-definite, its normalized eigenvectors are pairwise orthogonal, and that the smallest eigenvalue of L is zero, with an associated normalized eigenvector $u_1 = \frac{1}{\sqrt{n}}e$.

Thus the solution of the relaxation of the discrete optimization problem (10.2) is reduced to an eigenvalue computation of the Laplacian matrix of the graph, which can be performed quite efficiently using Lanczos' algorithm or some other polynomial time algorithm. Then, the relaxed solution must be mapped back into the feasible region of the discrete problem. Spectral bisection does this by finding the median of the vector x and assigning all nodes with x_i less than the median to one partition ($x_i = -1$) and all the other nodes to the other partition ($x_i = +1$). This integer point represents the point satisfying the constraints in (10.2) that is closest to x, and the hope is that it is an optimal solution of (10.2) or near-optimal. The above theory also shows that a lower bound on the number of cut edges of the best separator of the graph $\mathcal{G}(V, E)$ is given by $\frac{n\lambda_2}{4}$. By using the eigenvectors of the Laplacian matrix which represent properties of the whole graph, this method makes use of global information about the graph, as opposed to local refinement graph partitioning schemes like the Kernighan-Lin algorithm [Kernighan and Lin, 1970].

The spectral bisection algorithm and its extensions, namely recursive spectral bisection, or RSB for short, and recursive spectral quadrisection and octasection [Hendrickson and Leland, 1995b] that generalize bisection by considering two or three eigenvectors of the Laplacian matrix of the graph) have received a lot of attention because of their demonstrated excellent performance. However, they have two potential drawbacks; the first is the need to construct the Laplacian matrix of the graph, which in large problems can lead to memory problems. The second drawback is the fact that they cannot easily partition a graph into an arbitrary number of components. For this end, dummy nodes must be introduced, and the resulting partition is unlikely to be near-optimal (balancing issues might also arise). Furthermore, as the computational results show, these methods fail to find very good quality solutions for the class of large 5-point uniform grids by failing to take full advantage of the geometry and special properties of the grid.

1.2.2   Multilevel k-Way Partitioning.

Karypis and Kumar [Karypis and Kumar, 1995], developed a multilevel k-way graph partitioning scheme that first reduces the size of the graph by collapsing nodes and edges, solves the k-way partitioning problem on this collapsed graph, and then projects this partition on the original graph by uncollapsing the reduced graph. The first phase of the scheme, is called the coarsening phase, and the second, the uncoarsening phase (in which, certain refinements are performed so as to improve on the final solution). This scheme is mainly motivated by computations in dynamically adaptive meshes arising in finite element computations. The algorithms based on this scheme form the METIS family of graph partitioning algorithms. While the algorithms are very fast and can partition unstructured meshes of more than a million nodes in a few seconds, the overall partition more often than not, violates the load balancing constraints.

1.2.3   Geometric Mesh Partitioning.

In 1993, Miller et. al. proposed a new method for partitioning a d-dimensional mesh that used geometric information of the graph, namely the coordinates of its nodes in $\mathcal{R}^d$ to obtain provably good edge or vertex separators [Miller et al., 1993]. The idea underlying geometric partitioning is that once the coordinates of the nodes of a given graph are given, one can map the d-dimensional mesh onto a $(d+1)$-dimensional space using a stereographic projection which will map all the nodes of the graph into the unit sphere in $\mathcal{R}^{d+1}$. Any node v in $\mathcal{R}^d$ is projected to this sphere at the point where the line passing through v and the north pole $(0, \dots, 0, 1)$ and the sphere intersect. Then, when the centerpoint of the projected nodes has been computed, a rotation of all the projected nodes about the origin is performed until the centerpoint aligns itself somewhere on the $(d + 1)$-st axis. Then, the nodes on the surface of the sphere are dilated until the centerpoint becomes the origin. The method chooses a random circle on the unit sphere in $\mathcal{R}^{d+1}$, and maps it back to $\mathcal{R}^d$ by undoing the dilation, rotation, and stereographic projection operations. At that point, all the original nodes in $\mathcal{R}^d$ which are located in the "neighborhood" of the projected circle form a vertex separator of the graph, from which one can compute an edge separator as well.

The most striking property of this method is that it doesn't make use of edge information at all. Indeed, all that it requires to compute the vertex separator is the coordinates of the nodes of the graph. Its power comes from the fact that in most geometric meshes that arise from practical applications, nodes that are away from each other in $\mathcal{R}^d$ tend not to have edges connecting them.

The geometric mesh partitioner attacks the problem of graph partitioning using the idea of exploiting the position of the nodes in space to extract "good" cuts, but requires geometric information about the graph that might not be available. Furthermore, generalizing it to partition meshes into an arbitrary number of components that is not a power of two is not straightforward .

1.2.4 Greedy Randomized Adaptive Search Procedures for Graph Partitioning.

Laguna et. al. presented in [Laguna et al., 1994] a method for partitioning an arbitrary weighted graph into two balanced partitions while minimizing the sum of the weights of the cut edges. This GRASP heuristic chooses one node at a time from a list of candidate nodes and alternates insertion of the chosen node between the two partitions. The nodes in the candidate list are the ones that maximize a gain function, which is simply the sum of the weights of the arcs from the given node to the nodes in the partition that will accept the node minus the sum of the weights of the arcs from the given node to the nodes in the other partition.

This GRASP heuristic, when implemented with appropriate data structures, has been shown to outperform the Kernighan-Lin heuristic. Even though the authors do not provide an extension to the general k-way partitioning, it is conceptually easy to design a GRASP heuristic for partitioning a graph among any number of components.

1.2.5 Fair Binary Recursive Decomposition.

Fair Binary Recursive Decomposition (FBRD) is a method proposed by Crandall and Quinn in [Crandall and Quinn, 1995] for the partitioning of uniform and non-uniform 5-point grids; their discussion is focused on rectilinear grids. They allow different processor speeds, which implies possible different area sizes for each partition. Their method divides the processors into two sets of processors so that the sum of the processor speeds in each list is as balanced as possible. Then, they divide the grid along the longest dimension of the (rectilinear) grid so that the resulting subgrids can accommodate the processor sublists. Then, each of the subgrids is partitioned among the processors in the processor list assigned to it recursively.

The method, although mainly focused on rectilinear 5-point grids (uniform or non-uniform), allows partitions among an arbitrary number of processors with different speeds. The main problem with FBRD is that for large numbers of processors, this technique may lead to inefficient partitions due to the fact that the individual components no longer look near-rectangular.

1.2.6   The Kernighan-Lin Heuristic.   One of the most famous algorithms for graph partitioning, the K-L heuristic [Kernighan and Lin, 1970] is a very well established method for general graph partitioning that is used in many modern codes as a post-processing routine. It requires an arbitrary initial partition of the graph in two (balanced) sets and then until no more improvement can be made, it exchanges pairs of nodes between the two sets so as to improve the total cost of the partition. In order to effectively do these swaps, it maintains some heap data structures to help sort the gains of the nodes; the complexity of performing a swap (always the best swap is performed, i.e. the swap that maximizes the cost improvement of the partition) is then shown to be $O(n^2 \log(n))$ where n is the number of vertices in the graph. For an unweighted version of the graph partitioning problem, the minimum number of K-L iterations is $O(|E|)$ where E is the set of edges of the graph.

In general, K-L is a fast and efficient local refinement technique that can improve on the partitions found by other algorithms. However, it does need a relatively good partition to begin with. Extensions to the K-L algorithm are presented in [Hendrickson and Leland, 1993] where Hendrickson and Leland show how a generalized K-L heuristic (GKL) can partition a graph among an arbitrary number of nodes.

1.2.7   Genetic Algorithms Using Intelligent Structural Operators.   Genetic algorithms (GAs) have been proposed before for solving the graph partitioning problem. In [von Laszewski, 1991], Von Laszewksi uses a representation for a partition of a graph into k components that assigns one allele in the chromosome for each node in the graph. This means that the total length of the chromosome of each individual is at least $|V|$ where V is the node set of the graph. Simple crossover and mutation is likely to result in an unbalanced (and thus infeasible) partition, so a different strategy is used to create the offspring: first a component is selected from one of the parents, and is copied onto a temporary copy of the other parent. Other nodes assigned to the same component that was just copied are unassigned, and are assigned back to another component using a repair strategy so as to maintain the balance of the partition. Similarly, a mutation is defined in this context as the exchange of two numbers of the coding. Finally, a variant of the K-L heuristic is applied to the offspring partition. The GA is run on a network of transputers, and the selection process follows a neighborhood model (instead of the traditional panmictic model) where individuals chose their mates from a neighborhood scheme that is implemented as a ring.

As the author notes, the encoding used does not capture the solution of the problem at a level that allows the mechanisms of crossover to successfully transmit good building blocks of solutions to the next generations. Still, the experimental results showed the superiority of the GA versus K-L for partitioning a graph among any number of processors for small graphs (one randomly generated graph with 900 nodes and average node degree of 4 served as the test bed of the experiments reported in the paper). It is not clear how the algorithm will perform as the number of nodes increases (to the order of many thousands).

1.2.8   Classical Branch & Bound Methods for Integer Programming.
At the expense of introducing more variables and constraints, we can reformulate problem (10.1) as a mixed linear integer program; such a linear mixed integer programming formulation (MIP) is shown in [Christou and Meyer, 1996c].

Unfortunately, as the size of the grid increases, the number of binary variables needed in any of these formulations renders the problem intractable by means of classical Branch & Bound methods [Nemhauser and Wolsey, 1985] that solve subproblems with relaxed integrality constraints. In fact, graph partitioning presents extra difficulties due to the symmetry of the problem implying at least $P!$ algebraically distinct optimal solutions which force the branch-and-bound tree to grow extremely large before optimality can be proved (more details are given in [Christou et al., 1999a]). Experiments with the commercially available CPLEX solver [CPL, 1998], showed that even for very small problems and good branching strategies, the results were unsatisfactory both in quality and in solution time.

1.2.9   Stripe and Snake Decomposition.
In the next sections, we will present methods that employ a high level approach that follows the decomposition - coordination paradigm for large scale optimization without relaxing the integrality constraints of the problem as Branch & Bound methods do. The decomposition of the problem relaxes the *coupling assignment constraints* and results in P independent subproblems, each of which can be solved to optimality. Genetic Algorithms or knapsack techniques can be used to coordinate blocks of subproblem solutions, modifying them as little as possible so as to obtain a solution that observes the assignment constraints of the overall problem. This approach leads to the development of a theory that enables us to construct

high quality solutions very efficiently, having the property that as the problem parameters grow to infinity, the resulting partitions become asymptotically optimal in the sense that their relative distance from a computable lower bound approaches zero.

2.   THE DECOMPOSITION - COORDINATION FRAMEWORK

Large-scale optimization problems arising from industrial applications are formed by concatenating and linking blocks of data corresponding to major components of the problem. For example, in financial applications, each block may represent a distinct time period or investment scenario. The corresponding block-constraints involve only the decision variables for that period. Variables from a period relate to the variables of the next periods via *coupling* constraints. These constraints may involve variables from all periods. In network optimization, components of the problem may correspond to distinct commodities to be transferred through common routes. This is the multi-commodity network flow problem (formulated by Fulkerson in [Fulkerson, 1963]). Block-angular problems have the following structure:

$$
\begin{aligned}
min_x \quad & f_1(x_1) + f_2(x_2) + \ldots + f_P(x_P) \\
& A_1 x_1 && = b_1 \\
& A_2 x_2 && = b_2 \\
& \ddots \\
& A_P x_P && = b_P \\
& D_1 x_1 + D_2 x_2 + \ldots + D_P x_P && \leq d \\
& 0 \leq x_i \leq u_i, \quad i = 1, \ldots, P
\end{aligned}
$$
(10.3)

In this formulation, x_p represents a block of variables, and the matrices A_p and $D_p, p = 1 \ldots P$ correspond to P blocks of constraints. The last constraints $Dx \leq d$ represent the *coupling constraints* of the problem. Some or all of the variables may be integer. In large problems, there may be tens or hundreds of blocks. In each block, there may be several thousand variables and constraints.

2.1   BLOCK ANGULAR OPTIMIZATION

The classical Dantzig-Wolfe [Dantzig and Wolfe, 1960] method was the first to exploit the special structure of the *continuous* version of problem (10.3). The problem can be reformulated as an optimization problem that involves determining optimal convex combinations of the extreme points and stepsizes of the extreme rays of the feasible sets defined by the block constraints $B_i = \{x : A_i x = b_i, 0 \leq x \leq u_i\}$. Column generation techniques may now be used; columns with negative reduced costs can be obtained by solving P independent subproblems having an appropriate objective function, and subject to the block constraints only. When no more such columns can be generated from any of the independent subproblems, the overall problem is solved.

The emergence of parallel computers equipped with very powerful general-purpose processors interconnected with fast switches stimulated research in the area of "coarse-grained decomposition" (DeLeone et. al. [DeLeone et al., 1994]). Price-directive decomposition methods solve the general continuous multicommodity flow problem iteratively. They use techniques from the domains of interior-point logarithmic barrier functions (see Schultz and Meyer [Schultz and Meyer, 1991]), augmented Lagrangian techniques (see Zakarian's thesis [Zakarian, 1995]), and alternating directions. In each iteration, the problem is decomposed into P independent subproblems with a modified objective function. Each of these subproblems is solved, subject to the block constraints B_i only. Then, in a coordination phase, a master problem combines the subproblem solutions, to produce a feasible solution that satisfies the coupling constraints. These methods are well suited for implementation on multiple-instruction-multiple-data (MIMD) machines, such as the IBM SP/2, or other loosely-coupled networks of workstations (NOWs).

Notice that the MP problem can also be written in the form (10.3), by formulating it as

$$\min \sum_{p=1}^{P} \mathcal{P}(x^p) \tag{10.4}$$

$$s.t. \begin{cases} \sum_{i \in \mathcal{G}} x_i^p = a_p & p = 1 \dots P \\ \sum_{p=1}^{P} x_i^p = 1 & i \in \mathcal{G} \\ x_i^p \in \mathbf{B} = \{0, 1\} & i \in \mathcal{G} \ p = 1 \dots P \end{cases}$$

where $\mathcal{P}(x^p)$ is the perimeter of component p and x^p is the block of variables x_i^p with p fixed, and i ranging over the cells of the grid domain $\mathcal{G}$.

Following the decomposition - coordination approach described above, we eliminate the coupling constraints $\sum_{p=1}^{P} x_i^p = 1$ $i \in \mathcal{G}$ to obtain a relaxed problem of the form

$$\min \sum_{p=1}^{P} \mathcal{P}(x^p) \tag{10.5}$$

$$s.t. \begin{cases} \sum_{i \in \mathcal{G}} x_i^p = a_p & p = 1 \dots P \\ x_i^p \in \mathbf{B} = \{0, 1\} & i \in \mathcal{G} \ p = 1 \dots P \end{cases}$$

that decomposes into P independent sub-problems of the form:

$$\min \mathcal{P}(x^p) \tag{10.6}$$

$$s.t. \begin{cases} \sum_{i \in \mathcal{G}} x_i^p = a_p \\ x_i^p \in \mathbf{B} = \{0, 1\} \quad i \in \mathcal{G} \end{cases}$$

Each such subproblem requires the determination of the minimum perimeter of a collection of a_p cells in the domain. In the next section we develop the necessary tools for achieving this task.

Having obtained solutions from the P independent subproblems (10.6), we must then combine them, to obtain a feasible solution of the overall MP problem. As discussed previously, *for continuous optimization* problems, the coordination phase combines the solutions of appropriately modified independent subproblems using linear weights to obtain a feasible solution of the original problem. The approach works because these weights are continuous variables and the coordination problem is a tractable continuous optimization problem. In contrast, for discrete optimization problems, the use of continuous weights in the coordination phase would likely destroy the integrality of the subproblem solutions, so the coordination phase becomes a hard discrete optimization problem. Stripe and snake decomposition are techniques for solving the coordination problem utilizing Genetic Algorithms, as GAs excel in combining parts of solutions to obtain better solutions. Indeed, using a GA with an appropriate representation that encodes subproblem solutions (or collections of subproblem solutions) has allowed us to obtain optimal or near optimal solutions to problems with millions or even billions of binary variables in seconds. We report on these results in Section 6..

2.2   OPTIMAL SUBPROBLEM SOLUTIONS TO MP

We now present theoretical results on the nature of the subproblem solutions. In particular, we give details on the shape and the objective value of the optimal collections of a given number of cells in a domain. We use these results to construct asymptotically optimal solutions for large classes of the MP problem.

2.2.1   Lower Bounds and Optimal Configurations.
Yackel and Meyer showed in [Yackel and Meyer, 1992a], that for any given area $\mathcal{A}_p$, a lower bound on the perimeter of *any* configuration of $\mathcal{A}_p$ cells is given by

$$\Pi^*(\mathcal{A}_p) = 2 \left\lceil 2\sqrt{\mathcal{A}_p} \right\rceil. \tag{10.7}$$

Furthermore, this lower bound is tight, as there exists a library of configurations, called the *optimal shapes* for $\mathcal{A}_p$, that achieves this lower bound on the perimeter. We can generate these shapes as follows: we start with a rectangle with perimeter $\Pi^*(\mathcal{A}_p)$ and area at least $\mathcal{A}_p$. We then remove the corner cells of this rectangle until the area of the remaining object is exactly $\mathcal{A}_p$. The remaining object is an optimal shape for $\mathcal{A}_p$. All the optimal shapes for a given area size $\mathcal{A}_p$ can be constructed using this technique [Yackel, 1993]. This library, there-fore, is the complete set of optimal solutions to the subproblems (10.6). The optimal shapes for areas 5 and 7 are shown in Figure 10.2. All of these optimal

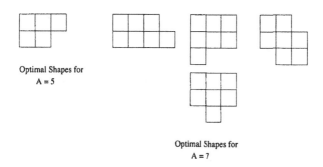

Optimal Shapes for
A = 5

Optimal Shapes for
A = 7

Figure 10.2 Optimal shapes for areas 5 and 7

shapes have a property called slice-convexity, which is the same as convexity in the "polyomino" literature. As a consequence of this property, the perimeter of any optimal shape is twice its semi-perimeter. (The semi-perimeter of a shape is the sum of the height and width of the smallest rectangle that contains it.) Many of these optimal shapes are rectangles with a "fringe" attached to one of their sides (Yackel and Meyer [Yackel and Meyer, 1992b]). Thus, they can be characterized by three numbers, namely the dimensions of the rectangle h, w and the size of the fringe f. We will denote such a near-rectangular shape by (h, w, f). In general, the number of such near-rectangular optimal shapes is of order $\mathcal{A}_p^{1/4}$ (see [Yackel et al., 1997]). However, this does not encompass all possible minimum perimeter configurations. There is a wealth of literature (see for example the papers by Lin and Bousquet-Melou [Lin, 1991, Melou, 1994]) on using generating functions to determine the exact number of "convex polyominoes" with various properties.

These near-rectangular optimal configurations play a key role in the analysis below. We prove a particular optimality test for these shapes:

Lemma 1 *Given an area $\mathcal{A}_p$ of unit cells, let $k := \lfloor \sqrt{\mathcal{A}_p} \rfloor$. Then*

- *if $k^2 = \mathcal{A}_p$ the shape $(k, k, 0)$ (a $k \times k$ square) is an optimal shape and its perimeter is $4k$.*

258 NONLINEAR ASSIGNMENT PROBLEMS

- *if $k^2 < A_p \le k(k+1)$ then the shapes (k, k, f) and $(k+1, k-1, f')$ with $f \le k$ and $f' \le k+1$ are optimal shapes and their perimeter is $4k+2$.*

- *else if $k(k+1) < A_p$ then the shapes $(k, k+1, f)$ and $(k+1, k, f)$ are optimal and their perimeter is $4(k+1)$.*

Proof:

- Consider $k^2 = A_p$. The lower bound on the perimeter of any configuration of A_p cells is $2 \lceil 2\sqrt{A_p} \rceil = 4k$. Since $(k, k, 0)$ has area A_p and perimeter $4k$ it is an optimal shape.

- Consider the case $k^2 < A_p \le k(k+1)$. Now the lower bound is $2 \lceil 2\sqrt{A_p} \rceil \ge 4\sqrt{A_p} > 4k \Rightarrow \Pi^*(A_p) \ge 4k+2$ (since Π^* is always an even number). Now, (k, k, f) with $f = A_p - k^2 \le k$ has perimeter $2(2k+1)$, and so is an optimal shape, as is the shape $(k+1, k-1, f')$ with $f' = A_p - k^2 + 1 \le k+1$ because its perimeter is also $2(2k+1)$.

- Finally, assume that $k(k+1) < A_p$. Then $2\sqrt{A_p} = 2(k+r)$ with $r \in (0, 1)$ and therefore $\Pi^*(A_p) = 2 \lceil 2k + 2r \rceil$ which is equal to $4(k+1)$ iff $r > \frac{1}{2}$. But we have that

$$\sqrt{A_p} = k + r \Rightarrow A_p = k^2 + 2kr + r^2 > k^2 + k \Leftrightarrow$$
$$A_p - k(k+1) = r^2 + 2kr - k > 0$$

If $r \le \frac{1}{2}$ then $r^2 \le \frac{1}{4}$ and we get $A_p - k(k+1) \le \frac{1}{4}$ which is a contradiction since $A_p - k(k+1)$ is an integer greater than 0. So it must be $r > \frac{1}{2}$, and therefore $\Pi^*(A_p) = 4(k+1)$. The shapes $(k, k+1, f)$ and $(k+1, k, f)$ with $f = A_p - k(k+1) < k+1 \Rightarrow f \le k$ (since $(k+1)^2 > A_p$) have perimeter equal to $4(k+1)$ and so are optimal configurations for A_p. ∎

When P, the number of processors divides the total area A of the grid, in the equi-partitioning case each processor is assigned $\frac{A}{P}$ cells. However, when P does not divide A, some processors will be assigned one more cell than others (to keep the loads as balanced as possible). Assuming that P_1 processors will get a lower load $a_i = A_1$ and that the rest $P_2 := P - P_1$ processors will get a heavier load $a_i = A_2 := A_1 + 1$, from the equations

$$A_1 P_1 + A_2 P_2 = A$$
$$P_1 + P_2 = P$$

we get that

$$P_1 A_1 + (P - P_1)(A_1 + 1) = A \implies PA_1 + P_2 = A.$$

Therefore,

$$\text{for } i = 1 \dots P_1 \quad a_i = A_1 = A \div P = \left\lfloor \frac{A}{P} \right\rfloor$$

$$P_1 = P - A \mod P$$

$$\text{and for } i = P_1 + 1 \dots P \quad a_i = A_2 = A \div P + 1 = \left\lceil \frac{A}{P} \right\rceil$$

$$P_2 = A \mod P$$

A lower bound on the solution of the MP($\mathcal{G}, P$) (equi-partition of a uniform 5-point grid of total area A among P processors) is then

$$L(A, P) = \begin{cases} P\Pi^*(\frac{A}{P}) & \text{if } A \mod P = 0 \\ P_1\Pi^*(A_1) + P_2\Pi^*(A_2) & \text{else} \end{cases} \tag{10.8}$$

Computational experience shows that even on the restricted class of rectangular grids, the lower bound in (10.8) is not always achievable. For an example, consider the rectangular grid of dimensions $1 \times N$ with N even, to be equi-partitioned among 2 processors. One reason that the lower bound (10.8) fails to be tight is that the dimensions of the grid may not be large enough to accommodate the relatively square optimal shapes [Christou and Meyer, 1996b]. In particular, we have the following lemma:

Lemma 2 *Assume that $M < N$ and that the following problem ($\mathcal{P}$) is feasible:*

$$\min_{h,w} h + w$$

$$s.t. \begin{cases} hw \geq A \\ h \leq M \\ w \leq N \\ h, w \in \mathcal{N}. \end{cases}$$

Let ($\mathcal{P}_{rel}$) denote the relaxed problem

$$\min_{h,w} h + w$$

$$s.t. \begin{cases} hw \geq A \\ h, w \in \mathcal{N}. \end{cases}$$

Assume that all optimal solutions of $(\mathcal{P}_{rel})$ violate at least one of the constraints of $(\mathcal{P})$. Then, an optimal solution of $(\mathcal{P})$ is $(h^*, w^*) = (M, \lceil \frac{A}{M} \rceil)$.

Proof: For each $h = 1 \ldots M$ that corresponds to a feasible point of $(\mathcal{P})$ (i. e. satisfies $\lceil \frac{A}{h} \rceil \leq N$), the w in the range $\lceil \frac{A}{h} \rceil \ldots N$ that yields the best objective is the value $\lceil \frac{A}{h} \rceil$. Thus, we need only find the number h in the range $1 \ldots M$ that corresponds to a feasible solution and minimizes the function $f(h) = h + \lceil \frac{A}{h} \rceil$. But the function $f(h)$ in the range $1 \ldots \lfloor \sqrt{A} \rfloor$ is non-increasing. To see this assume $i < \lfloor \sqrt{A} \rfloor$; we are going to show that $i + \lceil \frac{A}{i} \rceil \geq i + 1 + \lceil \frac{A}{i+1} \rceil$, or equivalently that

$$\left\lceil \frac{A}{i} \right\rceil - \left\lceil \frac{A}{i+1} \right\rceil \geq 1.$$

But the number $d := \frac{A}{i} - \frac{A}{i+1} = \frac{A}{i(i+1)} > 1$ as $i(i+1) < \lfloor \sqrt{A} \rfloor^2 \leq A$. Therefore, the ceilings of the two numbers $\frac{A}{i}$, $\frac{A}{i+1}$ are at a distance greater than, or equal to one, so $\lceil \frac{A}{i} \rceil - \lceil \frac{A}{i+1} \rceil \geq 1$.

As $M < \lfloor \sqrt{A} \rfloor$ (otherwise, there exists an optimal solution of $(\mathcal{P}_{rel})$ that does not violate the extra constraints of $(\mathcal{P})$ because it is shown in [Yackel and Meyer, 1992a] that there always exist an optimal solution of $(\mathcal{P}_{rel})$ that has $h^* = \lfloor \sqrt{A} \rfloor$.) an optimal solution of $(\mathcal{P})$ is $(h^*, w^*) = (M, \lceil \frac{A}{M} \rceil)$ and the optimal objective value of $(\mathcal{P})$ is $M + \lceil \frac{A}{M} \rceil$.   ∎

The above lemma 2 confirms that, when the domain is a sufficiently narrow horizontal band (for small enough M) so that no optimal shape from the collection of optimal shapes fits in the domain, then the optimal perimeter is $2(M + \lceil \frac{A}{M} \rceil)$.

3.   STRIPE DECOMPOSITION

3.1   ERROR BOUNDS FOR EQUI-PARTITIONS OF RECTANGULAR DOMAINS

We have presented a computable lower bound on the perimeter of the equipartition of *any uniform 5-point grid* among P processors. Now, we proceed to show that for rectangular domains, as long as the number of processors dominates the individual dimensions of the rectangle, partitions exist with a total perimeter whose relative distance from the lower bound goes to zero as the problem parameters tend to infinity. A very important aspect of these theorems is that the proofs are constructive, i.e. we provide a way to compute such solutions in polynomial time. Then, we extend the theorems to encompass more general grids (*irregular-boundary* grids) and show that as long as the grid contains a finite number of large rectangular areas to be partitioned among many processors, the above properties still hold, i.e. there exist partitions whose relative distance from the lower bound converges to zero as the problem parameters tend to infinity. These theoretical results will provide the basis of the algorithms to be discussed in the next sections.

Before we proceed to develop the theory, it is important to notice that there are solutions to the MP problem whose relative distance from the lower bound grows to infinity as the problem parameters grow to infinity. As an example, consider an $N \times N$ grid partitioned among N processors by having each processor assigned to one row of the grid. Simple algebra shows that the relative distance of this solution from the lower bound grows proportionately with $N^{\frac{1}{2}}$. The MP problem is therefore quite different from large classes of *random* QAP problems for which it has been shown that the distance between the best and worst solution goes to zero as the dimension of the problem goes to infinity.

In much of the analysis below, rather than dealing directly with perimeter, it is more convenient to use the concept of semi-perimeter (introduced earlier in the previous section). First, we consider the equi-partitioning problem MP($M \times N$, P) for rectangular grids of dimensions $M \times N$ and we assume perfect load balancing, i.e. that P divides MN and that each processor is therefore assigned $A_p = \frac{MN}{P}$ cells.

A key observation is that for many instances of the problem, a stripe decomposition of the domain is possible; that is, an optimal –or near optimal– partition exists, where the sub-domains form horizontal stripes of height approximately $\sqrt{A_p}$ that partition the rows of the grid; observe the stripes of Figure 10.3 which shows an optimal partition (i.e. one that matches the lower bound) of a 200×200 rectangular grid among 200 processors. To establish this claim, we are going to prove two lemmas. These two lemmas combined, guarantee the existence of such solutions for a large class of instances of the minimum perimeter problem (see also [Christou and Meyer, 1996c]).

Lemma 3 *Given two nonnegative integers* m, k *there exist natural numbers* a, b *such that*

$$m = ak + b(k+1) \tag{10.9}$$

iff $r_m = 0$ *or*

$$k \leq d_m + r_m$$
$$\text{where } d_m = m \div (k+1)$$
$$\text{and } r_m = m \mod (k+1).$$

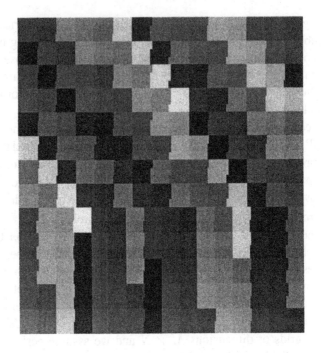

Figure 10.3   Optimal stripe partition of a 200x200 grid among 200 procs

Proof: The case $r_m = 0$ is trivial, so in the arguments below, we assume $r_m > 0$. From the definition of d_m and r_m we have

$$m = d_m(k+1) + r_m.$$

But this can be written as

$$m = (d_m - c)(k+1) + (r_m + c + kc)$$

for any c. Now, we can write m in the desired form (10.9) if k divides $r_m + c$ and $0 \leq c \leq d_m$. The smallest c that satisfies this requirement is $k - r_m$, and thus, if $c = k - r_m \leq d_m$ then simply set $b = d_m - (k - r_m)$ $a = \frac{r_m + (k+1)(k-r_m)}{k} = k + 1 - r_m$. For the other direction, observe that $k - r_m$ is the smallest c that allows k to divide the number $r_m + (k+1)c$ so if this c is greater than d_m, there exists no natural such that the required decomposition is possible.   ∎

A useful corollary of this lemma is the following:

Corollary 1 *Given two nonnegative integers* m, k *there exist natural numbers* a, b *such that equation (10.9) holds if* $m \geq k(k-1)$.

Proof: The corollary trivially holds for $k = 0$ or $k = 1$. Assume therefore $k \geq 2$. If $k(k-1) \leq m \leq k^2 - 1$, then $m = k^2 - r$ for some $r = 1 \ldots k$, and thus write $m = (r-1)k + (k-r)(k+1)$. Else $m \geq k^2$. For all m between k^2 and $k(k+1) - 1$ we have $d_m = k - 1$ and $r_m \geq 1$, so $k \leq d_m + r_m$ and the claim holds. For all m greater than or equal to $k(k+1)$ we have $d_m \geq k$ and so $k \leq d_m + r_m$ and the claim holds true again.   ∎

The next lemma states that the class of problems $MP(N \times N, N)$ is amenable to such decomposition. In other words, for all $N > 0$, we can partition the rows of the grid with a number of stripes, each of which has a height that is equal to the height of an optimal shape.

Lemma 4 *Given N, we can always find r near-rectangular optimal shapes* (h_i, w_i, f_i) *–not necessarily distinct– such that*

$$\sum_{i=1}^{r} h_i = N.$$

Proof: We are going to show that we can always find two optimal shapes (h_1, w_1, f_1) and (h_2, w_2, f_2) where $f_1 < h_1, f_2 < h_2$, such that $ah_1 + bh_2 = N$ for some natural numbers a, b. Let $k = \lfloor \sqrt{N} \rfloor$.

- Assume $k(k+1) > N$. The discussion in Section 2.2.1 implies that $(k, k, N - k^2)$ is an optimal shape and its semi-perimeter is $2k + 1$ (unless $N = k^2$ in which case the semi-perimeter is $2k$, and we can get a perfect partition using the optimal shape $(k, k, 0)$). Furthermore, trying the rectangle $(k+1, k-1)$ we get

$$(k+1)(k-1) = k^2 - 1 < N$$

  and $f = N - k^2 + 1 < k + 1$ because $N < k(k+1)$ so the shape $(k+1, k-1, N - k^2 + 1)$ is also an optimal shape. Both of these optimal shapes have fringe size less than the height of the corresponding block. Since $N \geq k^2$, by corollary 1 we can find two naturals a, b, such that $N = ak + b(k+1)$.

- Next assume that $k(k+1) = N$. This simply means that the $MP(N \times N, N)$ has an optimal shape that is a rectangle and thus we can obtain a perfect partition using N rectangles of dimensions $k \times (k+1)$ all oriented in the same way.

- Finally, assume $k(k+1) < N$. Observe that $(k+1)^2 > N$ from the definition of k. Now, the shapes $(k+1, k, f)$ and $(k, k+1, f)$ where $f = N - k(k+1) < k + 1$ are optimal. Again, because $N > k^2$ corollary 1 applies and the required decomposition of the rows of the grid is possible. Note that if $f = k$ the rectangle $k \times (k + 2)$ is an optimal shape, and a perfect partition using N such rectangles all oriented the same way is possible.

∎

Lemma 4 proves that for the MP($N \times N$, N), it is possible to partition the rows of the grid into r stripes of heights that correspond to the height of optimal shapes for area N. Such a stripe will have area hN and can be filled with h shapes of area N and height h using a *stripe-filling* process that fills the stripe with exactly h such shapes assigning them processor indices $1 \dots h$. We present this routine in the form of pseudocode and then describe it in more detail, as this routine forms the basis of the arguments to be presented in the next theorem.

```
algorithm stripeFill(h,A: integer; var str: Grid)
   /* input : h,A - the dimensions of the stripe */
   /* output: str - the processor index assignments of the cells */
begin
   proc := 1; area[proc] := 0;
   for col:=1 to A
      for row:=1 to h
         str[row,col] := proc;
         area[proc] := area[proc] + 1
         if (area[proc] = A)
            proc := proc + 1;
            area[proc] := 0;
         endif;
      endfor;
   endfor;
end;
```

In terms of placing optimal shapes, the effect of this process is to place the block-part of the current optimal shape so that its leftmost column occupies the first completely unassigned column of the stripe. If there exist any unassigned cells in the column to the left of this column, the routine places as much of the fringe as possible there. If there is some part of the fringe that does not fit there, the algorithm places this remainder in the immediate right neighboring column of the block. However, if all fringe cells are assigned to the left and there remain neighboring cells on the left that are not assigned, the algorithm alters the shape by removing cells from the rightmost column of its block and using them to fill the residual left neighbors of the block (in this latter case, the perimeter of the resulting shape remains optimal).

Using lemma 3 and 4 and corollary 1 together with the *stripeFill(...)* routine we can proceed to show that a large class of minimum perimeter problems is amenable to a decomposition that yields partitions that become asymptotically optimal as the area that each processor is assigned grows to infinity. This class of problems is the equi-partitioning problems MP($M \times N$, P) with $P \geq \max(M, N)$. We prove this fact in a series of constructive theorems starting with the case $M = N = P$. For this end, we define the term "deviation" as the distance from the lower bound of the perimeter of a collection of cells.

Theorem 1 *The MP($N \times N$, N) problem (equi-partition an $N \times N$ grid among N components) has a solution whose relative distance δ from the lower bound (10.8) satisfies*

$$\delta < \frac{1}{\left\lceil 2\sqrt{N} \right\rceil}. \tag{10.10}$$

Proof: Let (h_i, w_i, f_i) denote an optimal shape of height h_i, width w_i and fringe size f_i for area size $\mathcal{A}_p = \frac{N \times N}{N} = N$ where $f_i < h_i$.

As we have already shown in Lemma 4, we can always find r shapes – not necessarily distinct– such that $\sum_{i=1}^{r} h_i = N$. These numbers $h_1, \ldots, h_r$ induce a partition on the rows of the grid. The first h_1 rows of the grid are called the first stripe, the following h_2 rows are called the second stripe etc.

Now the i-th stripe can be filled with h_i components using the shape (h_i, w_i, f_i). In order to do this simply use the *stripeFill(...)* routine described earlier. In this manner, the stripe is filled using exactly h_i components, because the area of the stripe is $h_i N$, and the total area of h_i components is $h_i N$ also.

If $f_i = 0$ then no deviation from occurs in the i-th stripe. If $f_i > 0$, the deviation in this stripe can be no more than $f_i - 1$. To see this observe that each shape is either optimally placed (if it occupies $w_i + 1$ columns of the stripe) or its semi-perimeter is suboptimal by 1 (if the fringe part of the shape is split between the immediate left and right columns of the block). So, we can measure the deviation in the stripe by counting the number of the suboptimal shapes, or equivalently, by counting the "surplus" columns corresponding to regions that occupy $w_i + 2$ columns.

In the stripe, assume there are e_0 shapes that fill completely $w_i - 1$ columns of the stripe, and occupy part of their immediate left and right neighboring columns, e_0^+ shapes that fill w_i columns of the stripe and occupy part of one immediate neighboring column, and e_i shapes that are suboptimal, that is they fill w_i columns of the stripe, and they occupy part of both their immediate neighboring columns. Thus, letting t denote the number of columns containing more than one component index, we have

$$e_0 + e_0^+ + e_i = h_i \tag{10.11}$$

$$h_i w_i + f_i = N \tag{10.12}$$

$$e_0(w_i - 1) + e_0^+ w_i + e_i w_i + t = N \tag{10.13}$$

from which, after substitution, we conclude that

$$t = f_i + e_0.$$

Let us now associate each of the t doubly indexed columns with the component corresponding to the block to its left. Then the shapes corresponding to e_0 each contribute to t as do the e_i suboptimal shapes and the first e_0^+ shape at the left end of the stripe. Therefore, $e_0 + 1 + e_i \leq t$. Combining this with the preceding equation implies

$$e_i \leq f_i - 1.$$

Therefore, the semi-perimeter deviation in each stripe is not more than $f_i - 1$ and the stripes cover the grid completely and with no overlap using a total of $\sum_{i=1}^{r} h_i = N$ components, so the relative deviation is bounded by

$$\delta \leq \frac{1}{N \lceil 2\sqrt{N} \rceil} \sum_{i=1}^{r} (f_i - 1).$$

Defining $\pi_i \equiv \frac{f_i}{h_i}$, $\pi = \max_i \pi_i$, we have $\pi_i \in [0, 1)$, $\pi \in [0, 1)$ so substituting in the above we get

$$\delta \leq \frac{1}{N \lceil 2\sqrt{N} \rceil} \sum_{i=1}^{r} (f_i - 1) = \frac{1}{N \lceil 2\sqrt{N} \rceil} \sum_{i=1}^{r} (\pi_i h_i - 1)$$

$$\leq \frac{\pi N - r}{N \lceil 2\sqrt{N} \rceil} < \frac{1}{\lceil 2\sqrt{N} \rceil}$$

as stated.   ∎

It is worth mentioning that in the case when the fringe f_i of an optimal shape (h_i, w_i, f_i) divides its height h_i exactly, then there can be no surplus columns and therefore the deviation in a stripe using this shape is zero. The same zero deviation behavior of stripes occurs when $f_i \leq 1$. This implies that it is not unlikely in the best near-optimal solutions to observe a large number of stripes of zero total distance from the lower bound.

The techniques used in the proof of theorem 1 can be used to prove that similar quality solutions exist when the number of rows equals the number of processors and is greater than or equal to the number of columns of the grid. However, it now may become necessary to consider stripes of sub-optimal height (but by no more than 1).

Theorem 2 *The MP($M \times N, M$) with $M \geq N$ has a solution whose total perimeter possesses a relative distance δ from the lower bound that satisfies*

$$\delta < \frac{1}{\lceil 2\sqrt{N} \rceil}. \tag{10.14}$$

Proof: The proof is very similar to the proof of theorem 1. We are going to partition the M rows of the grid into $r_1 + r_2$ stripes having lengths
$h_1, \ldots, h_{r_1}, h_1^s, \ldots, h_{r_2}^s$. The first r_1 stripes will be filled with optimal shapes
(h_i, w_i, f_i) while the last r_2 stripes will use sub-optimal shapes (h_i^s, w_i^s, f_i^s).
These sub-optimal shapes have an area size equal to N, but their semi-perimeter
is 1 more than the semi-perimeter of an optimal shape for area N. The only
case in which we use these shapes is when N is a perfect square $N = k^2$. In
this case, the sub-optimal shape we are going to use is the shape $(k+1, k-1, 1)$
which has an area of $(k+1)(k-1) + 1 = N$ and a semi-perimeter equal to
$2k + 1$.

Let $k = \lfloor \sqrt{N} \rfloor$.

- Assume first $k(k+1) > N$. Furthermore, assume $N \neq k^2$. Under these
  assumptions, the shapes $(k, k, N - k^2)$ and $(k+1, k-1, N - k^2 + 1)$
  can be used to partition the grid. Applying the technique described in
  the previous theorem, in the i-th stripe we can place h_i shapes, and the
  deviation in each of them is

$$e_i \leq f_i - 1.$$

It only remains to prove that we can find nonnegative integers a, b such
that

$$M = ak + b(k+1). \tag{10.15}$$

But since $M \geq N \geq k^2$, from corollary 1, we have that equation (10.15)
holds.

Now, assume that $N = k^2$. In this case, the shape $(k+1, k-1, 1)$ is
sub-optimal by 1 as its semi-perimeter is $2k + 1$. Nevertheless, the area
size of this shape is N, and we can use $k + 1$ shapes to fill a stripe of
height $h_i^s = k + 1$. The absolute deviation in such a stripe will be

$$e_i = h_i^s = k + 1.$$

where the term h_i^s comes from the fact that each shape used in this stripe
has a semi-perimeter that is suboptimal by one. Note that since $f_i^s = 1$,
there exist no surplus columns in such a stripe. From the discussion
above, we have that $M = ak + b(k+1)$ for some $a, b \in \mathcal{N}$, so we can
partition the rows of the grid as desired. Now setting $r_1 = a$ and $r_2 = b$
we bound the total relative distance from above by

$$\delta \leq \frac{1}{M \lceil 2\sqrt{N} \rceil} \sum_{i=1}^{r_2} h_i^s$$

and since

$$\sum_{i=1}^{r_2} h_i^s = M - r_1 k$$

we get

$$\delta \le \frac{1}{M \left\lceil 2\sqrt{N} \right\rceil} \left[M - \left\lfloor \sqrt{N} \right\rfloor r_1 \right].$$

- Next, assume that $N = k(k+1)$. This means that $(k+1, k, 0)$ and $(k, k+1, 0)$ are optimal rectangles. Since $M \ge N \ge k^2$ by corollary 1 we can always write $M = ak + b(k+1)$ for some $a, b \in \mathcal{N}$. Note, that the deviation in each stripe is zero, which results in a perfect partition.

- Finally, in the case $N > k(k+1)$, the shapes $(k+1, k, f)$ and $(k, k+1, f)$ are optimal shapes for the MP($M \times N, M$). Note that $f = N - k(k+1)$, and if $f = k$ then the shape $(k, k+1, k)$ is really the optimal rectangle $(k, k+2, 0)$. Using the same arguments again, we can partition the rows of the grid by finding $a, b \in \mathcal{N}$ such that $M = ak + b(k+1)$. The deviation in the i-th stripe will be

$$e_i \le f_i - 1.$$

So we have shown that in all cases there exists a solution whose total perimeter has a relative distance from the theoretical lower bound that is

$$\delta \le \frac{\pi M - r}{M \left\lceil 2\sqrt{N} \right\rceil}$$

for some $\pi \in [0, 1)$ and $r \in \mathcal{N}$. ∎

The theorem we just proved will be used as an argument for the proof of the more general theorems proved in the following. The general idea is to reduce whatever grid is given into a series of rectangular grids of dimensions $M \times \mathcal{A}_p$ to be partitioned among M processors (with $M \ge \mathcal{A}_p$). Then if the remainder of the original grid is a small enough area compared to the whole of the grid, the relative distance of the resulting partition cannot grow too much if one partitions the remainder in a careful way. So, next we present the final and most general theorem for equi-partitioning a rectangular grid, when the number of processors divides the total area of the grid; the asymptotically optimal behavior of the class of grids to be presented was first established in [Christou and Meyer, 1996c]; the following result [Christou and Meyer, 1996b] is an improvement on the error bound, and the proof is more pleasing aesthetically.

Theorem 3 *Assuming P divides MN and that $P \geq max(M,N)$ the minimum perimeter problem $MP(M \times N, P)$ has a feasible solution whose relative distance δ from the lower bound satisfies*

$$\delta < \frac{1}{\sqrt{A_p}} + \frac{1}{A_p}. \tag{10.16}$$

Thus the error bound δ converges to zero as A_p (the area of each processor) tends to infinity.

Proof: The grid is shown in Figure 10.4. Note that $A_p \leq \min\{M,N\}$ and

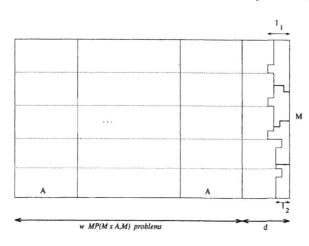

Figure 10.4 MP($M \times N, P$), $P \geq \max(M,N)$

write $N = wA_p + d$ for some naturals $w \geq 1$ and $d < A_p$. Define $k = \lfloor \sqrt{A_p} \rfloor$. Observe that the problem can be decomposed into w MP($M \times A_p, M$) problems, and a MP($M \times d, Md/A_p$). In each of the w problems MP($M \times A_p, M$), use the stripe decomposition method of theorem 2 to get a total absolute perimeter deviation $e < 2wM$. This striping technique (which partitions the rows of the grid into $r \leq M/k$ stripes of height $h_1, \ldots, h_r$) is continued over the last d columns in each stripe until no additional shape can be placed in the stripe. Let p denote the number of processors that have not been assigned. The stripe decomposition in the last d columns thus placed $\frac{Md}{A_p} - p$ processors, each of which may have a deviation in perimeter of no more than two.

The stripe decomposition for MP($M \times A_p, M$) uses at most two different shapes. Arrange the stripes of the grid so that all stripes that use the first shape are used in the top rows of the grid which we will refer to as area 1, and all the stripes that use the second shape are in the (remaining) bottom rows which we will refer to as area 2. Let l_i $i = 1, 2$ denote the maximum number of columns in area i that contain unassigned grid cells, and without any loss of generality, assume that $l_1 \geq l_2 \geq 0$.

We place the last p processors in the remaining area using the following "orthogonal stripe filling" algorithm that approximates the optimal shapes established in Lemma 2: starting from the top row of the grid, keep assigning the unassigned cells row-wise (interchanging left to right and then right to left) until the processor has $\mathcal{A}_p$ cells.

To compute the error bound in perimeter of the last p processors that were placed in the grid using this "orthogonal stripe filling" algorithm we compute the length of the boundary enclosing the region they occupy, plus the length of the border between processors, then subtract the lower bound. Thus, the maximum deviation in this region is

$$
\begin{aligned}
e < {}& (2M + l_1 + l_2) + (2r - 1 + l_1 - l_2) \\
& + 2\left[(p-1)l_1 + (p-1)\right] \\
& - 2(p\left\lceil 2\sqrt{\mathcal{A}_p}\right\rceil)
\end{aligned}
$$

The first six terms in the RHS of the inequality account for the left, right, top and bottom borders of this region, the next two terms account for the inner borders, and the last term is the lower bound. Note that the perimeter of the left border includes four terms $(M + (2r - 1) + l_1 - l_2)$ because it is not a straight line. Thus the total relative distance of the perimeter of the solution from the lower bound satisfies:

$$
\begin{aligned}
\delta < {}& \frac{2\left[wM + \frac{Md}{\mathcal{A}_p} - p + (p-1)l_1 + p - 1 - p\left\lceil 2\sqrt{\mathcal{A}_p}\right\rceil\right]}{2M\frac{w\mathcal{A}_p + d}{\mathcal{A}_p}\left\lceil 2\sqrt{\mathcal{A}_p}\right\rceil} \\
& + \frac{2M + l_1 + l_2 + (2r - 1) + l_1 - l_2}{2M\frac{w\mathcal{A}_p + d}{\mathcal{A}_p}\left\lceil 2\sqrt{\mathcal{A}_p}\right\rceil}
\end{aligned}
$$

or

$$
\delta < \frac{wM + \frac{Md}{\mathcal{A}_p} + (p-1)l_1 - 1 - p\left\lceil 2\sqrt{\mathcal{A}_p}\right\rceil + M + l_1 + r}{M\frac{w\mathcal{A}_p + d}{\mathcal{A}_p}\left\lceil 2\sqrt{\mathcal{A}_p}\right\rceil}.
$$

But for all $\mathcal{A}_p \geq 2$, we have $l_1 \leq \lfloor \sqrt{\mathcal{A}_p} \rfloor + 2 \leq \lceil 2\sqrt{\mathcal{A}_p} \rceil$ which implies that $(p-1)l_1 + l_1 = pl_1 \leq p\lceil 2\sqrt{\mathcal{A}_p} \rceil$, and since $r \leq \frac{M}{\lfloor \sqrt{\mathcal{A}_p} \rfloor}$ we get

$$\delta < \frac{M(w\mathcal{A}_p + d) + M\mathcal{A}_p + \frac{M\mathcal{A}_p}{\lfloor \sqrt{\mathcal{A}_p} \rfloor}}{M(w\mathcal{A}_p + d)\lceil 2\sqrt{\mathcal{A}_p} \rceil}$$

$$= \frac{1}{\lceil 2\sqrt{\mathcal{A}_p} \rceil} + \frac{\mathcal{A}_p}{(w\mathcal{A}_p + d)\lceil 2\sqrt{\mathcal{A}_p} \rceil} + \frac{\mathcal{A}_p}{(w\mathcal{A}_p + d)\lceil 2\sqrt{\mathcal{A}_p} \rceil \lfloor \sqrt{\mathcal{A}_p} \rfloor}.$$

It's easy to show that $\forall x \geq 1$ $x^2 \leq \lceil 2x \rceil \lfloor x \rfloor$ so (since $\mathcal{A}_p \leq M$, $\mathcal{A}_p \leq N$)

$$\delta < \frac{2}{\lceil 2\sqrt{\mathcal{A}_p} \rceil} + \frac{1}{N} < \frac{1}{\sqrt{\mathcal{A}_p}} + \frac{1}{\mathcal{A}_p}.$$

∎

It is easy to use the ideas behind the proof of theorem 3 to establish a similar result in the case where P does not divide MN. For a proof, see [Christou and Meyer, 1996b].

Theorem 4 *Assume P does not divide MN and that $P \geq max(M, N)$; the minimum perimeter problem MP($M \times N,P$) has a feasible solution whose relative distance δ from the lower bound satisfies*

$$\delta < \frac{1}{\sqrt{A_1}} + \frac{1}{\sqrt{A_2}} + \frac{1}{A_1} \qquad (10.17)$$

where $A_1 = \lfloor MN/P \rfloor$ and $A_2 = \lceil MN/P \rceil$. Thus the error bound δ converges to zero as A_1, A_2 (the areas of the processors) tend to infinity.

Theorems 3 and 4 guarantee the existence of good quality solutions for the MP($M \times N,P$) as long as the number of processors dominates the dimensions of the grid. These solutions become optimal in an asymptotic sense as the area that each processor is assigned to grows to infinity. The drawback of requiring too many processors in order to guarantee good partitioning schemes is real; parallel computing in networks of workstations has attracted a lot of attention because it has been shown to be a viable and much cheaper alternative to massively parallel supercomputers. In such an environment however, the number of available processors may not be as large as the domain (which is assumed big enough to require the use of parallel processing for the efficient solution of the problem in hand). Still, the computational results show that even when the number of processors is much less than the dimensions of the grid, stripe decomposition is able to find very good quality partitions. Furthermore, the technique used to fill the last d columns of the grid in Theorem 3 can be used to show that in the case where $N \leq P < M$ there exists a partition whose total

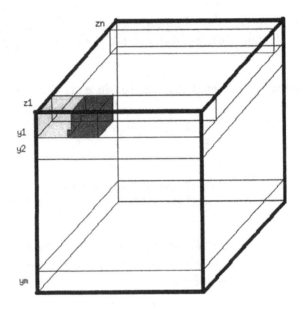

Figure 10.5   Partition of an $N \times N \times N$ cube among N^2 processors

It is easy to extend the theory of stripe decomposition for 5-point (2D) u-niform grids, to 7-point 3D uniform grids. In such domains, each cell must communicate with its immediate six neighbors, and the problem of minimizing communication is essentially a problem of minimizing total surface rather than perimeter of the resulting partition. The theory of optimal shapes developed in Section 2.2.1, extends to three dimensions via the notion of D-perimeter which for $D = 3$ corresponds to surface, to yield near-cubic optimal shapes (see Yackel et. al. [Yackel and Meyer, 1992b]). It is easy to see that for the 3D Minimum Surface problem MS($N \times N \times N$, N^2), one can perform stripe decomposition along two adjacent surfaces of the cube (by partitioning axes y and z as shown in Figure 10.5). This stripe decomposition of the surfaces of the cube is possible because each axis can be partitioned among segments of length $k = \lceil N^{1/3} \rceil$ and $\lceil N^{1/3} \rceil - 1$ since $N \geq k(k-1)$ (corollary 1). The result is a partition of the cube into horizontal "towers" (the 3D analog of a stripe) each of which can be filled exactly with $y_i z_j$ processors because the volume of each tunnel is $y_i z_j N$ and the volume each processor must get is N. Following an easy extension to the *stripeFill(...)* routine, we can easily partition the cube among N^2 processors with an error that should exhibit the same asymptotically optimal behavior. We have outlined a sketch of a proof of this claim. The formal proof and its extensions to arbitrary 3D rectilinear grids of dimensions $M \times N \times L$ to be partitioned among $P \geq \max(MN, ML, LN)$ processors is currently an open item for further investigation.

4. SNAKE DECOMPOSITION

The theorems developed in the previous discussion can be extended to show that *any* uniform 5-point grid can be equi-partitioned among P processors using stripe decomposition and as long as the grid contains a relatively large rectangular area the partition obtained will differ only slightly from the lower bound (10.8).

So, it becomes apparent that a 5-point general uniform grid can be decomposed very efficiently under certain assumptions. The *stripeFill(...)* routine that we presented can be extended so as to provide a general algorithm for partitioning *any* 5-point grid. *Snake decomposition* is an algorithm that extends the idea of partitioning a rectangular grid in stripes of optimal height to arbitrary 5-point uniform grids.

4.1 ERROR BOUNDS FOR EQUI-PARTITIONS OF IRREGULAR-BOUNDARY DOMAINS

Theorem 5 *Let $\mathcal{G}$ denote a 5-point uniform grid with total area $\mathcal{A}(\mathcal{G})$, to be equi-partitioned among P processors, each having an area $A_p = \frac{\mathcal{A}(\mathcal{G})}{P}$. Let (M_0, N_0) denote the dimensions of the largest rectangle that can fit in $\mathcal{G}$, and let P_0 denote the maximum number of processors that can fit in this rectangle. Then, if $P_0 \geq \max(M_0, N_0)$ the MP($\mathcal{G}, P$) possesses a solution whose relative distance δ_G from the lower bound (10.8) satisfies $\delta_G < \frac{1}{A_p} + \frac{1}{\sqrt{A_p}} + \left[1 - \frac{P_0 A_p}{\mathcal{A}(\mathcal{G})} \right] \sqrt{A_p}$.*

The proof of this theorem employs stripe decomposition on the largest embedded rectangles of the grid, and then scans all the cells of the grid and assigns every unassigned cell to any processor p that hasn't been assigned A_p cells yet. For details, see [Christou and Meyer, 1996a, Christou, 1996].

As an easy corollary we obtain the following:

Corollary 2 *Let $\mathcal{G}^k$ be a sequence of 5-point uniform grids. Assume that $\mathcal{G}^k$ has total area $\mathcal{A}(\mathcal{G}^k)$, (which contains a rectangle of dimensions (M_0^k, N_0^k)) to be equi-partitioned among P^k processors each having an area A_p^k. If $P_0^k :=$*

$\left\lfloor \frac{M_0^k N_0^k}{A_p^k} \right\rfloor \geq \max(M_0^k, N_0^k),$ *then if* $\left[1 - \frac{P_0^k A_p^k}{A(\mathcal{G}^k)}\right] \sqrt{A_p^k} \to 0$ *as* $A_p^k \to \infty$ *the sequence* $MP(\mathcal{G}^k, P^k)$ *has solutions with relative distances from the lower bound* $\delta_G^k \to 0$.

The above results show that asymptotically, grids that are near-rectangular can be optimally partitioned efficiently using stripe decomposition. A class of such grids that satisfies the conditions of corollary 2 are elongated trapezoids to be partitioned among a large number of processors (such grids arise from simulations in chemical engineering).

It is possible to extend the ideas behind the previous theorem, to include 5-point uniform grids that contain a finite number of relatively large disjoint rectangular areas (a proof of this claim appears in [Christou and Meyer, 1996a]).

4.2   THE SNAKE DECOMPOSITION ALGORITHM

To describe the snake decomposition method, observe that any 5-point uniform grid can be represented by an $M \times N$ rectangle with some of its cells having a certain value to indicate that they are not part of the grid, or "unavailable". This super-grid is the smallest rectangular grid that can accommodate our given grid; in the combinatorics literature [Melou, 1994], this rectangle is sometimes called the convex hull of the grid.

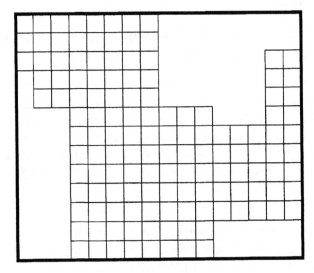

Figure 10.6   Rectangular hull representation of a grid

Snake decomposition accepts as input a partition of the rows of this super-grid (see the thick rectangle in Figure 10.6) into a set of near optimal heights, and then fills the grid with the required number of processors by filling the columns of each stripe consecutively in a way that resembles the movement of a snake. The first stripe's columns are filled left to right, then the second stripe's columns are filled going right to left, and then the process repeats. If the end of a stripe has been reached, but the processor that was used to fill the last part of it still hasn't been assigned $\mathcal{A}_p$ cells, then part of the stripe(s) immediately below the area that the current processor occupies are assigned to this processor in a row-by-row fashion. This is done in order to keep the perimeter small. A pseudo-code for this algorithm follows:

```
algorithm snake(P, stripe[S]: int; var G: Grid[M,N])
   /* Inputs: P is the number of processors */
   /* Inputs: grid G is an MxN array; */
   /* Inputs: stripe[S] is an array of S heights adding up to M */
   /* Output: An assignment of G among the P processors */
begin
   proc:=1; index_stripe:=1; start_row:=1; done:=FALSE;
   areas := computeAreas(area(G), P);
   while not(done)
      fin_row := start_row + stripe[index_stripe];
      fillLeftToRight(G, start_row, fin_row, proc, areas, new, done);
      if (done) return; endif;
      if not(new)
         end_next := fin_row+1+stripe[index_stripe+1];
         /* finish current */
         finish(G, fin_row+1, end_next, proc, areas);
      endif;
      start_row := start_row + stripe[index_stripe];
      index_stripe := index_stripe + 1;
      fin_row := fin_row + stripe[index_stripe];
      fillRightToLeft(G, start_row, fin_row, proc, areas, new, done);
      if (done) return; endif;
      if not(new)
         end_next := fin_row+1+stripe[index_stripe+1];
         /* finish current */
         finish(G, fin_row+1, M, proc, areas);
      endif;
   endwhile;
end;
```

Routine *fillLeftToRight(...)* accepts as input the grid $\mathcal{G}$ (with its partial assignment), the $start_{row}$ and fin_{row} rows of the stripe it will work on, and the current processor index proc; it outputs the variable done which indicates whether the assignment of processor indices to the grid cells has completed (in which case the algorithm terminates) and the variable new which indicates whether the current processor must continue on the next stripe. The routine starts from the leftmost column of the stripe and assigns unassigned cells in each column (scanning it top to bottom), with the current processor index and immediately decreasing the area of the processor by one. As soon as the area of the current processor becomes zero, the current processor index is increased by one. The routine *fillRightToLeft(...)* works identically except that it works its way from the rightmost column of the stripe and moves to the left end. A pseudocode for these routines follows:

```
algorithm fillLeftToRight(start, fin, p, areas[P]: int;
                    var new, done: Boolean;
                    var G: Grid[M,N])
   /* Inputs: The grid G is an MxN array; */
   /* Inputs: start, fin are stripe starting and ending rows */
   /* Inputs: p is the current processor index */
   /* Inputs: areas holds the remaining area for each proc */
   /* Output: new indicates whether a new processor is used */
   /* Output: done indicates termination */
   /* Output: An assignment of the stripe among processors */
begin
  for j := 1 to N
    for i := start to fin
      if (p < P and areas[p] = 0)
        p := p+1;                      /* increase current processor index */
        new := TRUE;
      endif;
      /* free cell? */
      if (G[i,j] = 0)
        G[i,j] := p;
        areas[p] := areas[p] - 1;
        new := FALSE;
      endif;
    endfor;
  endfor;
  if (areas[P] = 0) done := TRUE;  endif;
end;
```

The routine *fillRightToLeft(...)* is identical to the above routine except for the first statement which reads
```
for j := N down to 1
```

The routine *finish(...)* accepts as input the grid G, the starting and ending row of the next stripe and completes the assignment of the current processor by continuing into the immediate rows of the next stripe.

```
algorithm finish(start, end, proc, areas[P]: int,
                 var G: Grid[M,N])
   /* Inputs: The grid G is an MxN array; */
   /* Inputs: start is the beginning row of the next stripe */
   /* Inputs: end is the ending row of the next stripe */
   /* Inputs: proc is the current processor index */
   /* Inputs: areas[P] is the remaining areas for each proc */
   /* Output: Completes the assignment of the current proc */
begin
   s := startCol(proc, start-1, G);              /* get starting col */
   e := endCol(proc, start-1, G);               /* get ending col */
   for j := s to e
     for i := start to fin
       if (areas[proc] > 0 and G[i,j] = 0)
         G[i,j] := proc;                         /* assign free cell */
         areas[proc] := areas[proc] - 1;         /* decrease area */
       endif,
       else if (areas[proc] = 0)
          return;
       endif,
     endfor;
   endfor;
end;
```

The runtime of the algorithm is linear in the size of the grid since each grid cell may be examined at most twice; if a cell is set in the *fillLeftToRight(...)* or *fillRightToLeft(...)* routines then it is never examined again; otherwise, if it is set from within the body of the *finish(...)* routine, the next *fillRightToLeft(...)* or *fillLeftToRight(...)* call will examine it (and leave it intact). So the runtime of the algorithm, for any input array stripe[P], is $O(|G|)$ which makes it a very fast routine.

Figure 10.3 shows an optimal partition (i.e. one that matches the lower bound) of a 200×200 rectangular grid among 200 processors (obtainable by stripe or snake decomposition), while Figure 10.7 shows a partition of a circle into 64 equi-area subdomains obtained by snake decomposition. The horizontal lines that are formed in both figures at the boundaries of the partitions correspond to the various stripe heights. The partition of the circle has an associated relative gap of 5.87% from the lower bound, the best found by any method we have tried.

The main difference of the two methods lies in the fact that snake decomposition allows a stripe to "overflow", i.e. the last processor used in a stripe to "continue" over the stripe immediately below (see the shapes of the processors that occupy the ends of the stripes in Figure 10.7). The linear running time

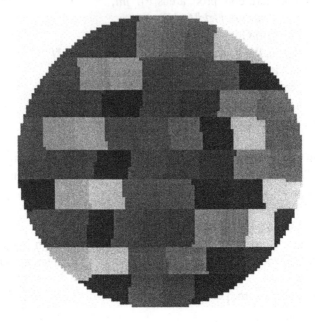

Figure 10.7   Snake partition of circle among 64 procs

of the procedure *snake(...)* indicates that we should expect very good response times of the algorithm in practical implementations. But our goal is twofold: to find very good quality solutions as fast as possible. The key to obtaining high-quality solutions is the availability of a partition of the rows of the grid that will produce stripes with minimal error in perimeter. The response time of the algorithm will be about the same for any such partition of rows, but the quality of the produced solutions can vary dramatically. In fact, solutions of high quality might actually be produced faster because of fewer double examinations of grid cells (in high quality solutions only a few processors "overflow" in the next stripe, so few double cell examinations are incurred).

4.3   EXTENSIONS TO 3D DOMAINS AND NON-UNIFORM GRIDS

In a similar manner as was done for the extension of stripe decomposition to 3D domains, one can extend snake decomposition for arbitrary 3D non-rectilinear 7-point grids by considering the 3D convex hull of the original grid, and performing snake decomposition along two surfaces. Extending the *s-nake(...)* routines to work in 3D domains is straightforward. Furthermore, if the theorems for rectangular 5-point grids can be extended to 3D 7-point grids, then so can the theorems for near-rectangular grids.

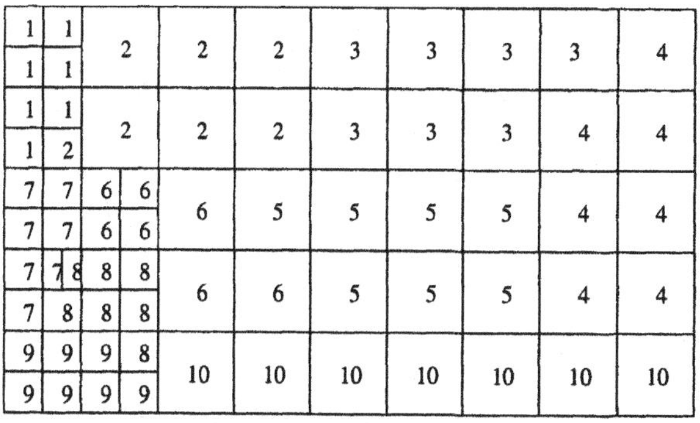

1 1 / 1 1	2	2	2	3	3	3	3	4
1 1 / 1 2	2	2	2	3	3	3	4	4
7 7 6 6 / 7 7 6 6		6	5	5	5	5	4	4
7 7 8 8 / 7 8 8 8		6	6	5	5	5	4	4
9 9 9 8 / 9 9 9 9		10	10	10	10	10	10	10

Figure 10.8   Non-uniform grid partitioned using snake decomposition

It is possible to extend the snake decomposition method to partition non-uniform domains by recursive application of the *snake(...)* routine. Non-uniform grids can be thought of as a hierarchy of grids that consist of grid cells, some of which are unit cells, and others of which are themselves grids. The snake process can start at the top level of the hierarchy and whenever it finds a "cell" that is itself a grid, it recursively applies itself to the next level of the hierarchy (in the grid contained within this grid-cell) starting with the current processor index. As an example, Figure 10.8 shows the resulting partition of a non-uniform grid among 10 processors. Lower bounds could be obtained by adding lower bounds for each level of the hierarchy, but such lower bounds will not have the simplicity of those considered so far, and will not be as close to the optimal value.

5.   PARALLEL GENETIC ALGORITHMS

Snake decomposition accepts as input a partition of the rows of the grid and subsequently computes a partition of the grid among the required number of processors. However, the number of possible partitions of the rows of the grid increases rapidly as M, the total number of rows increases. In particular, it is easy to show that the number of feasible partitions of M into k components is $n(k, M) := \binom{M-1}{k-1}$ where a partition of M into k components $(x_1, \dots, x_k)$ is feasible if $\sum_{i=1}^{k} x_i = M$, with $x_i \in \mathcal{N}^*$. Therefore, the total number of feasible partitions of M is

$$R(M) := \sum_{k=1}^{M} n(k, M) = 2^{M-1}$$

which grows exponentially fast.

To search the huge space of input partitions to the snake decomposition routine, we use the *Genetic Algorithm* paradigm. Genetic Algorithms (GAs for short) are a class of randomized algorithms that are inspired by the evolutionary processes of nature, and attempt to find optimum or near optimum solutions to a problem by breeding a population of solutions; this population is driven towards better solutions by a process that resembles "natural selection", the mechanism that is responsible for the evolution of species according to C. Darwin [Darwin, 1859]. Although the origins of this class of algorithms might be traced back to the early '50s (according to Michalewicz in [Michalewicz, 1994]) it was John Holland [Holland, 1992] who pioneered the field and developed a theoretical framework for the use of GAs. In the following, we briefly discuss traditional GAs and the models of generational replacement and the steady-state approach, and then describe in detail a new distributed, fully asynchronous GA (DGA) that is based on the *island* model. The results of this GA are superior to any other method we have tried for graph-partitioning.

5.1   THE GA MODEL

A GA is a randomized algorithm that breeds a –usually fixed– population of *individuals* which are represented by strings of an alphabet –traditionally the binary alphabet $\mathbf{B} = \{0, 1\}$– which themselves represent a possible solution to a given problem. Crucial to the success of the Genetic Algorithm is the existence of a *fitness function* that can be applied to any individual to produce a reasonable metric of the quality of the solution the individual represents for our problem. This metric is called the *fitness value* of the individual. Assuming the existence of such a function, the algorithm bootstraps its computations by creating a random initial population of individuals, and then proceeds in generations where at each generation some of the most fit individuals are selected for mating and form couples that exchange parts of their strings to form a new individual, a process called *crossover*. Other genetic operators such as mutation or inversion may be applied to the newly created individual and then this newborn offspring is evaluated using the fitness function. Then, according to a survival policy the individual might or might not replace one of its parents in the population. A pure survival policy will always replace one of the offspring's parents with the offspring no matter what the fitness value of the offspring is. More elitist strategies will discard the offspring in favor of the parents if they are more fit than their children. The algorithm stops when some termination criterion has been met, such as a provably optimal solution, or a maximum number of generations, or the population has converged to a point from which no further improvement can be made.

The above general procedure shown in Figure 10.9 leaves many decisions to be made about the specific workings of the GA. In fact, even the string representation of the solutions can be altered or generalized to include other more appropriate data structures for the original problem or simply more letters in the original alphabet, and this is some times the case in discrete optimization problems like the QAP or the traveling salesman (TSP) problem (see [Michalewicz, 1994] for more details on representation issues for GAs).

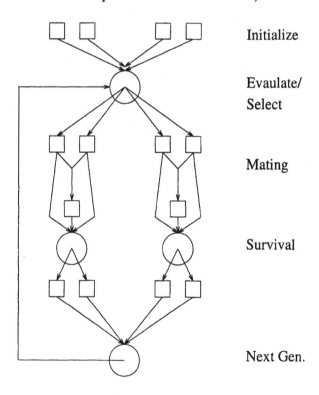

Figure 10.9   The GA model

- *Representation Issues:* The mapping between the strings manipulated by the GA and the actual solutions of the problem they represent, is referred to as the representation of the problem. It is now widely recognized [Falkenauer, 1998, Christou et al., 1999a, Hart, 1994, Michalewicz, 1994], that the efficiency of Genetic Algorithms heavily depends on an encoding scheme (i.e. representation) that facilitates the transmission good building blocks present in the current individuals to the next generation.

- *Genetic Operators:* Having decided on the representation of the solutions, the next task is to choose the genetic operators to be used for the creation of the next generation. The most widely used ones are one-point or two-point crossover (binary operators), mutation, and inversion (unary operators) but quite often special purpose operators might be needed to enhance performance or to ensure the feasibility of the solutions represented by the offspring.

- *Survival Policy:* The survival policy is a set of rules that determines if an offspring will replace one of its parents in the next generation, or it will be discarded. In general, the more elitist a policy is (keeping the best individuals in the population) the faster the whole population converges to a homogeneous state, where every individual is similar to every other, and with approximately the same fitness value. The problem is that fast convergence usually locates only a local optimum of moderate quality.

- *Selection:* This is the process that selects individuals for mating. Several such routines have been proposed. Among the most successful ones are tournament selection, roullete wheel based and remainder stochastic sampling with replacement.

- *Steady-State vs Generational Replacement:* Members of the current population might not be selected for mating but still continue their existence in the next generation. This policy is known as the steady-state approach, where in each generation only a small percentage of the population is s-elected for mating and their offspring replace the parent individuals. In the generational replacement model, the whole population is replaced by their offspring in the next generation.

- *Parameter Setting:* Finally, when all of the above decisions have been made, the appropriate parameters must be set (cross-over and mutation rate, population size etc). Choosing a set of parameters is usually done via extensive experimentation.

The GA paradigm offers an excellent compromise between the themes of exploration and exploitation. Problems with poor structure (such as the NP-hard problems) have to be tackled with search methods that are capable of exploiting promising regions of the search space as well as exploring in some reasonable way the rest of the space and if they are to be of any practical use, manage to accomplish both tasks without facing combinatorial explosion.

There is a lot of literature arguing why GAs indeed manage to strike a good balance between the two almost orthogonal goals of exploration (widening the

5.3   PARALLEL GA MODELS

Another very important aspect of GAs is the fact that they are by their nature, parallel algorithms. Working with whole populations of individuals rather than with a single individual at a time, parallel computing environments can be used to speed up the computations by assigning a number of individuals to each processor and letting the process of fitness evaluation proceed in parallel. Computationally time-consuming fitness evaluations (such as done for large equi-partition problems) are particularly well-suited for parallel computation.

Often, each processor, after computing the fitness values of its individuals, sends the results back to a coordinator master processor that decides (based on the selection policies it has) which individuals mate, and then communicates the answer back to the nodes. After the nodes receive the decision from the host and once they acquire the individuals they need for the mating to proceed, they begin another iteration (computing the offspring and evaluating them). In this scheme, coordination is needed, the master processor bases its decisions on global information about the whole population, and the programming style is a host-node synchronized communication-computation (see Figure 10.10). Similar versions of this model have been implemented by various researchers –see [von Laszewski, 1991, Chen et al., 1993, Christou and Meyer, 1996c].

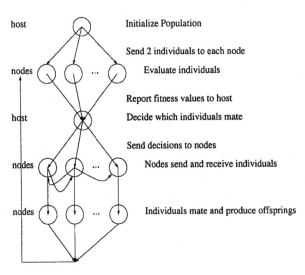

Figure 10.10   Host-node parallel GA

In another more recent approach proposed by Muhlenbein et. al. , Levine, and other researchers [Muhlenbein et al., 1991, Levine, 1995], each processor is considered an (almost isolated) island which periodically sends qualified individuals to another island according to some criterion. In this case, the whole GA process is carried out locally without global knowledge of what the populations in the other processors (islands) are doing (with the exception of a periodic *migration* of individuals). The advocates of this theory argue that the new model helps avoid the phenomenon of premature convergence, where early in the evolution process the population is driven to a local optimum and becomes nearly homogeneous; after this has happened, the evolution process cannot escape the local optimum because not enough diversity (genetic as well as schematic) is present to allow the genetic operators to find better quality solutions. Because premature convergence is often due to the early appearance of a super-individual, that is an individual whose fitness value is far above the average, which creates many offspring and thus drives the rest of the population towards it, having many islands that are isolated helps avoid this problem. Appearance of a super-individual no longer affects all of the population and genetic diversity is not lost. So the themes of exploration and exploitation may both continue. Furthermore, programming this model allows for a fully asynchronous distributed GA that periodically exchanges information between its processes (islands). A schematic figure of this model is shown in Figure 10.11.

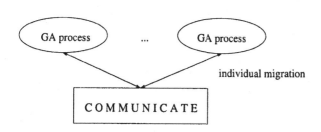

Figure 10.11   Asynchronous distributed GA communication

5.4   DGA: A DISTRIBUTED GA

Genetic Algorithms traditionally operate on a fixed population of individuals, using the general schemes described above. DGA is a new, fully asynchronous, distributed Genetic Algorithm following the island model of computation. When started, DGA spawns a certain number of processes (each of which is called an *island*) on the platform on which it is running, and each island is initialized with a certain number of individuals created randomly. It uses the steady state approach as the mechanism for creating the next generation of individuals, but it also includes a new feature, namely age (to be discussed below), in order to control the population and to maintain diversity of the population and prevent premature convergence.

Since our goal is to develop efficient algorithms and data structures for solving the minimum perimeter problem (MP), we have chosen the most natural representation of a partition of the rows of the grid for the GA: the string representation. However, each allele in the chromosome is not a binary number, but rather a natural number greater than zero that represents stripe height. This implies a modified design: each individual in the population may have different chromosome (DNA) length as long as it represents a valid solution, i.e. a partition of the rows of the grid. Furthermore, the age mechanism suggests a policy that allows each DGA island to have a variable population size. Figure 10.12 shows an illustration of the workings of DGA.

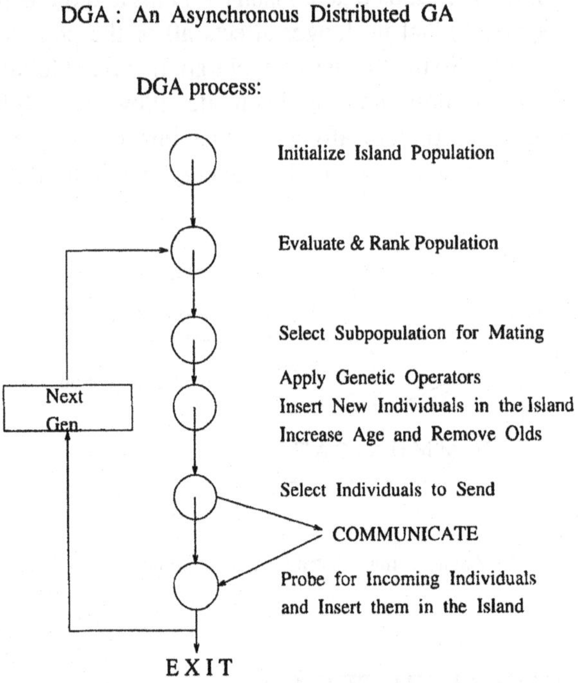

DGA : An Asynchronous Distributed GA

DGA process:

Figure 10.12   Asynchronous distributed GA

A main iteration in each island performs the following steps each of which will be explained in more detail below:

- For a certain number of times select two individuals, apply crossover and mutation to create a new offspring, evaluate it and place it in the island (replacing another individual following an elitist strategy if there is no empty space in the island).

- Increase the age of each individual and remove "olds" according to a probability distribution (specifically, the normal distribution).

- Broadcast the island's population.

- If migrating criteria are satisfied, select qualified individuals for migration to other islands and send them.

- Probe for incoming individuals and receive all that have arrived.

The algorithm is fully asynchronous in that all send and receives are done asynchronously, i.e. when a process sends a message to another process, it does not wait until the receiving process receives the message. Instead, it resumes computation as soon as the outgoing message is safely on its way on the network. The criteria for sending individuals to other islands are based on workload values: if an island's population has dropped too much then a decision is made to send the best individuals from another island to the depopulated island.

All processes during the probing step simply probe their buffer pool to check for pending messages, and if there are any, they are received, else computation is resumed from the point it was left. Thus, an individual sent to a process during an early generation might actually get there several generations later. However, the approach offers the advantage of not wasting network and CPU resources waiting for ACK/NACK signals, which in an environment like a network of workstations can make a difference. In the probing step, any incoming individuals are ranked according to their fitness value and inserted in the population.

Indirectly, the criterion for sending individuals to other islands helps increase the average fitness of the depopulated islands, because islands become deserted when individuals are deleted because of old age (migration should not affect an island's population too much because only islands that are very crowded may send individuals to other islands). But the age at which an individual is deleted is proportional to its fitness value and, thus, one would expect that islands with well fit individuals never become deserted while islands where the average fitness is low tend to shrink in population size, and thus become fertile ground for highly fit individuals to migrate. To the author's knowledge, such a use of an "age" mechanism in an island model of a GA has not been studied before.

The age mechanism also helps maintain diversity of the population within each island by eventually removing any individual no matter how well fit. Of course, this also means that the population will converge more slowly to a homogeneous state (if at all). But our goal is not to observe population convergence; we are mainly interested in finding high quality solutions to the MP problem, and thus we keep an incumbent individual for each island (which records the best individual ever found on this island) and our termination criterion is a fixed number of generations. With careful experimentation and parameter setting, we have been able to find near optimal –and often provably optimal– solutions to the MP problem.

6.   COMPUTATIONAL RESULTS

In this section we present computational results from runs of DGA having as fitness function the snake decomposition routine, and other well established codes for graph partitioning. We have used two classes of domains: rectangular grids, and non-rectangular grids (that is, uniform but irregular boundary 5-point grids). COW is a cluster of 40 high-performance Sun SPARC Server 20 workstations running Solaris at the Computer Sciences Dept. at the University of Wisconsin - Madison. Our experiments were performed on the nodes of this network (in [Christou and Meyer, 1996c] we report results from runs of another GA on a CM-5 with 32 nodes).

We compare the performance of DGA against that of recursive spectral bisection (RSB) and quadrisection (RSQ) as implemented in the Chaco package [Hendrickson and Leland, 1995a], a well-known package for graph partitioning developed at Sandia National Laboratories; on the class of rectangular grids we also make comparisons with the geometric mesh partitioner [Miller et al., 1993, Gilbert et al., 1995], another well known mesh partitioner and a state-of-the-art GRASP heuristic for the QAP [Li et al., 1994, Pardalos et al., 1995] that was written at AT&T Bell Labs and the University of Florida at Gainsville. Finally, we test DGA against R-SNAKE to check whether a Genetic Algorithm outperforms random selection in finding good inputs to the snake procedure. Comparisons with PERIX-GA [Christou and Meyer, 1996c], a GA following the host-node programming paradigm using generational replacement are made in [Christou, 1996].

6.1   SETTINGS

R-SNAKE was written in ANSI C, compiled with the -O2 optimization option and run on a Sun SPARC Server 20 workstation running the Solaris operating system. The same holds true for the Chaco package. All experiments with Chaco had the Kernighan-Lin post-processing phase option turned on. The implementation of the geometric mesh partitioner that was available to us was written in MATLAB, which partly explains the very long times of the method for many of the test problems. The GRASP code for the QAP was written in FORTRAN 77 and compiled with the -O optimization option; it run also on a Sun SPARC Server 20 workstation. It was allowed to run for 100 iterations. DGA was written in ANSI C, and utilizes the PVM 3.3 message passing interface for interprocess communication [Geist et al., 1994]. DGA runs on the COW.

R-SNAKE tests 100 random valid stripe arrays and it reports the best partition found. DGA requires more parameters. Each DGA island (process) maintains a maximum of 16 individuals per generation. There are 4 and 8 islands spawned (using the hostless programming paradigm). Whenever an individual arrives at an island, it replaces the worst individual in the population if there is no empty space in the island by the time it arrives. We use one-point crossover, with the rate set at 0.85. We set the mutation rate (defined here as the probability of mutating an allele of the chromosome) at 0.15. Since the product of the crossover and mutation operators may result in something that is not a legal individual, a *repair strategy* is used. Each new individual is given as input to a repair routine that modifies as few alleles as possible (sometimes even changing the length of the individual) so that the resulting individual is legal, i.e., represents an exact partition of the rows of the grid. Finally, as the algorithm follows the steady-state approach, approximately 70% of the populations new individuals are born in each generation (not necessarily replacing an equal number of old individuals, because whenever empty space exists, a new individual is inserted without replacing any existing individual). Each individual's lifespan is determined as a random variable that follows the Gaussian distribution $N(5 * fitness(individual), 0.09)$. DGA runs for 20 generations. Experimentation shows that the above parameters result in good overall GA behavior. More details on the experiments conducted with DGA can be found in [Christou and Meyer, 1996a].

6.2   RECTANGULAR GRIDS

In Table 10.1 we give the size of the rectangular problems in our test-suite using a QAP, linear MIP. In the QAP literature, the dimension of a problem is the number of rows (or columns) of the distance (or cost) matrix of the problem. Thus, for the QAP formulation of the MP we are using, the QAP dimension of a problem is equal to the number of grid cells in the graph.

In Table 10.2 we compare recursive spectral bisection and geometric mesh partitioner against GRASP for the QAP. The times on all tables are in seconds. The columns labeled "Gap" show the relative distance of the solution from the lower bound. An asterisk in Table 10.2 indicates the fact that the partition found was not balanced (i.e. there were components that had at least two more nodes than other components). For the first problem, since the number of partitions required is not a power of 2, neither RSB nor the Geometric Mesh Partitioner could solve it. Also, note that for the last problem, both the geometric and the spectral method run out of memory when trying to construct the adjacency matrix of the graph. It is apparent that even for small grid graphs, the QAP approach to solving the MP has difficulties because in the QAP literature, problems with dimension higher than 30 are considered challenging problems. (An experiment was performed to solve the MP(13 × 13, 13) problem on a Sun SPARC-Station 10 with 64MB of RAM, and it took GRASP more than 10,000

PROBLEM			QAP	MIP	
M	N	P	DIMENSION	VARS	CONSTR
7	7	7	49	427	3584
32	31	8	992	9857	108576
32	31	256	992	255873	125404128
32	30	64	960	63298	7492480
100	100	8	10000	99800	1118808
128	128	128	16384	2129664	5.285E+08
256	256	256	65536	1.690E+07	8.523E+09
512	512	512	262144	1.347E+08	1.369E+11

Table 10.1   Problem sizes under various formulations

PROBLEM		RSB		GEOMETRIC		GRASP	
$M \times N$	P	Time	Gap%	Time	Gap%	Time	Gap%
7 x 7	7	-	-	-	-	182.9	0.00
32 x 31	8	1.8	6.52	43.6	5.43	-	-
32 x 31	256	4.3	6.73	152.3	-2.73*	-	-
32 x 30	64	3.0	6.25	90.4	6.25	-	-
100 x 100	8	9.0	9.33	111.0	7.39	-	-
128 x 128	128	85.5	14.13	539.9	7.13	-	-
256 x 256	256	227.8	13.25	3304.2	4.15	-	-
512 x 512	512	-	-	-	-	-	-

Table 10.2   Spectral bisection, geometric and GRASP

seconds to come up with a solution that was more than 25% away from the lower bound which is actually attainable for this problem.)

PROBLEM		R-SNAKE		DGA 4procs		DGA 8procs	
$M \times N$	P	Time	Gap%	Time	Gap%	Time	Gap%
7 x 7	7	0.8	0.00	7.4	0.00	8.3	0.00
32 x 31	8	1.3	1.08	6.9	1.08	9.2	1.08
32 x 31	256	0.8	0.00	7.1	0.00	10.1	0.00
32 x 30	64	0.5	0.00	7.8	0.00	9.4	0.00
100 x 100	8	2.4	2.28	17.7	2.28	20.4	2.28
128 x 128	128	3.5	1.90	15.5	1.63	16.5	1.63
256 x 256	256	6.9	0.00	36.9	0.00	38.5	0.00
512 x 512	512	58.7	0.68	123.8	0.56	103.7	0.63

Table 10.3   R-SNAKE on 1 proc. vs DGA on 4 or 8 procs

As can be easily seen from tables 10.2 and 10.3, R-SNAKE and DGA significantly outperform recursive spectral bisection and quadrisection as well as the geometric mesh partitioner in solution quality, and in most cases in response time as well on all the rectangular domains we tested.

6.3   THE KNAPSACK APPROACH

For the rectangular domains, Martin's MSP algorithm [Martin, 1998] was applied to the class of problems MP($N \times N$, N), and MP($10N \times 10N$, N); the MSP algorithm determines all "valid" stripe heights (ones that result in stripes that can be covered exactly with an integer number of processors) and applies a knapsack routine to determine the best stripe form decomposition of the given problem. MSP is written in FORTRAN 77 and runs on a Sun SPARC Station 20 workstation running Solaris.

In Figure 10.13 we show the relative distance from the lower bound of the best stripe form partition of the problem MP($N \times N$, N) for $N = 5 \ldots 1000$. The average best stripe partition relative distance is 0.7%. It is interesting that 32.6% of the problems were solved to optimality. It is also interesting to see that the relative distance of these solutions from the lower bound decreases as N increases in the fashion predicted in Theorem 3.

Figure 10.14 on the other hand, shows the relative distance from the lower bound of the best stripe form partition of the problem MP($10N \times 10N$, N); here the area assigned to each processor is $100N$ and since the number of processors is 10 times smaller than each dimension of the grid, theorem 3 does

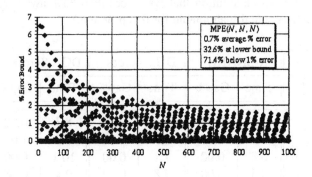

Figure 10.13   Distance from lower bound of stripe solutions (N,N,N)

not apply. However, one can clearly see that the resulting partitions are of excellent quality, so it is tempting to make the conjecture that even if P, the number of available processors does not dominate the dimensions of the grid, under very mild assumptions, the relative distance of the best solutions from the lower bound tends to zero as the problem parameters tend to infinity. For example, the solution obtained for the $10,000 \times 10,000$ grid partitioned among 1000 processors was within 0.042% of the lower bound.

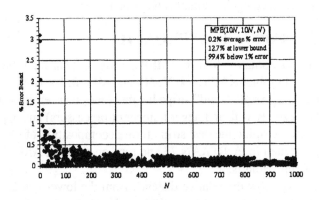

Figure 10.14   Distance from lower bound of stripe solutions (10N,10N,N)

6.4   NON-RECTANGULAR GRIDS

On the more difficult irregular boundary problems (for which the knapsack approach is not directly applicable), R-SNAKE and DGA run in times that are comparable to the ones given by RSB and RSQ, but the solution quality difference here is more dramatic, rising up to 22 percentage units for the diamond domain partitioned among 16 processors (see tables 10.4 and 10.5). The sole exception to this is the (small) elliptical domain partitioned among 64 processors, where RSB found a marginally better solution than R-SNAKE or DGA.

Comparing R-SNAKE with DGA, we observe that DGA finds better (but by no more than 1%) solutions but requires more time. Also, since the times for DGA with 8 islands running on 8 COW nodes (see Table 10.5) are almost the same as those for DGA running with 4 islands on 4 nodes, the communication penalties incurred by our method are minimal compared to other random network factors.

PROBLEM		RSB		RSQ		R-SNAKE	
Shape	P	Time	Gap%	Time	Gap%	Time	Gap%
circle	16	23.3	24.44	9.1	21.80	9.7	8.33
circle	64	34.7	16.87	14.5	28.34	11.7	6.35
ellipse	16	2.3	10.83	1.4	13.33	5.4	8.33
ellipse	64	3.5	5.16	2.2	15.10	4.9	5.56
torus	16	27.3	28.97	12.5	32.67	16.8	11.50
torus	64	36.5	22.86	18.5	34.3	9.9	11.08
diamond	16	14.0	38.67	6.5	35.74	14.6	17.70
diamond	64	18.7	29.78	9.0	28.80	13.2	14.60

Table 10.4   Spectral methods vs. R-SNAKE on non-rectangular grids

The size of the irregular boundary grids in our test suite varies; the circle has 7800 cells, the torus 7696 cells. The diamond domain is smaller, with 4019 cells, and the elliptical domain is the smallest in our test suite with only 823 cells. A partition of the torus among 64 processors as found by DGA running on 8 processors is shown in Figure 10.15. The torus is a disk of radius 50 with a hole of radius 7 in its center.

In Figure 10.16 we plot the response times of RSB, RSQ, and DGA with four (DGA4) and eight (DGA8) islands for the various irregular boundary grids in our test-suite. The labels on the x-axis indicate the problems shown in Table 10.6. Figure 10.17 then shows for the same problems and methods the quality of the solutions thus produced. Finally, we run DGA with one island only but leaving all the other parameters intact, disabling all communication (we don't even start PVM) so as to further check the effect of communication on response times of the algorithm. The results suggest that indeed the communication overhead of the algorithm is almost negligible. The quality of the resulting partitions is

PROBLEM		DGA (4 islands)		DGA (8 islands)	
Shape	P	Time	Gap	Time	Gap%
circle	16	19.8	8.33	20.2	8.47
circle	64	19.4	6.21	17.5	5.87
ellipsis	16	8.3	8.33	6.4	8.33
ellipsis	64	9.4	5.36	7.5	5.56
torus	16	18.8	11.50	16.7	11.50
torus	64	17.2	11.08	13.8	11.00
diamond	16	10.7	16.40	10.4	16.40
diamond	64	16.2	13.37	14.7	13.37

Table 10.5   DGA on 4 or 8 COW nodes on non-rectangular grids

Label	Problem	P
C16	circle	16
C64	circle	64
E16	ellipsis	16
E64	ellipsis	64
T16	torus	16
T64	torus	64
D16	diamond	16
D64	diamond	64

Table 10.6   Plot graph labels

Figure 10.15   Snake partition of a torus among 64 procs

comparable to that of R-SNAKE or the DGA with four (eight) islands, but not as good, and quite logically so, as more processes imply that more individuals are created and tested.

To isolate the effect of the island model on the evolution process, we also run DGA with 1 island but this time allowing a maximum of 64 or 128 individuals on the island (as opposed to the previous experiments where each island could maintain only up to 16 individuals). As expected, the solution times increase, but what is interesting is the fact that DGA running with a single population that is roughly 4 or 8 times bigger than the population on each of the islands of the previous experiments fails to find the same quality solutions. Occasionally, it beats DGA with one island and a maximum population size of 16, but it never finds the same quality solutions as DGA with 4 or 8 islands. This effect may be partly due to premature convergence.

7.   CONCLUSIONS AND FUTURE DIRECTIONS

We have presented stripe and snake decomposition, two powerful techniques for equi-partitioning grids with regular or irregular boundaries among any number of processors, a problem that is NP-hard. We have shown that the application of these techniques to a large class of domains produces solutions that are asymptotically optimal in the sense that as the problem parameters (dimensions

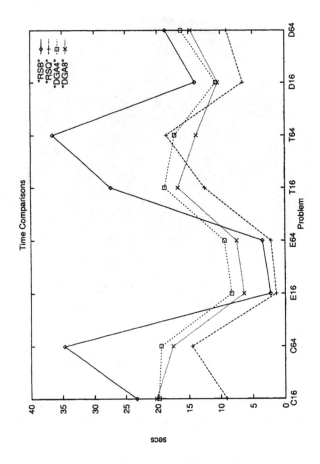

Figure 10.16   Time plots for various methods on non-rectangular grids

of the largest rectangles that fit in the grid and number of processors) grow larger, the relative distance of the perimeter of the corresponding solutions from a
theoretical lower bound tends to zero.

To generate even better solutions, we developed a new distributed Genetic
Algorithm (DGA) equipped with a snake-filling routine as the fitness function.
DGA follows the island model of computation using the steady-state approach
for generating the next population in each island. DGA is fully asynchronous
so as to avoid communication penalties from a possibly slow network, and to
allow for maximum interleaving of computation and communication. Migration
criteria are based on the workload of each processor, which is correlated to the
average fitness of the populations of each island because of a new mechanism,
namely *age*. This mechanism removes an individual from the population when

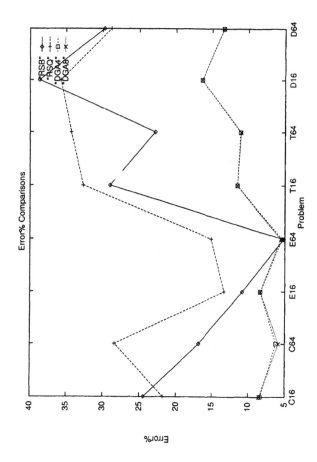

Figure 10.17 Error plots for various methods on non-Rectangular grids

this individual becomes "old". The age at which each individual is removed from the population depends upon its fitness.

The design and implementation of the DGA library functions proved so fast and flexible that it was recently embedded as the search engine for a system for solving large scale bidline generation problems which are an important and difficult scheduling problem in the airline industry [Christou et al., 1999b]. The computational results from DGA show that it compares well in response time with other popular graph partitioning methods and that –with one minor exception– produces superior quality partitions for all classes of domains tested.

Experimenting and evaluating DGA's mechanisms of age and asynchronous migration is another important topic for future research. In particular, it would be very interesting to assess the strength of the age mechanism in the avoidance of premature convergence, and to experiment with new –or variants of the existing– migration criteria that DGA employs.

Extensions of the results to triangulations or other data partitioning problems arising from parallel database design, and other types of fixed-charge networks also provide promising areas for further research. The same basic idea of fitting together, as well as possible, building blocks that represent optimal solutions to subproblems, applies to these domains as well. Recent results in telecommunications network design, pharmaceutical design, and crew rostering indicate that this paradigm of utilizing subproblem solutions within the context of genetic algorithms is very effective in those problem domains as well. With appropriately designed repair procedures for subproblem coordination, we can anticipate that the evolutionary progress associated with genetic algorithms will be effective in producing good solutions for broad classes of large-scale combinatorial network problems.

References

[TMC, 1991] (1991). *The Connection Machine CM-5 Technical Summary.* Thinking Machines Corporation, Cambridge, MA.

[CPL, 1998] (1998). *Using the CPLEX 6.0 Callable Library.* ILOG CPLEX Division, 889 Alder Ave. Incline Village, NV 89451.

[Ahuja et al., 1993] Ahuja, R. K., Magnanti, T. L., and Orlin, J. B. (1993). *Network Flows.* Prentice Hall, Englewood Cliffs, NJ.

[Chen et al., 1993] Chen, R., Meyer, R., and Yackel, J. (1993). A genetic algorithm for diversity minimization and its parallel implementation. In *Proceedings of the Fifth International Conference on Genetic Algorithms*, pages 163–170, Palo Alto, CA. Morgan Kaufman.

[Christou et al., 1999a] Christou, I., Martin, W., and Meyer, R. R. (1999a). Genetic algorithms as multi-coordinators in large scale optimization. In Davis, L. D., Whitley, L. D., DeJong, K., and Vose, M., editors, *Evolutionary Algorithms*, volume 111, pages 1–15, New York, NY. Springer-Verlag. The IMA Volumes in Mathematics and its Applications.

[Christou, 1996] Christou, I. T. (1996). *Distributed Genetic Algorithms for Partitioning Uniform Grids.* PhD thesis, Computer Sciences Dept. University of Wisconsin - Madison, Madison, WI.

[Christou and Meyer, 1996a] Christou, I. T. and Meyer, R. R. (1996a). Fast distributed genetic algorithms for partitioning uniform grids. In Rolim, H. and Yang, T., editors, *Lecture Notes in Computer Science*, volume 1117, New York, NY. Springer-Verlag. Proceedings of the Third International Workshop on Parallel Algorithms for Irregularly Structured Problems (IRREGULAR 96).

[Christou and Meyer, 1996b] Christou, I. T. and Meyer, R. R. (1996b). Optimal and asymptotically optimal equi-partition of rectangular domains via stripe decomposition. In Fischer, H., Riedmuller, B., and Schaffler, S., editors, *Applied Mathematics and Parallel Computing - Festschrift for Klaus Ritter*, pages 77–96, Berlin, Germany. Physica-Verlag.

[Christou and Meyer, 1996c] Christou, I. T. and Meyer, R. R. (1996c). Optimal equi-partition of rectangular domains for parallel computation. *Journal of Global Optimization*, 8:15–34.

[Christou et al., 1999b] Christou, I. T., Zakarian, A., Liu, J., and Carter, H. (1999b). A two-phase genetic algorithm for large scale bidline generation problems at delta air lines. *Interfaces*, 29(5).

[Crandall and Quinn, 1995] Crandall, P. and Quinn, M. (1995). Non-uniform 2-d grid partitioning for heterogeneous parallel architectures. In *Proceedings of the 9th International Symposium on Parallel Processing*, pages 428–435, Los Alamitos, CA. IEEE Computer Society Press.

[Dantzig and Wolfe, 1960] Dantzig, G. B. and Wolfe, P. (1960). Decomposition principle for linear programs. *Operations Research*, 8:101–111.

[Darwin, 1859] Darwin, C. (1859). *On the Origin of Species by Means of Natural Selection*. J. Murray, London, UK.

[DeLeone et al., 1994] DeLeone, R., Meyer, R. R., Kontogiorgis, S., Zakarian, A., and Zakeri, G. (1994). Coordination in coarse-grain decomposition. *SIAM Journal on Optimization*, 4(4):773–793.

[Falkenauer, 1998] Falkenauer, E. (1998). *Genetic Algorithms and Grouping Problems*. John Wiley & Sons, New York, NY.

[Fiedler, 1973] Fiedler, M. (1973). Algebraic connectivity of graphs. *Czechoslovak Math. Journal*, 23:298–305.

[Fulkerson, 1963] Fulkerson, D. R. (1963). Flows in networks. In *Recent Advances in Mathematical Programming*, pages 319–332, New York, NY. McGrow - Hill.

[Geist et al., 1994] Geist, A., Beguelin, A., Dongarra, J., Jiang, W., Manchek, R., and Sunderam, V. (1994). *PVM 3 User's Guide and Reference Manual*. Oak Ridge National Laboratory, Oak Ridge, TN.

[Gilbert et al., 1995] Gilbert, J. R., Miller, G. L., and Teng, S. H. (1995). Geometric mesh partitioning: Implementation and experiments. In *Proceedings of the 9th International Symposium on Parallel Processing*, pages 418–427, Los Alamitos, CA. IEEE Computer Society Press.

[Hart, 1994] Hart, W. E. (1994). *Adaptive global optimization with local search*. PhD thesis, Computer Science and Engineering Dept. University of California - San Diego, San Diego, CA.

[Hendrickson and Leland, 1993] Hendrickson, B. and Leland, R. (1993). Multidimensional spectral load balancing. In *Proceedings of the 6th SIAM Conference on Parallel Processing and Scientific Computing*, Philadelphia, PA. SIAM.

[Hendrickson and Leland, 1995a] Hendrickson, B. and Leland, R. (1995a). *The Chaco User's Guide Version 2.0*. Sandia National Laboratories, Sandia, NM.

[Hendrickson and Leland, 1995b] Hendrickson, B. and Leland, R. (1995b). An improved spectral graph partitioning algorithm for mapping parallel computations. *SIAM Journal on Scientific Computation*, 16:452–469.

[Holland, 1992] Holland, J. (1992). *Adaptation in Natural and Artificial Systems*. MIT Press, Cambridge, MA.

[Karypis and Kumar, 1995] Karypis, G. and Kumar, V. (1995). Multilevel k-way partitioning scheme for irregular graphs. Technical Report 95-064, Computer Science Dept. University of Minnesota.

[Kernighan and Lin, 1970] Kernighan, B. W. and Lin, S. (1970). An effective heuristic procedure for partitioning graphs. *Bell Systems Tech. Journal*, pages 291–308.

[Laguna et al., 1994] Laguna, M., Feo, T. A., and Elrod, H. C. (1994). A greedy randomized adaptive search procedure for the two - partition problem. *Operations Research*.

[Levine, 1995] Levine, D. (1995). *User's Guide to the PGAPack Parallel Genetic Algorithm Library Version 0.2*. Argonne National Laboratory, Argonne, IL.

[Li et al., 1994] Li, Y., Pardalos, P. M., and Resende, M. G. C. (1994). A greedy randomized adaptive search procedure for the quadratic assignment problem. In Pardalos, P. M. and Wolkowicz, H., editors, *Quadratic Assignment and Related Problems*, pages 237–262, Providence, RI. American Mathematical Society.

[Lin, 1991] Lin, K. Y. (1991). Exact solution of the convex polygon perimeter and area generating function. *J. Phys. A. Math Gen.*, 24:2411–2417.

[Martin, 1998] Martin, W. (1998). Fast equi-partitioning of rectangular domains using stripe decomposition. *Discrete Applied Mathematics*, 82(1–3):193–207.

[Melou, 1994] Melou, M. B. (1994). Codage des polyominos convexes et equation pour l'enumeration suivant l'aire. *Discrete Applied Mathematics*, 48:21–43.

[Michalewicz, 1994] Michalewicz, Z. (1994). *Genetic Algorithms + Data Structures = Evolution Programs*. Springer-Verlag, New York, NY.

[Miller et al., 1993] Miller, G. L., Teng, S. H., Thurston, W., and Vavasis, S. A. (1993). Automatic mesh partitioning. In George, A., Gilbert, J. R., and Liu, J. W. H., editors, *Graph Theory and Sparse Matrix Computation*, New York, NY. Springer-Verlag.

[Muhlenbein et al., 1991] Muhlenbein, H., Schomisch, M., and Born, J. (1991). The parallel genetic algorithm as function optimizer. In Belew, R. and Booker, L., editors, *Proceedings of the Fourth Intl. Conference on Genetic Algorithms*, pages 45–52, Los Altos, CA. Morgan Kaufmann Publishers.

[Nemhauser and Wolsey, 1985] Nemhauser, G. and Wolsey, L. (1985). *Integer and Combinatorial Optimization*. John Wiley & Sons, New York, NY.

[Pardalos et al., 1995] Pardalos, P., Pitsoulis, L., and Resende, M. (1995). A parallel grasp implementation for the quadratic assignment problem. In Ferreira, A. and Rolim, J., editors, *Parallel Algorithms for Irregularly Structured Problems*, pages 111–130, Dordrecht, The Netherlands. Kluwer Academic Publishers. Proceedings of the First International Workshop on Parallel Algorithms for Irregularly Structured Problems (IRREGULAR 94).

[Pardalos et al., 1993] Pardalos, P. M., Rendl, F., and Wolkowicz, H. (1993). The quadratic assignment problem: A survey and recent developments. In Pardalos, P. M. and Wolkowicz, H., editors, *Quadratic Assignment and Related Problems*, pages 1–42, Providence, RI. American Mathematical Society.

[Pothen et al., 1990] Pothen, A., Simon, H. D., and Liu, K. P. (1990). Partitioning sparse matrices with eigenvectors of graphs. *SIAM Journal on Matrix Analysis and Applications*, 11:430–452.

[Schalkoff, 1989] Schalkoff, R. J. (1989). *Digital Image Processing and Computer Vision*. John Wiley & Sons, New York, NY.

[Schultz and Meyer, 1991] Schultz, G. L. and Meyer, R. R. (1991). An interior point method for block angular optimization. *SIAM Journal on Optimization*, 1(4):583–602.

[Strikwerda, 1989] Strikwerda, J. (1989). *Finite Difference Schemes and Partial Differential Equations*. Wadsworth & Brooks Cole, Pacific Grove, CA.

[von Laszewski, 1991] von Laszewski, G. (1991). Intelligent structural operators for the k-way graph partitioning problem. In Belew, R. and Booker, L., editors, *Proceedings of the Fourth Intl. Conference on Genetic Algorithms*, pages 45–52, Los Altos, CA. Morgan Kaufmann Publishers.

[Yackel, 1993] Yackel, J. (1993). *Minimum Perimeter Tiling in Parallel Computation*. PhD thesis, Computer Sciences Dept. University of Wisconsin - Madison, Madison, WI.

[Yackel and Meyer, 1992a] Yackel, J. and Meyer, R. R. (1992a). Minimum perimeter decomposition. Technical Report 1078, University of Wisconsin - Madison, Madison, WI.

[Yackel and Meyer, 1992b] Yackel, J. and Meyer, R. R. (1992b). Optimal tilings for parallel database design. In Pardalos, P. M., editor, *Advances in Optimization and Parallel Computing*, pages 293–309, New York, NY. North - Holland.

[Yackel et al., 1997] Yackel, J., Meyer, R. R., and Christou, I. T. (1997). Minimum-perimeter domain assignment. *Mathematical Programming*, 78:283–303.

[Zakarian, 1995] Zakarian, A. (1995). *NonLinear Jacobi and epsilon - Relaxation Methods for Parallel Network Optimization*. PhD thesis, Computer Sciences Dept. University of Wisconsin - Madison.

Combinatorial Optimization

1. E. Çela: *The Quadratic Assignment Problem*. Theory and Algorithms. 1998
   ISBN 0-7923-4878-8
2. M.Sh. Levin: *Combinatorial Engineering of Decomposable Systems*. 1998
   ISBN 0-7923-4950-4
3. A.I. Barros: *Discrete and Fractional Programming Techniques for Location Models*.
   1998   ISBN 0-7923-5002-2
4. V. Boltyanski, H. Martini and V. Soltan: *Geometric Methods and Optimization Problems*. 1999   ISBN 0-7923-5454-0
5. P.M. Pardalos and S. Rajasekaran (eds.): *Advances in Randomized Parallel Computing*. 1999   ISBN 0-7923-5714-0
6. D.-Z. Du, J.M. Smith and J.H. Rubinstein (eds.): *Advances in Steiner Trees*. 2000
   ISBN 0-7923-6110-5
7. P.M. Pardalos and L.S. Pitsoulis (eds.): *Nonlinear Assignment Problems*. Algorithms
   and Applications. 2000   ISBN 0-7923-6646-8
8. C. Wynants: *Network Synthesis Problems*. 2000   ISBN 0-7923-6689-1

KLUWER ACADEMIC PUBLISHERS – DORDRECHT / BOSTON / LONDON